Environmental Studies

Anindita Basak

Reader in Chemistry, S.G. Women's College, Rourkela.

Pearson

ISBN 978-81-317-2118-6

First Impression

Published by Pearson India Education Services Pvt. Ltd, CIN: U72200TN2005PTC057128

Head Office: 15th Floor, Tower-B, World Trade Tower, Plot No. 1, Block-C, Sector 16, Noida 201 301, Uttar Pradesh, India.

Registered Office: 7th Floor, SDB2, ODC 7, 8 & 9, Survey No.01 ELCOT IT/ ITES SEZ, Sholinganallur, Chennai – 600119, Tamil Nadu, India. Website: in.pearson.com, Email: companysecretary.india@pearson.com

Digitally printed in India by Trinity Academy for Corporate Training Ltd, New Delhi in the year of 2022.

DEDICATED
TO MY

DEAR PARENTS

Contents

About the Author

Dr Anindita Basak, a Reader in Chemistry, has been teaching Chemistry, Polymer Science and Environmental Studies in the government colleges of Orissa like Fakir Mohan College, Balasore (presently Fakir Mohan University); Government Autonomous College, Rourkela; Government Autonomous Junior College, Rourkela; Government College, Sundargarh at the graduate and the postgraduate levels since 1987. She was also deputed as Visiting Scientist in National Institute of Technology, Rourkela during the period of 2004 to 2006.

Dr Basak is not only a popular teacher, a prolific author, an innovative research worker, an eminent administrator and social worker, but also has been a TV anchor. She did her M.Phil. in Chemistry from Ravenshaw College, Cuttack, Orissa (presently Ravenshaw University) and was a university topper. As a Junior Research Fellow at Bhabha Atomic Research Centre, Mumbai, she did advanced research work in the field of Polymer Science and obtained her Ph.D. from Utkal University in 1989. She has to her credit 16 technical papers published in national and international journals of repute. Dr Basak has been a National Scholarship holder throughout her career.

Dr Basak is closely associated with a number of social, cultural, educational and professional institutions like the Institute of Chemists, India; Professor Sukumar Maiti Polymer Award Foundation, Orissa; Chemical Society, The Uranium, a body propagating science education in Orissa; Rourkela Social and Cultural Trust; Orissa Gyan Vigyan Samiti; Saksharata Mission, Rourkela, etc.

At present, she is a Reader in Chemistry at Sushilavati Government Women's Junior College, Rourkela

Preface

The present plight of the world as a victim to a plethora of environmental setbacks ranging from global warming, ozone layer depletion to an alarming increase in world pollution levels is threatening the existence of the most intelligent species on earth. This has been enough for both environmentalists and laymen to wake up to the indisputable importance of environmental education. Almost every sphere that man has forged into has undergone a major demarcation from what it was before to suit the needs and greed of man better. Yet this Midas touch that came into major force in the last century was not quite the answer to the big question that it put on the sustainable development of our future generations. Rather, it proved to be a major hurdle in maintaining the ecological balance of nature, leading to the establishment of the United Nations in 1945 which was a natural and obvious decision in such a scenario.

Like all other developing nations, India too has been facing serious environmental deterioration and has suffered major destruction due to all kinds of natural calamities. This led the Supreme Court of India give a landmark judgment in October 2002 that made environmental education mandatory for all students up to postgraduation level irrespective of their major branch of study and, consequently, the introduction of environmental education in the UGC syllabi.

This book on environmental studies attempts to cover the concept of environment, the causes for its deterioration, the measures taken for its preservation and sustainable development. An insight into the contents of each of the eight units is as follows:

Unit-I is an introduction to the multifaceted subject and its importance to the society. Further, it elaborates the scope, significance, components and interdependence of the different components of the environment.

Unit-II describes natural resources, their use and various processes of extraction of resources such as water, mineral, food, land and forests. The consequences of overexploitation of natural resources have been depicted with supporting case studies. The Unit then moves on to the role that an individual can play towards conservation of nature through proper resource management and directs us towards a sustainable lifestyle by identification of the equitable use of resources.

In **Unit-III** the ecosystem and its functional components have been described and supported by updated data. A special thrust is given to ecological pyramids and facts such as balance of nature and biomagnifications. Different ecosystems such as forests, grasslands, deserts and water bodies have been described in great detail.

Unit-IV deals with biodiversity and the need for its conservation, supported by latest facts and figures in a very vivid and scientific manner. Details about the existing hotspots of biodiversity in addition to the recently discovered nine hotspots (exclusive of the book), the two hotspots present in India are discussed in this unit. Data showing the number of species found in India and its rank in the world for the same has also been described in detail. A list of endangered and endemic species of India and the threats posed to the present biodiversity is also included. Several case studies describing poaching of wildlife, conflicts between man and wildlife are incorporated to increase public awareness regarding protecting and conserving wildlife and thereby biodiversity. This Unit also brings up information regarding biopiracy.

Unit-V depicts environmental degradation with respect to air, water, soil, marine, noise and thermal pollution and nuclear hazards (accompanied by up-to-date statistics, information, graphs, photographs and tables) as per the UGC syllabus. The causes, effects and control measures of all the above types of pollution are specially highlighted in this unit. It also includes a description and disposal of the types of solid waste such as hazardous, non-hazardous, biomedical and e-waste. Recent concepts such as the generation of green electricity, recycling of plastic, mushroom cultivation, etc. are also explained lucidly and from a scientific standpoint.

Unit-VI focuses on the need for sustainable development for better understanding of social issues and the environment and then moves to contemporary subjects of importance such as rain water harvesting, watershed management and resettlement; that have been described after making an exhaustive study. Special emphasis is laid on elaborate discussion over a topic as essential and as sensitive as environmental ethics. A detailed account of climate change, global warming, with a reference to the phenomenon of ozone layer depletion, nuclear accidents and holocausts; ranging from the Hiroshima and Nagasaki incident to the Chernobyl accident have been included for the knowledge of today's students. Wasteland reclamation has been described with proper thrust and successful case studies like the establishment of the City of Dawn (Auroville). Issues such as consumerism, e-waste, and chemical waste which are a consequence of the present day technology boom also find a place in this Unit. The Environmental Protection Act, Wildlife Protection Act, Forest Protection Act, issues involved in enforcement of environmental legislations and public awareness for the same have been cited for reference. A brief discussion on EIA has also been included.

Unit-VII provides exhaustive coverage of recent up-to-date information complemented by appropriate discussions on human population growth in various nations, population explosion, family welfare programmes, human rights and human health in addition to suitable case studies makes up this Unit. HIV/AIDS, the most serious cause for concern for the youth in society and the student community in particular, has not only been discussed vividly but due importance has also been given to its remedial actions. With a view to encourage the nation's future—the youth, to become actively involved in the upliftment of our social environment, suitable case studies have been employed to illustrate the need for the welfare of women and children and various programmes involved in the above cause. It also underlines the role of IT in improving the environment as well as human health.

The uniqueness of the book also lies in **Unit-VIII** which is dedicated to providing guidelines for field study through important experiments so as to aid the practical understanding of environmental study.

Five appendices at the end of the book facilitate the understanding of the subject matter to readers of all disciplines. The first one discusses National River Action Plans for the abatement of river pollution in India. The global environmental summits held till date make up the second appendix, aptly followed by important environmental dates, important abbreviations relevant to the subject, keywords (short notes) and a glossary of frequently used terms.

Keeping in mind the above facts, *Environmental Studies* is presented in a very simple and lucid manner covering the module syllabus designed by UGC for undergraduate students. It also includes up-to-date facts for inquisitive readers. Enough photographs, pie-charts, graphs, tables, suitable case studies and appropriate discussions have been included to make the learning process enjoyable for readers. Above all, the book is well-equipped to assess the comprehension of students by way of a treasure of long, short and multiple choice questions in addition to model question papers and self assessment questions that will prepare students for term ends and for competitive examinations alike.

ANINDITA BASAK

Acknowledgments

It was a daunting task for me to write a book on environmental studies when there is a plethora of books on this topic. I must acknowledge that my students' inquisitiveness and thirst for a comprehensive book on this topic for the last few years sowed the seed for the present work.

I thank the Department of Higher Education, Government of Orissa; principals of government colleges where I have been working; NIT, Rourkela, where I worked as a Visiting Scientist during 2004 to 2006; UGC; Sambalpur University, and other institutions for giving me the opportunity to turn into an author from a reader.

I am also grateful to my parents, Nabin Chandra Basak and Anima Basak, for their constant inspiration, to Chitta Ranjan Patnaik and Dr Sandhya Patnaik for their support and guidance, to Dr Mukesh Rawal for his valuable advice and also to my daughters, Dipanwita and Nandita, for their support throughout the preparation of the manuscript.

My sincere thanks goes to the team of Pearson Education, specially Anita Yadav and Bhavna Chawla. I also thank D. K. Basak for his ready cooperation and Rajesh Singh for attending to my IT need even at odd hours.

Finally, I would like to thank my husband, Sudip Kumar Basak, without whom this book would have probably remained a mere dream.

ANINDITA BASAK

Core Module Syllabus for Environmental Studies

Unit – I Definition, Scope and Importance, Need for Public Awareness **2 lectures**

Unit – II Natural Resources: Renewable and Non-renewable resources **8 lectures**

Natural Resources and Associated Problems

(a) **Forest Resources**: Use and overexploitation. Deforestation case studies. Timber extraction, mining. Dams and their effects on forests and tribal people.

(b) **Water Resources:** Use and overexploitation of surface and ground water, floods, drought, conflicts over water. Dam: benefits and problems.

(c) **Mineral Resources:** Use and exploitation. Environmental effect of extracting and using mineral resource. Case studies.

(d) **Food Resources:** World food problems. Changes caused by agriculture and overgrazing. Effects of modern agriculture, fertilizer pesticide problems, water-logging, salinity. Case studies.

(e) **Energy Resources:** Growing energy needs, renewable and non-renewable energy sources, use of alternate energy source. Case Studies.

(f) **Land Resources:** Land as a resource. Land degradation. Man-induced landslides, soil erosion and desertification.

Role of individual in conservation of natural resources. Equitable use of resources for sustainable lifestyle.

Unit – III Ecosystems **6 lectures**

Concept of Ecosystems. Structure and function of an ecosystem. Producers, consumers and decomposers. Energy Flow in the Ecosystem, Ecological Succession, Food Chain, Food Web and Ecological Pyramids. Introduction, Types, Characteristic features, Structure and function of the following ecosystems:

(a) Forest Ecosystem

(b) Grassland Ecosystem

(c) Desert Ecosystem

(d) Aquatic Ecosystems including ponds, streams, lakes, oceans, estuaries.

Unit – IV Biodiversity and its conservation **4 lectures**

Introduction — Definition, Genetics, Species and Ecosystem diversity, Biogeographical classification of India, Values of diversity, Consumptive use, Productive use, social, ethical, aesthetic and optional values. Biodiversity and global, national and local levels. India as a megadiversity nation. Hotspots in diversity. Threats to diversity: Habitat loss, poaching of wildlife, man–wild life conflicts, endangered and endemic species of India. Conservation of Biodiversity, In-situ and Ex-situ conservation of Biodiversity.

Unit – V Environmental Pollution **8 lectures**

Definition, causes, effects and control measure of:

(a) Air Pollution.

(b) Water Pollution.

(c) Soil Pollution.
(d) Marine Pollution.
(e) Noise Pollution.
(f) Thermal Pollution.
(g) Nuclear Hazards.

Solid Waste Management: Causes, effects and control measure of urban and industrial wastes. Role of individuals in prevention of pollution. Case studies.
Disaster Management: Flood, Earthquakes, Cyclones and Landslides.

Unit – VI Social Issues and the Environment 7 lectures

From unsustainable to sustainable development. Urban problems related to energy, measures of water conservation including rain water harvesting, watershed management, resettlement and rehabilitation of people, their problems and concerns, case studies, environmental ethics: issues and possible solutions, climatic changes, global warming, acid rain, Ozone layer depletion, Nuclear accidents and holocaust, case studies, wasteland reclamation, consumerism and waste products, Environment (Protection) Act (Prevention and Control of Pollution) Act, Water (Prevention and control of pollution) Act, Wildlife Protection Act, Forest Conservation Act, Issues involved in enforcement of environmental legislation, Public awareness.

Unit- VII Human population and the Environment 5 lectures

Population growth variation among nations, Population explosion, family welfare programme, Environmental and Human Health, Human Rights, value education, HIV/AIDS women and child welfare, Role of Information Technology in Environment and Human Health, Case Studies.

Unit – VIII Field Work 5 lectures

Visit to a local area to document environment problems in forest, grassland, hill, mountain. Visit to a local polluted site-urban/rural/industrial/agricultural. Study of common plants/insects/birds. Study of simple ecosystems–pond, river, hill slopes etc.

UNIT I

Definition, Scope and Importance of Environmental Studies

LEARNING OBJECTIVES

After studying Unit I, students will be able to:

- Define the scope and importance of environmental studies.
- Describe the need for public awareness.
- Identify the different components of the environment.
- Describe the history of environmental laws, and myths and misconceptions about the environment.
- Explain the importance of environmental studies in promoting sustainable development.
- Analyse the components of natural, man-made environment and social environment of the Earth – and explain the interdependence of these components on each other.

1.1 Definition, Scope and Importance of Environmental Studies

Environmental Studies is a multidisciplinary subject. It employs subjects like chemistry, physics, botany, zoology, physiology, geography, geology, geophysics, and metrology to describe the biological and physical nature of our environment. In order to understand how people function separately and also in a group, **Environmental Studies** involves an understanding of philosophy, ethics, psychology, anthropology, demography, archaeology, economics and political science. Laid against an ecological format, the information gleaned through the study of all these varied disciplines gives us a holistic view of the environment for sustaining life on earth on an infinite time scale.

The unlimited exploitation of nature (environment) by mankind for the sake of development has threatened the survival of not just human beings but also all other living organisms. The number of living species has decreased, a large number are threatened, and many are even extinct. Human beings too, are suffering from various health problems. Today India is one of the top 10 industrialized countries in the world and the ever-increasing pollution levels in its environment are affecting all living organisms. People around the world are enjoying economic growth at the cost of 'quality of human life'. So the need of the hour is to save our environment by following a suitable developmental policy. This necessitates the knowledge of our environment, its components and the different issues affecting the environment.

Education for environmental awareness is required not only for environmental scientists, engineers, policymakers or NGOs, but also for every one of us. Only environmental study can make us conscious and careful about the environment. Environmental education is aimed at developing environmental ethics in people. It teaches them the importance of conservation of life and biodiversity of the environment. **Environmental study** also teaches people to understand their role in the environment and learn to live with limited natural resources so as to avoid future disasters. The casual attitude of human beings towards the environment and its conservation is the root cause of all environmental problems.

Therefore, proper education and public awareness are necessary to tackle environmental problems. Towards this end, **environmental studies** will provide sufficient knowledge about the philosophy, genesis and consequences of local and global environmental problems and the necessary knowledge for their abatement and control. Thus, for a sustainable environment and for the survival of the present and future generations, environmental education is necessary.

1.2 The Need for Public Awareness

As soon as man acquired the ability to generate fire in the Paleolithic era, the environment began to get adversely affected. In the Iron Age, man discovered the use of tools which he used to grind metals. This led to minor accumulations of discarded materials which were easily dispersed. The vast extent of natural vegetation made up for the miniscule pollution that the activities of man would have caused to air and water.

The end of the Middle Ages saw a growth in population, concentrated within cities. This created areas that suffered from contamination. Air pollution began to be recognized as a health issue, while polluted water in densely populated areas served as a conducive medium for transmission of diseases.

The earliest known writings concerned with pollution were Arabic medical treatises written between the 9th and 13th centuries. They covered a number of pollution related subjects, such as the contamination of air, water and soil. The writings mentioned about the mishandling of solid waste, and environmental assessments of certain localities.

In 1272, King Edward I of England banned the burning of sea coal in London by proclamation. This was done after its smoke had become a problem. Every person who disobeyed the law was executed. However, during the industrial revolution, air pollution continued to be a problem there. London also recorded one of the earliest cases of water quality problems with the Great Stink on the Thames of 1858. This led to the construction of the London sewerage system soon afterward.

These instances prove that environmental consciousness has been prevalent among people since ancient times. All religions too preach environmental conservation. However, rapid industrialization and several developmental projects have disturbed the ecological balance in such a way that the survival of life on earth has been seriously threatened. People are now suffering from contamination in gases, liquids, and solids as well as phase transfer of contaminants.

The London Smog, Los Angeles Smog, Bhopal Gas Tragedy, Chernobyl disaster, DDT Episode, Minamata disease, Itai-Itai disease, oil pollution, and solid waste disposal are all examples of the far reaching effects of pollution. The repeated occurrence of such accidents has drawn the attention of the world towards environmental awareness. The union government of India has also enacted a comprehensive Environment (Protection) Act, 1986 to safeguard and protect our environment. This shows the increased awareness in both the government and the common people about the environment.

Environmental protection requires both preventive and curative measures. This can be achieved through a policy of sustainable development. The increased awareness was also reflected at the Earth Summit at Rio de Janeiro in June 1972, where Agenda–21, a Global Action Plan was adopted with the aim of integrating environmental imperatives with developmental aspirations.

The Indian government has taken some important steps towards environmental protection. For example, no development work can progress without a proper Environmental Impact Assessment (EIA). The government has from time to time set up State Pollution Control Boards for controlling pollution from different industries, and established Green Benches in different High Courts of India. In addition to the government's responsibility, social awareness is a most important factor for conservation and regeneration of environmental resources.

Social awareness can encourage people to use biogas and solar energy or non-conventional energy instead of fossil fuels. An informed citizen can raise his voice against using unfit air and water resources and force the responsible agencies (industry) to treat the particular resources before they are released into the natural environment. Global environmental issues like ozone layer depletion, climatic changes, acid rain, and biodiversity losses are the result of rapid industrialization, different developmental projects, and fast urbanization leading to deforestation among other problems. In India, we also have problems of solid waste disposal, sanitation, and air and water pollution.

However, the major problems of developing countries like India are their poverty and illiteracy. Moreover, decisions about different environment management programmes are taken by the elite groups of society. People, who cannot fulfil their daily needs, cannot think about the environment. This ignorance and illiteracy has generated a number of misconceptions and superstitions such as:

(i) Diseases are caused by God's curse and demons. They have nothing to do with infection or hygiene.

(ii) Famines, floods, droughts are God's punishment for the sins of men and have no relationship with environmental management.

(iii) Rainfall is dependent on God's grace.

(iv) Cities are polluted not villages. (In reality, villages are more polluted due to the burning of fossil fuels [such as cow dung cakes and wood] while cities are polluted due to industrial effluents and automobiles exhaust).

(v) Deforestation is the result of the industrial revolution but growing urbanization is not responsible for this.

To get rid of these superstitions and misconceptions about nature, environmental awareness is imperative in India. This awareness ensures that everyone, from an environment conscious farmer in the village to a policy planner in the government knows about the consequences of his activities on nature. Agricultural production can get a boost only if people know the patterns of land use, of water resources and irrigation, if they use fertilizers and pesticides judiciously and above all if they are aware of the renewable resources of energy.

In the industrial sector too, people should know about the impact of industrial activity on nature. The demand for more food from more or less the same land has resulted in the use of science and technology for the development of industry. Industries such as fertilizers, chemicals, and pesticides and industries that manufacture different components of the same have come into existence. While it is true that this industrial revolution supports the green revolution, the fact remains that it is responsible for the speedy consumption and pollution of natural resources. However, industrialization is likely to be less of a health hazard if growing green belts around industrial areas, and maintaining wild life sanctuaries and national parks are made compulsory. This will help in maintaining the ecological balance and result in higher agricultural productivity.

Industrial development, along with urbanization, can generate health problems. However, awareness about healthy living conditions and the consequences of ill-health will automatically make people conscious about their health and hygienic living conditions.

From these discussions, we can conclude that public awareness is a must for sustainable environmental development.

1.3 Environment and its Components

Environment: The environment encompasses virtually everything that surrounds an organism in a holistic ecological approach.

Out of all the nine planets, meteorites, and satellites in our solar system, the earth is the only planet known to support life. Life on earth experiences different types of surroundings. These surroundings may be living or non-living. Each living organism constantly interacts with its surroundings and adapts to it. These surroundings are our environment. The physical environment, which consists of soil, air, water, sunlight among others, provides favourable conditions for the existence and growth of different life forms. Living beings constitute the biological environment.

Both the physical and the biological environments closely interact with each other to form a stable self-perpetuating system. Everything that influences an organism and its living processes from outside is collectively known as 'environment.' The living component of the environment is known as the biotic component and the non living component (things) as the abiotic component. Hence, the term 'environment' can be defined as the sum total of living and non-living components, their influences and events surrounding an organism.

No organism can live without interacting with the environment. Animals depend on green plants for food and oxygen, whereas plants depend on animals for pollination of flowers and dispersal of seed or fruit. Therefore, for the survival of human civilization, the protection of its environment is very important. For this, some fundamental principles have to be followed:

(i) Maintenance of biodiversity.

(ii) Maintenance of all gaseous and material cycles and interdependence of living organisms among themselves and with abiotic environments.

(iii) Maintenance of ecological order and natural balance, which depend on the food chain relationship, sustainable productivity and biotic interaction.

These principles were known to early human beings, who lived in harmony with nature. However, in the course of evolution, man has developed a new type of environment, the man-made environment. A large chunk of the imbalance in nature is caused by this man-made environment. This man-made imbalance has forced us to put restraints on the use of natural resources.

1.3.1 Segments of the Environment

Our environment can be broadly classified into natural and man-made environment.

Natural Environment

Each living organism has a specific surrounding with which it interacts and to which it is adapted. This surrounding is its natural environment. The natural environment can be broadly classified into two categories.

(i) The non-living or abiotic component, which includes:

  (a) Climatic factors such as solar radiation, temperature, wind, water current, and rainfall.

  (b) Physical factors such as light, air, pressure and geomagnetism.

  (c) Clinical factors such as oxygen, carbon dioxide, acidity, salinity, availability of inorganic nutrients and so on.

(ii) **Living or biotic factors** such as microbes, plants, animals and all living organisms and their organic by-products.

Man-made Environment

With the development of science and technology, human beings have begun to alter the environment to suit their requirements. This has led to the evolution of a man-made environment. Hence, the environment—which earlier comprised just air, land, and water—now also includes crop fields, urban areas, industrial space, vehicles, power plants, telecommunications, and much more.

The basic needs of human beings are shelter, followed by potable water and sanitation. The houses of people in the city are made of brick and cement and not of mud with a thatched roof. The resources for urban housing are transported from rural areas in cars, buses, trucks and trains, which consume a large amount of energy and pollute the atmosphere. The ever-increasing demand for comfort has resulted in the migration of people from villages to urban areas. Urban areas, on the other hand, are unable to meet the demands of basic civic amenities. As a result, they are becoming hovels of dirt, disease and crime. This has resulted in the paradox of concrete skyscrapers coexisting with slums and the atmosphere being polluted with exhaust from traffic, factories and domestic smoke.

Figure 1.1 Park: Man-made Environment

Social Environment

Human beings are social animals. This is why the socio-cultural environment plays an important role in their lives. The social environment is formed by the network of social institutions, which include political, religious and economic institutions. Family is one of the basic institutions of the social environment. It is here that human beings perform various activities, including socialization of children, and the transference of cultural heritage and morals from one generation to the next. Groups of families form communities which are classified according to their occupation, religious faith, and other parameters.

Environment of the Earth

The environment of the earth has been studied with various modern and scientific instruments, such as satellites, rockets, and balloons. The results indicate that the environment of our planet comprises mainly of three segments. These are:

(i) Air or atmosphere
(ii) Water or hydrosphere
(iii) Land or lithosphere

Atmosphere: The earth's atmosphere is an envelope of gases extending up to 2000 feet above the ground level. The gases include nitrogen, oxygen, argon, carbon dioxide, traces of carbon monoxide, oxides of nitrogen, sulphur and hydrocarbon, and very little amount of water vapour. The concentration of these gases decreases with an increase in altitude. The bulk of these gases are present within the atmospheric band that stretches up to 5 km above the earth.

The atmosphere protects the earth's biosphere by absorbing a major portion of the electromagnetic radiation and most of the cosmic rays. The atmosphere also absorbs infra-red radiation and thereby maintains the temperature of the earth at life sustaining levels. It also helps nature in maintaining its balance through different biochemical cycles, namely the oxygen cycle, nitrogen cycle, carbon cycle, and hydrological cycle.

However, scientific advancements of the modern man are polluting this protective blanket by dumping waste materials like carbon emissions and smoke into the atmosphere.

Layers of the atmosphere: The earth's atmosphere is broadly divided into five regions:

Troposphere: The lower portion of the atmosphere is called troposphere. It contains 70 per cent of the atmosphere's mass. The density of the troposphere decreases with altitude. The air near ground level is heated by the radiation from the earth, but the temperature decreases uniformly with altitude. This decrease of temperature with altitude is known as lapse rate. The cold layer (56°C) at the top of the troposphere, which shows a temperature inversion, that is, a negative to positive lapse rate, is known as tropopause. The global energy flow, resulting from the difference in heating and cooling rates between the equator and the poles, makes the troposphere a turbulent region.

Stratosphere: Above the troposphere, the quiescent layer with a positive lapse rate is known as the stratosphere. Very little water vapour is found here. The ozone molecule, present in the layer, absorbs the Sun's ultraviolet radiation, and decomposes into oxygen molecules and an oxygen atom. When these particles combine, energy is released as heat radiation which causes a positive lapse rate.

$$O_3 \underset{\text{Heat released}}{\overset{\text{UV radiation } 220\text{–}330 \text{ nm}}{\rightleftharpoons}} O_2 + O$$

The stratosphere not only shields life on earth from the injurious effects of the Sun's ultraviolet rays, it also supplies heat for separating the quiescent stratosphere from the turbulent troposphere. The stratopause separates the stratosphere from the mesosphere.

Mesosphere: In the mesosphere, the lapse rate is negative again due to low levels of ozone that absorbs ultraviolet radiation. The mesopause separates the mesosphere from the thermosphere.

Thermosphere: In thermosphere, the positive lapse rate raises the temperature to a maximum of about 1200°C. Hence, atmospheric gases such as oxygen and nitric oxide split into atoms, which absorb solar radiation in the far ultraviolet region and then undergo ionization. That is why this layer is called ionosphere.

Exosphere: The uppermost layer of the atmosphere is called the exosphere. This extends up to a height of about 1600 km and gives way to interplanetary space. It is extremely rarefied. The upper layers of the atmosphere are continuously pressing down on the lower ones. Hence, the density of the lower layers is higher and it decreases as we move upwards.

Having described the layers of the atmosphere, now we shall deal with the hydrosphere which forms the second of the three main segments of the earth's environment.

Hydrosphere: All types of water resources, namely the oceans, seas, rivers, lakes, ponds, polar ice caps, streams, glaciers, ground water, and water vapour are collectively known as the hydrosphere. Water being the elixir of life, all ancient civilizations were linked to major sources of water, be it the Egyptian Civilization along the River Nile, the Indus Valley Civilization along the River Indus, or the Mesopotamian Civilization between the Tigris and Euphrates rivers.

The hydrosphere is an important part of the earth's surface. About 70 per cent of the earth's surface is covered with water. The northern hemisphere is dominated by land surface, while the southern hemisphere is almost entirely occupied by water bodies (oceans).

Water is the most essential component of life for all living organisms. The hydrosphere is of immense importance to mankind. It maintains the availability of fresh water to the biosphere through the hydrological cycle. A major component of the hydrological cycle is the ocean. The oceans are great reservoirs of water and they also regulate carbon dioxide. The oceans can absorb more carbon dioxide than the atmosphere. Oceans are also the storehouses of vast resources, such as, water, salt, minerals, and food. The oceans are the largest sinks (pollutant receptor) of the planet. Thus, the role of the hydrosphere is critical to the sustenance of life on the earth. This is underlined by the fact that life on the earth originated under marine conditions.

Lithosphere: The lithosphere is the outermost mantle of the rocks constituting the earth's crust. Rocks are subjected to continuous physical, chemical and biological (attack by lichens) weathering. Plants grow and decay on the soil covering the rocks. Soil is the major component of the lithosphere. The organic matter in soil is decomposed by micro-organisms, thus forming biomass. This biomass is mixed with the soil fauna. The major components of soil are air, water, minerals, and inorganic matter obtained from weathering of the parent rock. Organic matter of soil comprises plant biomass that is in various stages of decay. It also includes a high population of bacteria, fungi and animals such as nematodes, micro arthropods, termites and earthworms. Soil plays a vital role in supplying nutrients to the plant kingdom. The nutrient supply power of soil is a measure of its fertility, while the productivity of the soil is a function of crop and animal biomass per unit area. Thus, the yield of crop depends solely on soil and crop management strategies. Therefore, this dynamic balance between the soil and the crop needs to be preserved to maintain the interrelationship between the two.

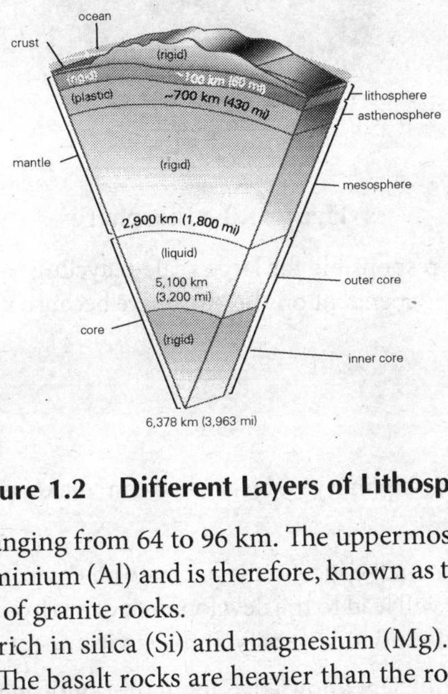

Figure 1.2 Different Layers of Lithosphere

The lithosphere has a thickness ranging from 64 to 96 km. The uppermost part of the lithosphere (the earth's crust) is rich in silica (Si) and aluminium (Al) and is therefore, known as the SiAl layer. The continents belong to the SiAl layer and are made up of granite rocks.

The zone next to the SiAl is rich in silica (Si) and magnesium (Mg). This layer is formed of basalt rocks and constitutes the ocean floors. The basalt rocks are heavier than the rocks formed by SiAl layer. Below the SiMg layer, the density of the layers increases with depth. Such differences in density cause the constituting layers to float, one over the other. The continents are basically large segments or 'plates' of the earth's crust floating on top of this heavier layer. These floating plates are responsible for the tectonic movement of the earth's surface during an earthquake.

Below the lithosphere lies the mantle, which has a thickness of about 2400 km. The upper part of the mantle is known as the Asthenosphere, while the lower mantle is called the Mesosphere. The interior-most part of the earth is called the Core, which consists of minerals such as iron, nickel, cobalt mixed with sulphur, and silica. The thickness of the core extends to about 3500 km. The Core consists of the outer core and the inner core. The inner core appears to be solid, while the outer core is molten and metallic. The temperature of the core ranges between 5000 and 5500°C.

The direct interaction between the atmosphere, hydrosphere and lithosphere for millions of years has made the earth suitable for life and has formed the biosphere.

Biosphere: Life on earth occupies a 'thin skin' extending more than a few kilometres below and above its surface. This is commonly known as the biosphere. Both the biosphere and environment influence each other a lot. The oxygen and carbon dioxide levels of the atmosphere depend entirely on the plant kingdom. All the different biogeochemical cycles are essential for the continuous circulation of constituents necessary for supporting life. This is possible due to the interaction of the biosphere and the environment

It is in the biosphere that radiant energy is converted to chemical form (carbohydrates) through the process of photosynthesis. Only then does energy transfer take place from chemical to mechanical, and heat forms during cellular metabolism.

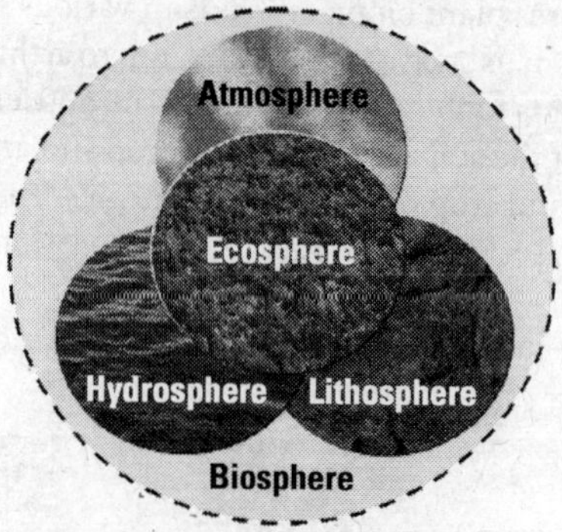

Figure 1.3 Biosphere

Thus, it is the biosphere which is responsible for large scale recycling of matter and energy. Even today, the existence of life on earth is closely dependent on the biosphere because it constitutes an essential life support system for all living beings.

SUMMARY

Environmental study is a multidisciplinary subject that can make us conscious and caring about our environment.

The primary objective of environmental education is to create an environment-friendly mindset and attitude among all sections of society. This will lead to the development of environmental ethics and increase the value people place on conservation of life and biodiversity in the environment.

Environmental studies provide sufficient knowledge about the philosophy, genesis and consequences of local and global environmental problems and the necessary knowledge for their abatement and control.

All religions preach environmental protection.

The Government of India enacted a comprehensive 'Environment (Protection) Act, 1986' to safeguard and protect our environment.

This shows increased environmental awareness on part of both the government and the people. This was also reflected at the Earth Summit at Rio de Janeiro in June 1972, where Agenda-21, a Global Action Plan was adopted with the aim of integrating environmental imperatives with developmental aspirations.

Environmental protection requires both preventive and curative measures.

ESSAY TYPE QUESTIONS

1. Write a note on the fundamentals of the environment.
2. Why are environmental studies necessary for all?
3. How will environmental awareness support sustainable development of the environment?
4. Why is environmental awareness mandatory for all the people in the society?
5. Write a short note on the environment of the earth.
6. Define atmosphere with special emphasis on its layers.
7. What is the Lapse Rate? Describe its nature in different layers of the atmosphere, giving reasons.

SHORT-ANSWER TYPE QUESTIONS

1. Define the term 'environment'.
2. What are the different components of the environment?
3. Write three major objectives of environmental studies.
4. Name the fundamental principles of the environment.
5. Why is social awareness about the environment necessary?
6. Why is the man-made environment different from the natural environment?
7. What is meant by hydrosphere?
8. Define the term 'lithosphere'.
9. Write a short note on biosphere.
10. Write a short note on the movement of tectonic plates.

MULTIPLE CHOICE QUESTIONS

1. Environmental Study is a subject that encompasses
   (a) all branches of science only.
   (b) only social science.
   (c) branches of science and social science.
   (d) all branches of science, arts and social science.
2. The unlimited exploitation of nature by human being resulted in
   (a) environmental pollution.
   (b) health problems.
   (c) decrease in biodiversity.
   (d) All of the above.

3. The primary objective of Environmental Study is to
   (a) generate an environment-friendly mindset among all classes of people.
   (b) enjoy economic growth at the cost of quality of human life.
   (c) realize the value of biodiversity by increased environmental ethics among all classes of people.
   (d) get a holistic view for the sustenance of life on the earth on an infinite scale.

4. The importance of environment
   (a) was only recognized in recent years.
   (b) was recognized in ancient time.
   (c) was not recognized in ancient time.
   (d) None of the above.

5. The conservation of natural resources
   (a) was not encouraged in ancient India.
   (b) was encouraged in ancient India.
   (c) is recently being used in India.
   (d) None of these.

6. Life on earth constantly interacts with its surroundings called environments which essentially consists of
   (a) biotic components only.
   (b) only abiotic components.
   (c) both biotic and abiotic components.
   (d) None of these.

7. The conservation of environment requires
   (a) maintenance of biodiversity.
   (b) maintenance of gaseous and material cycles.
   (c) maintenance of ecological order and natural balance.
   (d) All of these.

8. Biosphere is made up of
   (a) atmosphere and lithosphere.
   (b) hydrosphere and atmosphere.
   (c) lithosphere and hydrosphere.
   (d) lithosphere, hydrosphere and atmosphere.

9. The segment/segments of the environment, responsible for conservation of radiant energy to chemical energy (carbohydrate) through photosynthesis is/are
   (a) hydrosphere.
   (b) lithosphere.
   (c) atmosphere and lithosphere.
   (d) biosphere.

10. The largest sink (pollutant receptor) of the planet is
    (a) atmosphere.
    (b) hydrosphere.
    (c) lithosphere.
    (d) biosphere.

11. Which of the following components of the earth has least pollutant-receptor capability?
    (a) Atmosphere.
    (b) Hydrosphere.
    (c) Lithosphere.
    (d) Biosphere.

12. Biosphere is the
    (a) envelope of gases surrounding the earth.
    (b) the outermost layer of earth's crust.
    (c) the thin layer of organic matter containing all living things on the surface of the earth.
    (d) None of the above.

13. The Earth Summit at Rio de Janeiro was held with the aim of integrating environmental imperatives with developmental aspirations in
    (a) June 1972.
    (b) December 1972.
    (c) December 1993.
    (d) June 1993.

14. The Global Action Plan adopted at Earth Summit held at Rio de Janeiro in June 1972 is also known as
    (a) Agenda 20.
    (b) Agenda 22.
    (c) Agenda 23.
    (d) Agenda 21.

15. Environmental degradation is due to
    (a) rapid industrialization.
    (b) onset of different developmental works.
    (c) fast urbanization leading to deforestation.
    (d) All of the above.

ANSWERS

1 (d)	2 (d)	3 (d)	4 (b)	5 (b)	6 (c)	7 (d)	8 (d)	9 (b)	10 (b)
11 (a)	12 (c)	13 (a)	14 (d)	15 (d)					

UNIT II
Natural Resources

LEARNING OBJECTIVES

After studying Unit II, students will be able to:

- Define natural resources and identify problems associated with the use and extraction of natural resources in the context of environmental sciences.
- Describe the various natural resources such as forest resources, water resources, mineral resources, food resources, energy resources and land resources.
- Identify the uses and overexploitation of different resources.
- Describe case studies that underline examples of damage and exploitation of natural resources and highlight people's struggles to protect their environmental resources.
- Explain different processes of resource extraction of water, mineral, food or forest etc.
- Identify alternate resources and processes that can be less damaging to the environment.
- Define the role of the individual in conservation of natural resources.
- Identify the equitable use of resources as a way towards a sustainable lifestyle.

2.1 Natural Resources

A resource can be defined as any material that can be transformed into a more valuable and useful product or service. Ever since his appearance on earth, man has been dependent on the resources that nature provides. While air, water, soil, minerals, wind, solar energy and so on are non-living or abiotic resources of nature, plants, animals and other organisms are biotic or living forms of natural resources.

Everything that nature provides has some utility for mankind but its utilization is possible only with the help of appropriate technology. For example, coal and mineral oil have been present beneath the earth's surface for centuries but earlier we had no technology to help us make proper use of them. Thus, naturally found materials can be converted into a natural resource only after a suitable technology is discovered for its utilization and conversion to a more valuable product. Therefore, depending on the origin, natural resources are either biotic or abiotic. Similarly, based on availability and utility, natural resources may be classified into renewable resources and non-renewable resources.

(i) **Renewable Resources:** Resources which can be renewed even after use are known as renewable resources. For example, forest cover, solar energy and hydroelectricity are all renewable resources. Since these can be easily replaced they can be termed renewable or non-exhaustible resources.

(ii) **Non-Renewable Resources:** There are some resources in nature that have taken thousands of years in combination with highly favourable climatic conditions to form. If exploited recklessly they will be exhausted. Such resources that take too long to form and cannot be replaced easily are called exhaustible or non-renewable resources. Fossil fuels, minerals and so on come under this category.

Apart from these two types of resources, there are some natural resources that can be used continuously.

Water is one such natural resource. After using it for domestic, industrial and other purposes, water can be recycled and used repeatedly.

2.1.1 Natural Resources and Associated Problems

Due to the continuous increase in population, demand for natural resources has also increased. This is catalysed by scientific progress followed by technological advancement for the utilization of natural resources. These two factors are responsible for the overutilization of natural resources, particularly non-renewable resources.

Overexploitation of non-renewable resources will not only lead to economic imbalance between developed and developing countries, but also to environmental and ecological imbalance between nature and population growth. Today, developed countries are consuming up to 50 per cent more natural resources than developing countries and they are producing more than 75 per cent of global waste and Greenhouse gases. Overutilization of underground water has resulted in the depletion of the groundwater table and the drying of streams and rivers and that is why conservation of natural resources is very important for the sustenance of life on earth. The health of the ecosystem is an indicator of the quality of human life. For the maintenance of a quality ecosystem, it is necessary to have:

(i) control over the use of natural resources;
(ii) protection of the environment from pollution;
(iii) conservation of biodiversity and
(iv) controlled growth of human population.

2.2 Forest Resources

Forests are an important renewable natural resource. In fact, a major part of the earth's lithosphere is covered with forests. Forests which can be considered the centers of biodiversity are complex and they constantly change their environment to house wildlife, trees, shrubs, fauna and flora, microscopic soil organisms and much more. They are not only valuable from the economic, historical, cultural, recreational, aesthetic and religious points of view, but their resources too, are extremely important for mankind. Wood is still a source of fuel for one-third of the human population. While tribals living in and around forests directly depend on forests for their lives and livelihood, the remaining human population also indirectly depends on forests.

Houses, furniture, paper, clothes, dyes, gums, resins and lac are largely made from wood and other forest products. In fact, the photosynthesis of plants regulates the amount of the oxygen we inhale and the carbon dioxide we exhale into the atmosphere.

Harvesting, collecting and processing important forest products like timber generate employment. So does ecotourism including bird watching, outdoor adventures, nature study activities, hiking, camping and so on. Some of the uses of forest resources are listed in Table 2.1.

Table 2.1 Useful Functions of Forest Resources

Function	Types of Benefits
Centre of biodiversity	• Forests provide home to unique plants, animals and wildlife.
Watershed protection	• Forests reduce the pace of surface run-off of water and thus encourage its percolation. • They stop soil erosion and prevent flash floods. • They help in preventing droughts through prolonged, gradual run-off.

Function	Types of Benefits
Clean water	• Trees cleanse the ground because their root system filters water and clears toxins and impurities. • Trees facilitate storing of clean water and maintain the availability of water during summer, when it is most needed.
Land erosion control	• Forests hold soil by preventing rain from directly washing away soil. • They reinforce soil to avoid landslides in mountainous areas. • They maintain soil nutrients and structures.
Clean air	• Trees absorb solar energy and cool and refresh the air we breathe. • Theymaintain local climatic conditions. • They absorb carbon dioxide and harmful pollutants and release oxygen into the atmosphere.
Economic benefits	• Timber production and other wood-based industries constitute an important part of the national economy. • Forests supply wood for fuel. • They supply fodder for cattle. • Trees provide fibres for weaving baskets, ropes, nets, strings and so on. • Sericulture for silk, apiculture for honey and pollinating crops, medicinal plants for medicines are economically beneficial. • Forests provide shelter for tribal people and building materials for others. • They provide foods, fruits, nuts, flowers, fish and meat. • Forests facilitate ecotourism.
Environment benefits	• Trees provide clean air. • They provide clean water. • They prevent global climate change through absorption of carbon dioxide, a leading Greenhouse gas, to produce wood and leaf matter, known as carbon sequestration. • They help in controlling soil erosion. • They absorb noise and reduce stress. • They provide an aesthetic place for mental peace and healing qualities. • They help in controlling climate and heat island effects resulting from city environments. • Forests also help in global recycling of water, carbon and nitrogen.

The overexploitation of forest resources has resulted in a serious threat to mankind. This was the cause of heated discussions at the Earth Summit in June 1972 at Rio de Janeiro on the topic of global transition to sustainable forest management. Forest cover all over the world is depleting fast, thus endangering rare varieties of plants, wildlife and other natural resources. India itself is losing about 15 lakh hectares of good forest land annually, which is estimated to be equivalent to the country's total consumption of oil, coal and electricity.

2.3 Deforestation

A demographic explosion has resulted in a steadily increasing demand for food and fodder for livestock, firewood and industrial raw materials. Croplands are being increasingly used for non-agricultural purposes due to rapid urbanization and industrialization; new croplands are being created by clearing forests. This wanton cutting of trees including lopping, felling, removal of forest litter, browsing, grazing and so on is known as deforestation.

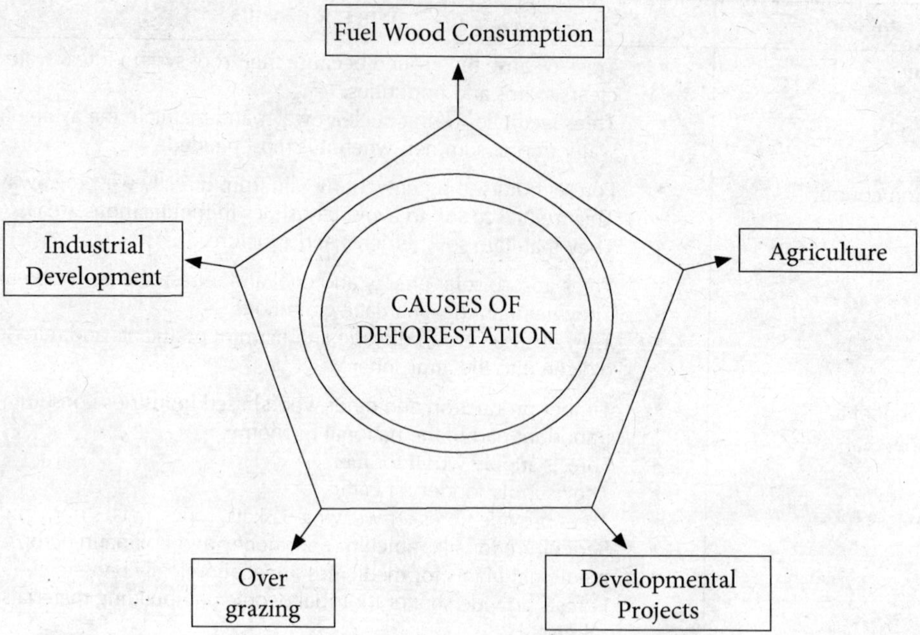

Figure 2.1 Causes of Deforestation

The main causes of deforestation (as shown in Fig. 2.1) are:

(i) Fuel wood consumption;

(ii) Agriculture;

(iii) Industrial development;

(iv) Overgrazing and

(v) Developmental projects.

The causes of deforestation are fairly complex and range from a competitive global economy, compelling international debt payment by developing countries, population growth and consequent increase in demands of agriculture to commercial logging, a shift to a cash crop economy and developmental projects like dams and mines.

Around the world, most of the forests are cleared for grazing cattle, planting crops and so on. Poor farmers cut down village common lands and forests, burn down trees to make the land fit for agriculture. Commercial logging of timber also enables the local tribals to enter the forests thus causing further deforestation.

In developing countries, cash crops like rubber, tobacco, oil and palm need marginal lands which results in the clearing of rain forests or dense forests. Mining, industrialization, dams and hydroelectric projects also cause deforestation. The fact is that rich mineral resources are available mostly in forest areas. Similarly, steep embankments of river valleys suitable for dams, hydro and irrigation projects are usually fertile lands sustaining dense forests and biodiversity. Therefore, the construction of dams for hydroelectric projects or irrigation usually submerges forests, displacing tribals from their natural habitat.

The massive cutting of trees for firewood, developmental projects, raw materials for industry and/or other various reasons have resulted in the shrinking of forest cover thereby adversely affecting precious wildlife. A recent study shows that 600 different species of animals have become extinct on earth over the last 2,000 years.

2.3.1 Timber Extraction

Timber extraction results in deforestation and in the fragmentation of the last remaining forests. It harms valuable species of trees, birds and wild animals. In spite of this, it is sometimes necessary to extract timber, so as to meet the needs of a developing country. During the extraction of timber, cutting, felling and handling should be done selectively, carefully and in a planned manner, in order to save the remaining forests and biodiversity. The United Nations Food and Agriculture Organization (UNFAO) has framed certain guidelines for the same.

Timber extraction can be classified as follows:

(i) Clear felling;
(ii) Selective logging;
(iii) Mechanized logging;
(iv) Handlogging and
(v) Reduced-impact logging.

Clear felling generally means the complete destruction of native forest, modifying it by harvesting commercial trees to create an even aged group and removing non-commercial trees, if required. Now, industrial timber logging is being done through clear felling all over the world.

In selective logging, only large individual trees of a few economically marketable species are harvested. The other trees are left untouched till the next harvesting. Although in selective logging only three to 10 of the tallest trees are targeted per hectare, the damage done may be as high as 50 per cent of the total forest areas because of the need to create access routes, dragging cut trees on the forest floor and lack of planning.

In mechanized logging, heavy machineries are used to pull, lift and transport the trees. This process can be used in clear felling or selective logging operations.

Local people use handlogging for non-commercial felling of timber or clearing of forests for agriculture. This labour-intensive, non-mechanized means involves the felling of trees by hand-held chain saws and then transporting the logs manually. This method is used in peat swamp forests which are regularly water logged and where heavy machinery movement is not possible.

Reduced-impact logging is now a common feature in industrialized nations where environmental damage can be minimized through the selection of site-sensitive techniques of harvesting and logging.

2.3.2 Dams and their Effects on Forests and Tribal People

Post–Independence, a series of large dams were planned and built on some of the major rivers in India. In fact, India is one of the largest dam-building nations in the world, with as many as 4,291 dams either built or under construction. Though the estimates vary significantly the fact is that dams are the single largest cause of human displacement in India and account for 75 per cent to 80 per cent of displacement of about four to five crore people.

The impact of large dams on forests and on the lifestyle and identity of tribal people is extremely high. Almost 40 per cent of those displaced by dams belong to scheduled tribes and 20 per cent to schedule castes. Only 25 per cent of the displaced people have been rehabilitated so far.

Of the 32 dams of a height of more than 50 m that were completed between 1951 and 1970, only nine (or less than 30 per cent) were in tribal areas. That figure doubled between 1971 and 1990, a period when another 85 dams of a height of more than 50 m were under construction. These were taller and more sophisticated than the earlier dams and about 60 per cent of these dams were in tribal areas. A recent government report based on a study of 110 projects stated that more than 50 per cent of the total 1.69 million people displaced by these projects were tribals. This means that the tribal communities which account for just 8 per cent of

India's total population constitute about 40 per cent of the displaced persons. About 92 per cent of the tribal people in India live in rural areas which are dry, forested or hilly. Most of them depend on agriculture and minor forest produce for sustenance. These largely self-sufficient tribal communities live in close proximity to forests, rivers and mountains. Since these areas are rich in natural resources they are most likely to be developed for dams, mines, industries and so on.

Immediately after Independence, only a few dams were built in tribal areas. However, by the 1970s, when the resources in more accessible areas were exhausted, more dams were planned in tribal areas thereby displacing a large number of tribal people.

Tribals are socially, economically and politically the weakest and the most deprived community in India. They have been evicted from their ancestral homes and are either forced to migrate to urban slums in search of employment or become landless labourers in rural areas to pay the price of 'development'. Unfortunately, tribal people hardly get to share the benefits of development projects that cause their displacement. They are always forced to live without the basic amenities like roads, electricity, transport, communication, healthcare, drinking water or sanitation. On the contrary, a majority of them end up with less income than before, less work opportunities, inferior houses, less access to the resources of the common people such as fuel wood and fodder, poor nutrition and poor physical and mental health. Developmental projects have invariably led to the dispersal of communities, the breakdown of traditional support systems and the devaluation of their cultural identity. Therefore, the government should devise a strategy to minimize tribal displacement. It must ensure 100 per cent rehabilitation and make sure that the fruits of development are shared with the dispersed people as well. It must augment the rehabilitation of the displaced persons of previous projects, protect the customary rights of the tribal people over natural resources and take their opinions into consideration for future projects.

CASE STUDY

***Chipko* Movement**: The world famous *Chipko* Movement, pioneered by *Dasohli Gram Swarajya Mandal* in Gopeshwar brought about a general awareness about conservation of forests. The first *Chipko* Movement dates back to 1731, when a village woman named Amrita Bai led the *Bishnoi* women against the *Maharaja's* men to prevent them from cutting trees. In this attempt to save the trees, she sacrificed her life along with the lives of her husband, three daughters and 363 people. The movement was given this name because the village women embraced or hugged the trees to stop them from being cut. In 1972, in Uttar Pradesh, the *Chipko* Movement was led by Bachnoi Devi of Advani who protected the hill forests from the contractor's axe men.

2.4 Water Resources

Water is the most important factor of life. History reveals that life began with water. Human civilization too, flourished wherever water was available in abundance. In fact, several cities and civilizations disappeared due to the shortage of water.

2.4.1 Hydrosphere—As a Source of Water on Earth

The hydrosphere includes all sources of water such as oceans, lakes, rivers and underground water. Earth is known as the 'watery planet' because it is the only planet in the solar system with an abundant source of water. About 97 per cent of the earth's total water supply lies in the oceans although it is unsuitable for human consumption due to its saline content. About 2 per cent of the earth's water that is in the polar ice caps is frozen. The remaining 1 per cent of the earth's water is available as fresh water flowing in rivers, lakes, streams and groundwater which is suitable for human consumption.

The composition of surface water differs widely from groundwater. Surface water contains a lot of organic matter and mineral nutrients, whereas groundwater contains dissolved minerals. When water seeps into the ground, a number of micro-organisms originally present in the surface water get depleted. Hence, groundwater is the most suitable domestic source of water.

Table 2.2 shows proportions of different types of water in the hydrosphere

Table 2.2 Proportion of Different Types of Water

Form	Relative proportion (%)
Water in the oceans	97.390
Water in the rivers and lakes	00.020
Water in the atmosphere as vapour	00.001
Water in glaciers and polar ice caps	02.010
Underground water and soil moisture	00.580
Total hydrospheric fresh water	02.601

Water is also the most important component of protoplasm. In different metabolic processes, water is the only source of hydrogen and one among the several sources of oxygen. All living organisms draw water from the hydrosphere. Water consumed during different metabolic processes by organisms is partly given back into the environment in different ways. For instance, solar energy causes water to evaporate from different sources. These vapours cool down after gaining altitude and condense to form clouds which then precipitate as rain or snow and thereby return to the parent water system – the hydrosphere.

Water is an indispensable ecological factor for both plant and animal life. It is abundantly distributed in all segments of the environment, namely atmosphere, hydrosphere and lithosphere. In the atmosphere, water is present both as invisible and visible water vapour. Humidity is a form of water vapour present in the atmosphere which can be felt but not seen. However, clouds or fog are visible forms of atmospheric water vapour.

In soil water exists as gravitational water, capillary water, hygroscopic water, combined water and water vapour. When water moves downwards through moist soil under the force of gravity till it reaches a deep saturated zone of soil, it is called gravitational water. The upper surface of this is called the water table and it is available to the plant kingdom only after frequent showers. However, this pool of water sometimes gushes out onto the earth's surface through any weak point, when it is known as spring water. This water can also be made available artificially by digging wells and through lift irrigation. This water is generally free from impurities and may even contain dissolved minerals.

Wells are a major source of drinking water in villages and often well water has medicinal values too. After the draining of gravitational water, the water that is left over around the soil particles is known as capillary water. The sun dries this capillary water. The water that remains in the soil after the drying of capillary water is known as hygroscopic water. This water cannot be usually dried completely by the sun. Only if the soil is heated to about 105°C the hygroscopic water gets lost.

After the soil loses its hygroscopic water, the remaining water is present in the soil as oxides of iron, silicon and aluminium and is therefore called combined water. Both hygroscopic and combined water are of no use to plants. On the contrary, it is water vapour or moisture present in the soil particles that helps in the germination of seeds and the growth of the root and so on.

2.4.2 Use and Overexploitation of Surface and Groundwater

As discussed earlier, the need for water, land, and energy has increased enormously with the increase in population. The demand for more land and energy has resulted in deforestation and the destruction of wetlands and flood plains. The demand for more water has forced human beings to use and overuse the natural water storage systems (aquifers). This has resulted in the destruction of nature's ecosystem. The water table has also receded because of overuse. Further, the loss of vegetative cover is not only a deterrent in the filling of these natural reservoirs but also causes floods and droughts.

The abundant availability of water has led to the overuse and abuse of water in every area of our lives. Water is overused in household chores during brushing, bathing, washing and cleaning. Water is overused in agricultural fields too. One can get the same or even more yield with less amount of water by using the drip irrigation method. Water is overused by industries too. The situation worsens when untreated water is released by these industries into nearby water bodies, leading to water pollution.

Dams and irrigation canals have been constructed on rivers for the production of hydroelectricity and year-round supply of water. Since these dams and irrigation canals alter the flow of the river, they are the cause of overexploitation of water. The large landholders get the maximum share of water from canals, while, unfortunately the conditions of the poor, weak cultivators remain the same.

Water is wasted in urban India too. For instance, water can often be seen flowing incessantly from municipal taps that are left open on the roadside or from pipes that leak. Although water is overused, the cost of water has remained constant.

2.4.3 Floods

A river overflowing its banks has always been a great threat to human civilization. Ancient civilizations have flourished along the banks of all of the world's greatest rivers. As population density has increased, the need for more land has risen. To create more space, the catchment areas are deforested and wetlands filled up. The advancement of science and technology has strangled the rivers with bunds, dams and so on. As a result, the wetlands of the flood plains which were natural flood controllers have been lost. No longer can they act as sponges to hold the excess water from the rivers. Deforestation of hills and mountains, which are the sources of rivers, has added to the problem. Water does percolate underground but due to deforestation it runs downward, carrying with it a great load of top soil. This not only washes away the fertile top soil but temporarily blocks the rivers only to suddenly allow an enormous amount of water to move downstream towards the plains either through the same path or a changed path. In both the cases, the river swells and due to the lack of floodplains and wetlands the water overflows, submerging cities, towns, villages, agricultural lands and much more. A detailed report on floods and their mitigation has been included in Unit-V.

2.4.4 Drought

Drought which is mainly the consequence of climatic conditions that lead to a serious scarcity of water is one of the major problems in India. It occurs due to the failure of more than one successive monsoon in the area. Usually arid and semi-arid regions are drought-prone and they have very little or no vegetative cover which explains the condition of groundwater in the area. During a drought, water scarcity becomes so acute that there is no water for farms, industries, households or even drinking. Moreover, the intensity of the drought depends on how long the drought lasts in that area. A prolonged drought may result in famine.

Thus, droughts occur mainly due to the failure of the monsoon, supplemented by the non-availability of groundwater due to a receding water table. Although the failure of the monsoon cannot be prevented the intensity of a drought can be reduced with the help of proper watershed management.

One of the main causes of water scarcity during a drought is deforestation. Due to the wanton cutting of trees in forests and hill slopes, rain water cannot percolate the subsoil and recharge the natural aquifers. However, afforestation can recharge the natural aquifers.

We also need to control the overuse of water. During a drought, this stored water can be used by the animal and plant kingdom to reduce the effects of drought. In a water-scarce region, the financially stronger section of society draws groundwater with the aid of pumps. As a result, members of the poorer and weaker sections are deprived of their supply of adequate water. Moreover, the release of industrial waste water into nearby water bodies makes the water unfit for human consumption besides turning it into a death trap for all aquatic life. Hence, strict measures have to be taken to compel industries to release liquid waste into water bodies only after adequate treatment to prevent pollution.

2.4.5 Conflicts over Water

Nearly three-fourths of the earth's surface is covered with water. There is abundant sea water on earth but it is not fit for human consumption, agriculture or manufacturing industries. Fresh water is essential to sustain life, development and the environment. It has been estimated that 1 billion people do not have access to clean water and 1.7 billion do not have access to sanitation.

As water becomes increasingly scarce, national conflicts over water have risen. Industry, agriculture and citizens compete with each other for the water they require for sustenance and development. National tension over the distribution of water can quickly escalate into discord between groups dependant on a shared resource. Over 200 water bodies are shared by two or more countries or areas. There are water disputes in the Middle East, eastern Europe and South East Asia. In the Indian subcontinent, the Indus Water Dispute between India and Pakistan, the Ganga river controversy between India and Bangladesh and the inter-state water disputes, namely the Cauvery Water Dispute, Yamuna Water Dispute, Krishna–Godavari Water Dispute, Ravi–Beas Water Dispute and Bansadhara Water Dispute are some major ones.

Indus Water Dispute between India and Pakistan: The dispute started with the Partition in 1947 when the Indus water basin was divided between India and Pakistan. Pakistan disputed India's share of irrigation water from the Indus river. The dispute ended in 1960 when the Indus Water Agreement was signed by the two countries after 12 years of World Bank-led negotiations.

Ganga Water Dispute between India and Bangladesh: This dispute too, started with the Partition in 1947 between India and East Pakistan (present Bangladesh). In 1962, India built and operated the Farakka Barrage 40 km away from the Indo-Bangladesh border for navigability of the Calcutta Port. This resulted in the escalation of the dispute regarding allocation of the flow of the Ganga River and its tributaries between India and Bangladesh and in the development of a rational plan for an integrated watershed development, including supplementation of the flow of the Ganga. During the long conflict, short term agreements were reached in 1977, 1982, and 1985 that temporarily settled the dispute between 1977 and 1982, and 1982 to 1984 and 1985 to 1988. In December 1996, a 30-year treaty was signed between the two riparian based on the 1985 accord to help reduce the regional tensions.

Cauvery Water Dispute: The Cauvery river is one of the most contentious sources of water in Southern India. Its watershed is divided between Karnataka (former princely state of Mysore) and Tamil Nadu (former Madras Presidency). Although Tamil Nadu does not control any of the Cauvery head waters, it is in possession of two of its tributaries, Bhavana and Moyar. Similar to the other divided watersheds (Mekong in Southeast Asia and Colorado in Western United States) there is peace in times of good rains.

The Cauvery Water Dispute has been a serious issue since 1974 when a 50-year-old agreement, signed in 1924 between Madras Presidency and Mysore State expired, compounding a century-old dispute over the vital interests of farmers in Tamil Nadu and Karnataka. According to Karnataka, the 1924 agreement entailed a discontinuation of the water supply to Tamil Nadu after 50 years. In 1991, the Supreme Court reassigned a tribunal to settle the dispute. The tribunal gave the decision that Karnataka must release 205 TMC of water from the Cauvery reservoirs to Tamil Nadu on a monthly basis. Karnataka declined to implement the order, arguing that if more than 100 TMC of water is released to Tamil Nadu, it would cause distress to its people.

Yamuna Water Dispute: The sharing of the waters of the Yamuna river between the three states of Uttar Pradesh, Delhi and Haryana comes with its own share of conflicts. The case was earlier referred to the Supreme Court for intervention and solution. When that failed, the recent political intervention resolved the issue. The Chief Ministers of the three states and the Central Government held conferences and resolved the issue.

Krishna–Godavari Water Dispute: This dispute was mainly about the inter-state utilization of untapped surplus water between the states of Maharashtra, Karnataka, Andhra Pradesh, Madhya Pradesh and Orissa. After failed negotiations, the dispute was resolved by a tribunal judge who ordered equitable apportioning of water. The Krishna Tribunal reached its decision in 1973 and the verdict was published in 1976, while the Godavari Tribunal commenced hearing in 1974 and gave its final verdict in 1979. Meanwhile, the states continued negotiations among themselves and reached agreements on all disputed issues and the tribunal only endorsed the same.

Ravi–Beas Water Dispute: The present Ravi–Beas Water Dispute between Punjab and Haryana started in 1966 when Haryana became a separate state with a part of east Punjab. Ravi, Beas, Sutlej and Yamuna flow through both the states and the two states are highly dependant on irrigated agriculture. The two states disputed over the allocation of water. This resulted in an agreement in 1976 which Punjab has been disputing till now. A new agreement was made in 1981. Later, a tribunal was formed in 1986 to resolve the issue. The verdict of the tribunal has not yet been published or implemented.

Dams: Benefits and Problems

With more than 4,200 dams, India is one of the largest dam-building nations of the world. India has invested enormous capital, effort and resources in building large dams and irrigation projects amounting to more than Rs 80,000 crore, which is the total irrigation budget in the 50 years since Independence.

Large dams are meant to solve problems of hunger and starvation by providing irrigation, boosting food production, controlling floods and providing much-needed electricity for industrial development. Dams are therefore one of the major causes of development in India, as they result in enormous benefits to society in the areas of economy, agriculture, food production, irrigation and water supply, electricity generation, industrial development and so on.

It is these benefits that prompted Pandit Jawaharlal Nehru, the country's first Prime Minister, to term dams as '**secular temples of modern India**.' Along with the stupendous benefits, dams also pose great problems to society. The problems that include environmental degradation, development-induced displacement and consequent aggravation of an already skewed social structure are massive. The utilitarian logic of 'few people having to sacrifice for the greater national good' for dam building has often grossly underestimated the environmental and social costs of dam building.

The impact of dams on forests and tribal communities has also been discussed in the section titled Forest Resources in this Unit. Dams are the single largest cause of 'displacement for development'. In fact, approximately 80 per cent of about five crore people, who have been estimated to be displaced in India since Independence for all development projects, are due to dams. Displacement caused by large dams has actually resulted in the transfer of resources from the weaker sections of the society to the more privileged ones. Large dams do little to change the existing social inequalities. On the contrary, they further aggravate the already skewed social structure. The largely self-sufficient, nature-dependant tribal societies generally live near forests, rivers, and mountains. These are resource-rich areas likely to be mined or turned into a dam. The impact of displacement of people affected by large dams has been overwhelmingly negative in India. The majority of displaced persons in most of the resettlement operations have ended up with lower incomes, less land than before, less work opportunities, inferior housing and less access to common resources such as fuel wood and fodder. They also suffer from lack of nutrition and poor physical and mental health.

Moreover, even a relatively liberal compensation and rehabilitation package cannot compensate for the loss of forests, rivers and ancestral land which is intricately woven into the social, cultural, religious and economic practices of a community.

After Independence, large dams were built as symbols of India's national development. All human and ecological costs were justified in its name. On this premise, people's movements such as the *Narmada Bachao Andolan* (NBA) have raised several fundamental questions such as: Who constitutes the nation? Who benefits from dams? Who pays the cost? Who decides and who obeys? Why? Can development that impoverishes the poor and the marginalized communities and destroys the ecological balance for the sake of the privileged classes be called development? Our society needs just and sustainable development, which in turn needs dams to be built selectively without inflicting massive human and social exploitation by the state.

Narmada Bachao Andolan (NBA)

The Narmada Valley Development Project is the single largest river development scheme in India. The Government of India plans to build some 3,200 dams on the Narmada River, 90 per cent of which flows through Madhya Pradesh, skirts the northern border of Maharashtra and flows through Gujarat prior to its confluence in the Arabian Sea. It is one of the largest hydroelectric projects in the world and has displaced approximately 1.5 million people from their homes in these three states.

Building a dam over the Narmada River will not only erode fertile agricultural soil due to continuous irrigation and salination, it will also make the soil toxic. Sardar Sarovar, the largest dam under construction, will flood more than 37,000 hectares of forest and agricultural land on completion and will displace about half a million people from their homes, land and livelihood. But the government's Resettlement and Rehabilitation (R&R) Programme will benefit only a certain percentage of the displaced persons. According to the World Commission on Dams, the impact assessment includes all people in the reservoir, upstream, downstream and in catchment areas whose properties, livelihood and non-material resources are affected and also those affected by the construction of dam-related infrastructure like canals and transmission lines. But the people affected by the canal system are not considered as Project Affected People (PAP).

Dissatisfied with the government's R&R Programme, the people's organizations of each state came forward in 1986 to form *Narmada Bachao Andolan* (Save the Narmada River Movement). In 1988, they called for a stop to all work of the Narmada Valley Development Projects. In September 1989, more than 50,000 people from across India gathered in the valley and pledged to fight against this destructive development programme. A year later again, thousands of villagers congregated in a small town in Madhya Pradesh to reiterate their pledge to drown rather than move from their homes. Under this immense pressure, the World Bank withdrew its loan for the Sardar Sarovar Project in 1993, thereby endorsing all the main concerns raised by the NBA. But the Gujarat Government decided to raise $200 million and go ahead with the project.

In response to the continuous protests by the NBA, the Government formed a five-member team to review the project. But this team too, endorsed the World Bank view. Following a writ petition by the NBA in 1994 which called for a comprehensive review of the project, the Supreme Court of India stopped the construction of the Sardar Sarovar Dam in 1995. After this, the NBA shifted its attention to the other two big dams in Madhya Pradesh, the Narmada Sagar and Maheshwar. Although the height of these dams was less than their project-approved height, their impact on the environment was already apparent. However, in 1999, the Supreme Court ordered the height of those dams to be raised from 80 m to 88 m, followed by a judgment in October 2000 for immediate construction of the Sardar Sarovar Dam to a height of 90 m with permission to increase its height to the originally planned 138 m. These court decrees have come despite major unresolved issues on resettlement, the environment's biodiversity and the cost and benefit of the project. According to the government, the project will provide water to 20 to 40 million people, irrigate 1.8 to 1.9 million hectares of land and produce 1,450 MW of power. But the social, environmental and human costs of the projects were obviously not evaluated properly by the planners.

2.5 Mineral Resources

A mineral is a natural substance that forms in the earth's crust over a period of millions of years. It has a definite chemical composition and identifiable physical properties. On extraction from the earth's interior, with or without technological and economic benefits, minerals can be used as raw materials for industries or for domestic purposes. They are of the following categories:

(i) metals and metallic compounds like iron, aluminum, zinc, copper, manganese, limestone, gypsum and dolomite;

(ii) rare earth metals such as uranium and niobium;

(iii) non-metals such as silica;

(iv) building materials such as granite, marble and mica;

(v) gems such as diamonds, rubies and emeralds;

(vi) noble metals such as gold, silver and platinum and

(vii) fossil fuels such as oil, gas and coal.

Some of these minerals are ores. An ore is a mineral or combination of minerals from which a useful substance such as a metal can be extracted and used to manufacture a useful product. For example, hematite and magnetite are ores used for steel manufacture; bauxite is an ore used for aluminum extraction and rutile is used for the manufacture of titanium.

2.5.1 Mining

The process of extraction of these underground minerals is known as mining. Mining operations follow four phases:

(i) Prospecting: Searching for the mineral.

(ii) Exploration: Assessing the size, shape, locations, qualities and economic value of the deposits.

(iii) Development: Preparing for the extraction of the mineral

(iv) Exploitation: Extracting the mineral from the mines.

Two types of procedures are followed for mining: Opencast or open-pit mining and deep or shaft mining. Of the two, an appropriate method is selected to get maximum yield at minimum cost and one that provides least danger to the mining personnel.

Mine Safety: Open-pit (surface) mining is always less hazardous than underground or shaft mining and metal mining is safer than coal mining. Since these processes are used when mining is done on an industrial scale, it is undoubtedly a hazardous task. During the process, mining personnel are posed with dangers such as roof felling, insufficient ventilation, release of poisonous gases such as methane (specially in coal mines), flooding and radiation and also long term effects like lung diseases (black lung disease or pneumoconiosis) and bronchitis.

Environmental Problems: Mining also causes irreparable environmental damage. During the prospecting and exploration phases, modern sophisticated and heavy machineries are used by geologists, mining engineers, geochemists, geophysicists and others. Such activities increase access to remote forest areas and disturb the local habitat. The most significant degradation of the natural environment occurs during the next two phases of mining, that is, development, exploitation and also long after exploitation.
Some of the adverse effects of mining on forests and the environment are:

(i) Minerals are non-renewable natural resources. Their unplanned extraction and exploitation poses a threat to raw material resources for future generations.

(ii) Disposal of the waste which is generated during the process of mining is a problem as it causes land, soil and water pollution.

(iii) Land erosion which is the transportation of fine soil particles carried away by water and deposited in the tailing dam causes a problem as it leads to sedimentation.

(iv) Mining causes damage to forest cover as well as felling of trees.

(v) Dust generated during mining causes air pollution, which in turn causes respiratory problems and asphyxia of plants and trees.

(vi) A large quantity of timber is used in underground mining.

(vii) A lot of water is used in hydro-metallurgical the beneficiation of minerals and the discharge of the same, contaminates water.

(viii) Movement of heavy machinery used during mining and the process of blasting causes heavy noise pollution.

(ix) Mining causes large disturbances to the environment adversely affecting terrestrial habitats.

(x) Some of the adverse social impacts of mining include:

(a) Loss of land of local people.

(b) Impact on health.

(c) Destruction of forms of community subsistence and life.

(d) Alteration of social relationships and social disintegration.

Mineral Resources of India

India has a large number of economically useful minerals. About 25 per cent of the world's known mineral reserves are found in India. The availability of various ores and minerals in India is given in Table 2.3.

Table 2.3 Mineral Resources of India

Minerals	Main sources	Other available sources	Remarks
Iron ore	Orissa, Jharkhand	Madhya Pradesh, Maharashtra, Chhattisgarh, Karnataka, Goa, Tamil Nadu, Himachal Pradesh	2nd largest reserve after Brazil.
Coal	Orissa, Jharkhand, West Bengal,	Chhattisgarh, Maharashtra, Tamil Nadu	
Manganese	Madhya Pradesh, Maharashtra	Orissa, Jharkhand	2nd largest reserve after Russia.
Chromites	Orissa, Jharkhand,	Andhra Pradesh, Karnataka,	
Bauxite	Orissa, Jharkhand, Maharashtra	Tamil Nadu, Karnataka	India produces 75 per cent of the world's total production.
Mica	Bihar, Andhra Pradesh, Rajasthan		
Gypsum	Tamil Nadu, Rajasthan		
Nickel	Orissa		
Ilmenite	Kerala		

Minerals	Main sources	Other available sources	Remarks
Silmanite	Madhya Pradesh,Meghalaya		
Copper	Andhra Pradesh, Jharkhand, Rajasthan, Sikkim, Karnataka		
Gold	Andhra Pradesh, Karnataka		
Diamond	Madhya Pradesh,		
Uranium, rare earths	Jharkhand		
Petroleum	Assam, Gujarat, Maharashtra	Tripura, West Bengal, Punjab, Himachal Pradesh Andaman & Nicobar Islands	

CASE STUDY

1. Two-thirds of India's iron ore reserves lie in Orissa and Jharkhand. Due to the growth of the steel industry worldwide, many international steel players have been attracted to the iron ore reserves in these two states and have been coming to Orissa and Jharkhand to mine iron ore reserves and, if possible, to export the ore. They also sign memoranda of understanding, promising to set up steel plants in the states. For example, POSCO signed an agreement with the Orissa Government to set up a Rs 53,000 crore steel plant with a capacity of 12 million tonnes per annum. They want to export iron ore to Brazil also. Similarly Arcelor Mittal, the number one steel maker in the world and Tata also signed agreements with the Jharkhand Government to set up steel plants with a capacity of 10 million tonnes each per annum at an estimated investment of Rs 42,000 crore and Rs 39,000 crore, respectively.

   Arcelor Mittal had to withdraw their proposal of right to export iron ore before signing the agreement. If all these recent developments materialize, the states of Orissa and Jharkhand will be leaders in the industrial and economic map of India.

2. A classic example of the local population's right to mineral resources leading to the birth of a new state is Bougainville Island which is a part of Papua New Guinea. On May 17, 1990 Bougainville Island spread across an area of 10,000 sq km with a population of about 1, 50,000 people, declared itself an independent nation. The island was a part of Solomon Islands under British Rule till 1899. Then, it was a part of New Guinea under German rule till 1914. In 1947, Australian Papua merged with Guinea and came under Australian administration under the United Nations (UN) trusteeship. Papua New Guinea (PNG) became independent on September 6, 1975. Bougainville's CRA (Conzine Rio-Tinto Australia) copper mine started its operations in 1972 and accounted for 40 per cent of PNG's export and 20 per cent of government revenue.

   Bougainville Island leaders wanted independence from PNG since the 1960s and submitted their demand to the UN's Decolonization Committee. Displacement, environmental pollution and inadequate compensation made Bougainville Islanders fight against the PNG administration which ultimately culminated in the closure of the PANGUA mine in May 1989. After almost a decade of ethno-nationalist conflicts, which started with the Bougainville Revolutionary Army's demand for independence and PNG's resistance and oppression; Bougainville Island headed towards peace in 1997–1998 due to international intervention.

2.6 Food Resources

Food is essential for the survival of all living beings on earth. Equilibrium of the food chain is therefore necessary for the sustenance of the environment. More so because it is estimated that starvation and malnutrition take the lives of 18 million people per annum. Most of the victims are children and women and those who survive suffer from hunger and dietary deficiencies.

An assurance of a constant food supply or the ability of all people to earn their meals at all times is therefore absolutely essential for the maintenance of peace and harmony in the society. This depends on the following factors:

(i) Availability of enough food for all (sufficient food production for the total population).

(ii) Accessibility of food for all persons (buying power and freedom for every person to purchase food).

(iii) Adequacy of food for all (food utilized by everybody should meet the nutritional requirement).

2.6.1 World Food Problems

As the world population continues to rise, great pressure is being placed on arable land, water, energy and biological resources to provide an adequate supply of food while maintaining the integrity of our ecosystem. According to the World Bank and the United Nations, between one to two billion human beings on earth are malnourished, indicating a combination of insufficient food, low incomes and inadequate distribution of food. This is the largest number of hungry humans ever recorded in history. Based on current statistics, the world population is projected to double from roughly six billion to more than 12 billion in less than 50 years. Reports from the Food and Agricultural Organization (FAO) of the United Nations numerous other international organizations and scientific research also confirm the existence of a serious food problem owing to the phenomenal growth in population. For example, the per capita availability of food grains which make up 80 per cent of the world's food has been declining for the past 15 years. With a quarter million people being added to the world population each day, the need for grains and all other foods will certainly reach unprecedented levels.

More than 99 per cent of the world's food supply comes from land while less than 1 per cent comes from oceans and other aquatic habitats. The continued production of an adequate food supply is directly dependant on ample fertile land, fresh water, energy, along with the maintenance of biodiversity. As the human population grows, the requirement for these resources is also growing. Even if these resources do not get depleted on a per capita basis, they will decline significantly because they must be divided among more people.

At present, fertile cropland is getting lost at an alarming rate. For instance, nearly one-thirds of the world's cropland (1.5 billion hectares) has been abandoned during the past 40 years because erosion has made it unproductive (Pimentel et al., 1995). Solving the problem of erosion is difficult because it takes 500 years to form 25 mm of soil under agricultural conditions.

Eroded agricultural land is now being replaced by marginal lands and forestland. The pressing need for agricultural land accounts for 60 per cent to 80 per cent of the world's deforestation. Despite such land replacement strategies, per capita world cropland has been declining and is now only 0.27 hectare per capita; in China only 0.08 hectare is available at present. This is only 15 per cent of the 0.5 hectare per capita cropland considered minimal for a diet as diverse as that in the United States and Europe. The shortage of productive cropland combined with decreasing land productivity is partially the cause of the current food shortages and associated human malnutrition. Other factors such as political unrest, economic insecurity and unequal food distribution patterns also contribute to food shortages.

Competition for water resources among individuals, regions and countries and associated human activities are already occurring because of the current world population. About 40 per cent of the people live in regions that directly compete for shared water resources. Worldwide water shortages are reflected in the per capita decline in irrigation over the past 20 years. A major threat in maintaining future water supplies is the continuing over-draft of surface and groundwater resources.

Diseases associated with water, rob people of health, nutrients and livelihood. The problem is serious in developing countries. For example, about 90 per cent of the diseases occurring in developing countries result from the lack of clean water. Worldwide, about four billion people are contracting water-borne diseases and approximately six million deaths are being caused by the same each year. When a person is ill with diarrhoea, malaria or some other serious disease, anywhere from 5 to 20 per cent of an individual's food intake offsets the stress of the disease.

Disease and malnutrition problems in the Third World are as serious in rural areas as they are in urban areas, especially among the poor. This will intensify in the future. Furthermore, the number of people living in urban areas is doubling every 10 to 20 years, creating major environmental problems including water and air pollution, increase in the number of diseases, and food shortage.

Fossil energy is another prime resource used for food production. The intensive farming technologies of developed countries use massive amounts of fossil energy for fertilizers, pesticides, irrigation and for machines as a substitute for human labour.

In developing countries fossil energy has been used primarily for fertilizers and irrigation to help maintain yields. Economic analyses often overlook the biological and physical constraints that exist in all food production systems. The assumption is that market mechanisms and international trade are effective insurances against future food shortages. A rich economy is expected to guarantee a food supply adequate enough to meet a country's demand despite the existing local ecological constraints. In fact, it is the opposite. When global biological and physical limits to domestic food production are reached, food import will no longer be a viable option for any country. At that point, import of food for the rich can only be sustained by starvation of the powerless poor.

Improved technology will certainly assist in more effective management and use of resources but it cannot produce an unlimited flow of those vital natural resources that are the raw material for sustained agricultural production. For instance, fertilizers enhance the fertility of eroded soils but humans cannot make topsoil. However, fertilizers made from finite fossil fuels are presently being used to compensate for eroded topsoil.

Per capita fish catch has not increased even though the size and speed of fishing vessels have improved. Consider also the supplies of fresh water that are available not only for agriculture but also for industry and public use. No available technologies can double the flow of rivers, although effective water conservation would help. Similarly, the shrinking groundwater resources stored in vast aquifers cannot be refilled by human technology. Rainfall is the only source of supply.

Strategies for the future must be based first and foremost on the conservation and careful management of land, water, energy and biological resources needed for food production. Our usage of world resources must change and the basic needs of people must be balanced with those resources that sustain human life. The conservation of these resources will require coordinated efforts and incentives from individuals and countries. Once these finite resources are exhausted they cannot be replaced by human technology. Further, more efficient and environmentally sound agricultural technologies must be developed and put into practice to support the continued productivity of agriculture.

Yet none of these measures will be sufficient to ensure adequate food supplies for future generations unless the growth in human population is simultaneously curtailed.

2.6.2 Changes Caused by Agriculture and Overgrazing

Productive lands for sustainable agriculture have been systematically damaged by overexploitation, mainly through agriculture and overgrazing. The Human Development Report, 1998 states that nearly one-sixths of the world's productive land, amounting to two billion hectares has been degraded since 1945. The loss has been spectacular in developing countries where two-thirds of the world's poorest people live. Modern agriculture is mainly responsible for the pollution of land through unskilled irrigation, non-judicious use of chemical fertilizers and pesticides and the practice of shifting cultivation and so on. Unskilled irrigation causes serious problems of salination and water-logging. This converts healthy land into a wet desert. These days, inorganic fertilizers are being used on a large scale to increase crop yield. This increases contamination in run-off water and groundwater. Tribal people all over the world use the shifting-cultivation technique which involves slashing and burning forests to create cropland.

The increase in livestock population with a simultaneous decrease in grazing land has resulted in overexploitation of natural resources. The removal of vegetative cover on the soil and the gradual depletion of soil organisms turn productive land into wasteland.

2.6.3 Effects of Modern Agriculture

The effects of modern agriculture, as discussed in world food problems and elsewhere, are not very good. The adverse impacts are:

(i) Soil pollution;
(ii) Contamination of water;
(iii) Water scarcity;
(iv) Global climate change;
(v) Water-logging;
(vi) Soil salinity and
(vii) Loss of genetic diversity.

Soil pollution can be traced to wind or water erosion of exposed top soil, compacting of soil, depletion of organic matter in the soil, loss of water retention capacity, reduction in biological activity, salination of soil, accumulation of irrigation water in irrigated farming area due to poor absorption or poor drainage and desertification due to overgrazing.

The use of pesticides and chemical fertilizers results in water contamination. Water is becoming increasingly scarce due to its overuse for irrigation and increase in domestic and industrial requirements.

Deforestation and loss of vegetation cover are the consequences of modern agriculture. This may cause climatic change in the area. Overirrigation and agriculture cause water-logging and salinity. Water-logging is caused by the excessive use of water in a land, mostly clayish land, due to overirrigation. The surplus water in the overirrigated land evaporates, resulting in an increase in the salt content in the soil which leads to a loss of crop productivity. Modern agricultural practices have also resulted in a serious loss of genetic variety of crops.

Fertilizer Pesticide Problems

The problems that arise from the usage of fertilizers and pesticides are commonly known. However, this has been discussed in the context of soil pollution in Unit-V.

Water-logging and Soil Salinity

Water-logging and soil salinity, as mentioned earlier, are caused by overirrigation.

When there is water-logging due to excessive rainwater, irrigation, flood in clay-type soil or in soil containing an impermeable layer of clay, plants decay because their roots cannot get enough oxygen to breathe and grow. Soil salinity is also due to overirrigation. When crops are over irrigated surplus water evaporates and the dissolved salts are left behind in the soil thereby increasing the salinity of both the soil and the remaining water. Increased soil salinity affects water intake by plants and thus their productivity. The effects are most evident in fruits, followed by vegetables and then crops.

CASE STUDY

1. In India, some traditional communities in urban and semi-urban areas cultivate vegetables in their backyards with waste water from their own homes.
2. The Kolkata Municipal Corporation releases its waste water into surrounding lagoons in the eastern side (Dhapa). Fish farming is done in these waters and this water is used to grow vegetables.
3. Israel is short of water. So, it started the drip irrigation system to increase crop production. Over a period of 20 years, irrigation efficiency improved by 95 per cent and food production has doubled in that country.
4. Desertification and salination of the Aral Sea basin, covering about 36,000 sq m, in some of the independent states of Kazakhstan, Kyrgyzstan, Tajikistan, Turkmenistan and Uzbekistan of the former Soviet Union resulted in the poisoning of food and drinking water due to accumulated salts and contamination by harmful pesticides and fertilizers. These spread by wind to other areas. This resulted in a food crisis and diseases affecting the lives of about five million people. Former Soviet planners are responsible for this disaster. The water of the Aral Sea receded because the water of the Amu Darya and Syr Darya rivers was diverted into deserts to grow cotton. Cotton production shot up by the 1980s. However, the Aral Sea basin which once covered an area of the size of Lithuania became a desert exemplifying the harm caused by modern agriculture.

2.7 Energy Resources

Energy is the capacity of doing work. It can neither be created nor destroyed but can be converted from one form to another. The advancement of science and technology has increased the energy consumption of the human civilization. The rate of energy consumption by a country has now become a measure of the pace of its development. The higher the rate of energy consumption, the faster is the pace of development. To meet this increasing energy demand, efforts are being made to improve existing technologies and to develop new energy resources as well.

Energy can be obtained from various sources. The sun is an inexpensive and unlimited, therefore ideal source of energy. Besides energy from the sun (solar energy) can also be obtained from different resources such as fossil fuels, biomass, biogas, nuclear power, hydroelectricity, wind energy, tidal energy and geothermal energy.

Depending on the source, energy can be broadly classified into renewable and non-renewable energy.

Renewable or non-exhaustible sources of energy are those that are present in nature and are continuously produced, irrespective of human activity. Solar, wind, tidal, geothermal and biogas energy are examples of renewable resources of energy.

Non-renewable resources are also present in nature as deposits. Once they are used, these deposits get exhausted (depleted) and cannot be replaced. Fossil fuels like coal, petroleum, natural gas and timber, different ores and electricity are all non-renewable sources of energy.

Due to constant use, not only is the stock of non-renewable resources reducing but their increased use is also creating a number of socio-economic and environmental problems like forest cover depletion, global warming and pollution

2.7.1 Growing Energy Needs

The economic growth and development of a nation constantly needs more energy to be generated and consumed, irrespective of the increase in energy efficiency due to technological development. The demand

and consumption of energy is therefore an index of a nation's development. For example, according to the World Resources Institute, in 1998, the per capita energy consumption in USA was 24 times the per capita energy consumption in India thus establishing the superiority of economic development of the United States of America. Demand and consumption of energy increases with the passage of time and as human civilization grows in stature. For example, the world's energy consumption increased four-fold in the 40 years from 1950 to 1990. The form of energy consumed also changes with the passage of time, geographic location and technological development. Electrical energy is the most widely used. Coal was the primary energy source in the world till the 19th century. Oil was the most widely consumed energy source in the world at the end of 20th century, accounting for 39 per cent, with coal, natural gas, nuclear energy and hydro/renewable energy accounting for 24 per cent, 24 per cent, 7 per cent and 6 per cent, respectively. In India, as estimated in 2001, the predominant commercial energy source was coal accounting for 55 per cent, followed by oil (31 per cent), natural gas (8 per cent), hydro-energy (5 per cent) and nuclear energy (1 per cent). However, rural India uses biomass, fuel wood, cow dung and so on which account for 40 per cent of the primary energy supply of India.

Coal, oil, natural gas, forests and others are non-renewable and exhaustible energy resources which account for 87 per cent and 94 per cent of the energy sources of the world and India, respectively. They need to be preserved for our future generations keeping in mind the increasing population load, economic development and consumption demand. We have to harness other forms of energy such as solar, nuclear and renewable energy, enhance our energy efficiency with technological development and restrict the overuse of energy. However, our energy demands are bound to grow as long as the earth exists and we have to sustain energy sources innovatively and find alternative sources of energy.

2.7.2 Non-renewable Sources

Non-renewable sources of energy include fossil fuels and ores such as uranium and so on. These sources of energy get exhausted with use and cannot be replaced.

About 90 per cent of present-day energy consumption is based on non-renewable sources. Since they deplete fast, the demand for renewable sources of energy has become more important in recent times.

Fossil Fuels: Fossil fuels include coal, petroleum and natural gas. About 90 per cent of the world's energy requirements are met by burning fossil fuels. Burning these not only produces energy but also many harmful gases such as carbon dioxide, carbon monoxide, sulphur dioxide and oxides of nitrogen. About 60 per cent of air pollution occurs due to the burning of petrol and diesel in automobiles.

The presence of carbon dioxide within a certain limit is not harmful since it is used by plants during photosynthesis but excess deposit of carbon dioxide in the earth's atmosphere causes global warming due to the Greenhouse Effect. Carbon monoxide, if combined with haemoglobin of blood can cause death in humans within seconds. Oxides of sulphur cause irritation of the respiratory system and prevent the growth of plants. Sulphur dioxide combines with atmospheric moisture to cause acid rain. Oxides of nitrogen affect the respiratory system, pulmonary function and cause irritation in the eyes and nose. Photochemical smog is a result of the increase in oxides of nitrogen in the atmosphere. Therefore, in order to check air pollution the rampant burning of fossil fuels has to be banned.

Coal: Due to its high abundance and easy availability, coal is the most widely used fossil fuel. It is a solid fossil fuel formed by partial decomposition of plants deposited in layers at varying depths.

Coal was formed about 300 million years ago. At that time the earth's atmosphere was hot, damp and rich in nitrogen and carbon dioxide. This favoured the growth of huge non-flowering plants which in due course of time died and fell into swampy water. Layers of lush growth formed a layer several feet thick of decaying vegetation. Volcanic eruptions and earthquakes facilitated the burning process. Gradually, the high pressure of mud and sand squeezed out water and oxygen from the layers and converted them to a pasty mass which slowly hardened to form coal. The conversion of wood to coal usually takes thousands of years.

Depending upon the depth, pressure and quality of plant materials, the quality of coal varies. Chemical coal is a complex material of approximate composition $C_{100}H_{85}S_2N_{2.1}O_{1.5}$.

Depending on the carbon content coal has different grades.

Lignite Coal: It is the lowest grade of coal and is also known as brown coal. It contains about 70 per cent carbon and 27 per cent volatile materials.

Bituminous Coal: It is the widely used solid fuel all over the world. When lignite coal is buried deep, the pressure drives out more volatile matters present in it and its carbon content increases. Bituminous coal contains 75 to 85 per cent of carbon. It is also known as soft coal.

Anthracite Coal: This is the best quality of coal with 95 per cent carbon content and 5 per cent volatile matter. So, on burning it gives off very little smoke. It is hard and jet black in colour. It is found deep inside the earth's crust. Anthracite coal produces 6,000 to 7,000 kcal of heat per kilogram. The high cost and less availability has restricted the use of this good quality coal.

Coal is used mainly for cooking and heating purposes. It is used as a fuel for steam power plants and for running locomotive engines and industries. It is also used for the generation of electricity in thermal power plants. Coal can be transformed to gas, liquid or low-sulphur, low-ash solid fuels and used as a substitute for petroleum.

Many chemical products are also made from it. On destructive distillation, bituminous coal produces gas, coke, tar, and ammonia and sulphur compounds. After purification, this gas is also used as a substitute for petroleum. From tar, many synthetic products like dyes, drugs and plastics are manufactured. Ammonia is used as a refrigerating substance, while sulphur compounds are used in fertilizers, explosives and dyes. The residual coke is used for the manufacture of iron and steel.

Natural Gas: It is formed by the decomposition of organic matter buried in the interiors of the earth and is recovered by a process of compression and cooling. Natural gas is a mixture of helium and a number of hydrocarbons like methane, propane and butane. It burns with a hot, blue, non-luminous flame. For its easy transportation and cleanliness it is used as a cooking gas in almost all households.

Petroleum: Petroleum is a mineral oil found between the rocks under the earth's surface. It is mostly covered with compressed natural gas. The availability of petroleum and natural gas governs the energy growth and status of a country. Although the Industrial Revolution was initially fuelled by coal subsequently the emphasis shifted to cleaner fuels like oil and gas.

According to scientists, petroleum is formed in the same way as other fossil fuels such as coal and natural gas, from the remains of plants and animals, which were buried under sand and mud due to earthquakes. The high pressure and temperature of the earths crust change these remains to petroleum. Depending upon the depth, pressure and quality of buried material, the nature and texture of fossil fuel vary from gas to coal and finally to petroleum (solid, gas or liquid).

These underground mineral oils are explored through seismic methods by exploding dynamite in shallow holes drilled in the earth's crust. The shock waves produced by the explosion travel down to various layers of rock. The vibrations reflected through various layers are recorded on seismographs. The time gap between the detonation and reception of the reflected signal gives information regarding the depth of the layer. The strength of the vibration provides information about the nature of the layers of rocks. More sophisticated tests are performed on soft rocks to get information about the presence of oil.

Underground crude oil is always associated with natural gas. When the oil is drilled, the compressed natural gas forces the crude oil up the well. However, pumping becomes necessary when gas pressure is less than sufficient.

Petroleum is obtained by refining crude oil. It is a mixture of hydrocarbons of different chain-lengths. Small molecules with one to four carbon atoms such as methane, ethane, propane and butane form gases, while larger molecules with up to 10 carbon atoms form gasoline. Light fuels contain hydrocarbon chains with at least 50 carbon atoms while giant molecules up to several hundred carbon atoms constitute heavy fuels and waxes. Different components of the crude oil are separated by fractional distillation. Gasoline is used as a fuel in automobiles and airplanes while diesel is also used as a fuel in automobiles and in locomotives. About 70 per cent of these oils are used in power stations in producing steam, lubricating oils, gear oil and greases, which are used as lubricators to reduce friction.

2.7.3 Renewable Sources

The non-exhaustible sources of energy like solar energy, hydroelectricity, wave and tidal energy, wind energy and bioenergy from biomass or biogas are known as renewable sources of energy.

Biomass: All types of biological substances like plant products (wood, crop, algae and aquatic plants), their residues (straw, husk, sawdust, cow dung, animal droppings) and also waste materials like garbage and night soil are collectively known as biomass.

Biomass is used as a source of energy in many parts of the world. Burning of the biomass produces heat energy known as bioenergy. It is obtained through the oxidation of biomass. The residue left after burning of the biomass is used as manure in agriculture fields. Burning one kilo of wood produces about 4,000 to 5,000 kcal of heat. Wood is inexpensive and lights up easily, but the burning of wood causes depletion of forest cover, which in turn leads to the accumulation of carbon dioxide in the atmosphere, the Greenhouse Effect and soil erosion.

These problems are being tackled now-a-days by a wide range of social forestry or energy plantation schemes. Plants like eucalyptus, acacia and cassia are planted widely to be used as fuel wood.

Biogas: India is rich in cattle, sheep, goat, horses and camels and they produce approximately 1,000 tonnes of dung every year. Besides dung, water hyacinth, hydrilla, duck weed, salvinia and so on are also used to produce biogas. First, the dung and other materials are allowed to ferment in the absence of air in a slurry form. The produced gas, known as *gobar gas* or biogas, contains 55 per cent methane and 45 per cent carbon dioxide by volume. The residue left in a *gobar gas* plant after gasification makes for good manure and is used in fields. A *gobar gas* plant can be operated with a maximum output of two cubic metres per day, which requires collection from just four heads of cattle per day. This amount of gas is enough for cooking and lighting the house of a family of six. In fact, one cubic metre of gas produces 4,713 kcal of heat. *Gobar gas* (biogas) is used for cooking, lighting and running pump sets, engines and machines in factories and turbines to generate electricity.

The construction of a *gobar gas* plant is very simple. The main unit is a well called a digestive well. The floors and walls of the digestive well are made up of brick and cement and it is covered with a metal dome that prevents the entry of external air into the well, thereby helping in aerobic fermentation of the dung by microbes. The *gobar gas* plant also has an inlet and an outlet pipe. A mixture of dung and water in the ratio of 1: 5, is fed into the well through the inlet pipe. As the gas forms inside the well by fermentation, the metal dome gradually pushes up due to the pressure. The gas is then released through the outlet pipe provided at the top of the dome. The residue left (slurry) is then taken out and dried in the sun. Drying in the sun increases the nitrogen content of the slurry from 0.25 to 1.5 per cent, thus, making it a better fertilizer than cow dung. Presently, there are more than 6.5 lakh biogas plants in India. This number will increase rapidly in the near future as the potential for the production of biogas and slurry in India is very high.

Solar Energy: The sun is a storehouse of enormous energy formed by nuclear fusion reactions in the interior of the sun. In fusion reactions, light nuclei fuse together to yield heavier nuclei, where the mass of the end-product is less than the total masses of the reacting species. The difference of masses between the reacting and product species is usually converted to energy. When heavy elements like radium and plutonium undergo nuclear fusion reaction, they produce huge amounts of energy.

Scientists have predicted the central temperature of the Sun to be nearly 20 million degree Celsius. They have also predicted that the sun contains mainly hydrogen. About 56.4 million tonnes of hydrogen nuclei fuse together per minute to produce 56 million tonnes of helium. So the sun is losing about 400,000 tonnes of matter per minute which is converted to energy.

$$_1H^2 + _1H^2 \longrightarrow _2He^3 + _0n^1 + 1\text{MeV} \text{ where n is a neutron}$$

This energy reaches the earth as electromagnetic radiation, commonly known as solar energy. The solar energy consists of mainly ultraviolet, visible and infrared radiation. Of these, the harmful ultraviolet radiation is mostly absorbed by the ozonosphere. The visible and infrared radiation reaches the earth as heat radiation.

The sun is 150 million kilometres away from the earth and its rays travel this path and lose most of their energy while travelling. Only 4 per cent of the total solar energy reaches the earth, which is approximately 1 kW per sq m of flat land on the earth's surface.

Approximately 35 per cent of the total solar energy is reflected back, 18 per cent is absorbed by the atmosphere and the solar energy reaching the earth depends upon atmospheric conditions.

Since it is non-polluting, solar energy helps in maintaining the changes in the atmosphere and the climate cycle. It is a life sustaining source of energy. Solar energy is used by plants for photosynthesis to produce food energy. Human beings use solar energy for evaporation. But for its effective use, solar energy has to be gathered and concentrated. For this, solar mirrors are used. Solar energy is mainly used for heating purposes. The availability of solar energy depends on seasons, length of days, and the latitude of the place and so on. For commercial use, this energy is collected and stored, so that it can be used on days when the sunlight is not adequate.

Electrical energy can be generated from solar energy through solar power plants. The largest solar power plant is in the deserts of California in USA and it consists of a power tower 90 metres high, encircled by 1,800 giant movable mirrors known as Heliostats. These Heliostats reflect the sunlight in the tower to boil water which turns to steam. The steam, in turn, rotates the turbines and produces electricity. In western countries, solar power is gathered during the day to light streetlamps at night. Now-a-days photo-electric cells are used for calculators and in small machines. Solar cells are also used to provide power to satellites, water pumping, communication and radio and television receivers in remote areas. The Oil and Natural Gas Commission (ONGC) is using photocells for drilling in Kerala. Solar energy is also used for cooking in solar cookers. A solar cooker is a simple hot box-type device provided with a reflector. It saves about 40 to 50 per cent of fuel. However, it suffers from some disadvantages:

(i) More time is required for cooking.

(ii) It becomes inactive in the absence of adequate sunlight.

Solar energy is also used in refrigerators for cooling purposes by liquefying ammonia which vaporizes to run the refrigerator.

Nuclear Power: All matter, whether an element or a compound, consists of an elementary particle called an atom. An atom has a tiny nucleus and electrons revolve around the nucleus.

Inside the nucleus, protons and neutrons are held together by a strong nuclear force of attraction. Hence, when an atomic nucleus splits, a tremendous amount of energy is released. This energy is known as nuclear energy or atomic energy. The atom bombs, which were dropped on **Hiroshima** and **Nagasaki** in August 1945, are examples of the tremendous power of atomic energy.

The splitting of the atomic nucleus is caused either by nuclear fusion (as it occurs in the sun) or nuclear fission.

Nuclear Fusion: In this process, two light atoms combine, to form a heavier one. But the atomic mass of the product is less than the sum total of the atomic mass of the reactants. Hence the difference of masses is released as nuclear energy. For example, the fusion reaction of deuterium and deuterium releases:

$${}_1H^2 + {}_1H^2 \longrightarrow {}_2He^3 + {}_0n^1 + 1\ \text{MeV}$$ about 1 MeV of energy.

The deuterium and tritium reaction is energetically more viable, since it promises an endless source of energy without any radioactive waste.

$${}_1H^2 + {}_1H^3 \longrightarrow {}_2He^4 + {}_0n^1 + 17.6\ \text{MeV}$$

Due to technological problems, scientists have not been able to harness energy from a nuclear fusion reaction.

Nuclear Fission: The principle of nuclear fission reaction is the opposite of nuclear fusion reaction. In this method, a heavy nucleus with a large mass is bombarded with fast moving neutrons. As a result, an atom with a heavy nucleus splits into smaller, lighter atoms. The mass lost in the process, is converted to huge amounts of energy.

Atoms with a large mass and heavy nucleus used for the purpose are uranium, thorium, plutonium, radium and so on. Of these, uranium is the ideal atom for fission reaction. Uranium is always found as a mixture of its three isotopes, namely U^{238}, U^{235} and U^{234}. Of these, U^{235} is more fissionable and is separated from the mixture by a process called 'enrichment.' All nuclear power (fission) reactors are based on fission of U^{235}.

$${}_{92}U^{235} + {}_0n^1 \longrightarrow {}_{51}Sb^{133} + {}_{41}Nb^{99} + 2.5\ {}_0n^1 + 200\ \text{MeV}.$$

The released energy is used to heat water in a reactor to produce steam to drive a steam turbine. The released 2.5 neutrons, on an average, per fission are utilized to bring about further nuclear fission reactions leading to a chain reaction. This reaction can become uncontrollable if proper care is not taken.

Nuclear Reactors: A nuclear reactor is a nuclear reaction chamber in which nuclear fission reaction is allowed to occur at a controlled rate by controlling the speed and number of neutrons. It consists of (a) a fuel rod of U^{235}, (b) control rod for absorbing excess neutrons made of Boron (B) and Aluminium (Al), (c) moderator like heavy water or granite to slow-down the speed of the neutron, (d) coolants like heavy water to carry the heat and (e) shield to prevent heat loss. The large amount of energy which is released in the form of heat due to the controlled reaction rate is used to produce steam which can run turbines and produce electricity.

Hydropower: The energy we get from flowing water is known as hydropower. India is rich in rivers like the Ganges, Narmada, Godavari, Krishna, Mahanadi, Brahmani and others. As dams are constructed over big rivers in addition to generation of hydropower, they also help in (i) irrigation, (ii) flood control, (iii) generation of electricity, (iv) employment generation, (v) transportation, (vi) inland fisheries and (vii) recreation. At present, India has 1,550 major hydel power projects and many thousand minor projects. An estimate shows that the cost of one unit of hydel energy is about 17 to 30 *paise* against 9 *paise* in the case of wood or 7 *paise* for getting the same amount of energy from coal. This high cost of hydropower is due to the high rate of transmission and losses in distribution.

Dams also control floods and provide water for irrigation. These projects also help in local transportation, recreation and development of fisheries. However, recently it has been found that the large hydel projects are ecologically disruptive as (i) construction of big dams changes the lifestyle of the *Adivasi* community which traditionally depends on forests, (ii) construction of dams lead to destruction of forest, which includes felling of trees and destruction of wild animals, (iii) big dams destroy the riverine system, thereby affecting a large number of people who depend on rivers, (iv) big dams cause soil pollution leading to infertile land due to extensive water-logging and salination and (v) big dams increase the risk of earthquakes because the huge amount of standing water puts enormous pressure on land.

Hence, it is both economically and ecologically profitable to construct medium-level hydel projects.

Wind Energy: Wind energy is another alternative source of renewable energy. It can be converted into mechanical and electrical energies and made to run turbines to generate electricity. Mechanical and electrical energies, are particularly useful for lighting, pumping and devising the change of barriers in remote and windy areas and where the aerodynamics of the area provide suitable conditions. Wind powered mills of approximately 2 MW capacity have been installed in a number of places in India where the wind density is as high as 10 KWM/m^3/day in winter and 4 KWM/m^3/day in other 5–7 months of the year in peninsular and central areas of the country.

According to the Department of Non-Conventional Energy Source (DNES) in India, 20,000 MW of electricity can be generated from wind. Wind farms have already been set up in Mandvi, Okha, Deogarh, Tuticorin and Puri.

Ocean (Tidal) Energy: In some countries, tidal energy is also used to rotate turbines and generate electricity. The energy derived from the rise and fall of the sea tide is converted into electricity in small, tide-driven turbines at sea shores. France was the first country to construct a major tidal electricity generation plant.

In India, places like the Gulf of Kutch, Cambay and Sundarbans are exploited for the most prospective tidal energy harnessing sites. Sea shores near Lakshadweep and the Andaman and Nicobar Islands are also found to be most suitable for the generation of electricity from the ocean. At both these places, cold water is available at a depth of 1,000 mm near the shore. In fact, a one MW plant for Lakshadweep Island has already been designed.

Tidal power potential of 9,000 MW has been identified in India. In Ocean Thermal Energy Conversion (OTEC) and wave energy, the temperature difference between warm, surface sea water (28–30°C) and cold, deep sea water (5–7°C) at 800 to 1,000 m depth in tropical waters is utilized to drive turbines to generate electricity. A floating OTEC plant can generate electricity even in the middle of the sea for offshore mining operations and so on.

Geothermal Energy: This is the heat of the interior of the earth present at volcanic regions, geysers or hot springs. This is utilized to generate electricity. Exploration, for the identification of prospective sites of geothermal energy is to be undertaken. Presently, nearly 350 geothermal springs have been identified in India.

Dendrothermal Energy: Wastelands can be used for the plantation of fast growing, high calorific shrubs and trees, which can be used for both fuel and fodder. This method is widely practiced in Philippines. In 1987, the energy plant produced 1.5 MW power through gasification of over 8,000 hectares. In India, this method is yet to become popular. The wastelands of Rajasthan and Madhya Pradesh can be exploited for the same.

Energy from Urban Waste: In Delhi, a pilot plant has already been set up where municipal sewage and solid waste is used for energy generation. Bagasse and other farm waste can also be used to generate electricity, which will not only make the industry energy-sufficient, but the surplus power that is generated can also be used for improvement in neighbouring areas. Energy generation from solid waste has been discussed in detail in Unit-V, under the heading Solid Waste Management.

2.7.4 Land Resources

Land is a critical natural resource but it constitutes only about 29 per cent of the total earth's surface. Yet, all life (both animals and plants) flourish on land. Be it the prehistoric 'hunter gatherer' society or today's 'technocrat', human beings depend on land for their food (hunting and agriculture), occupation (roads, mining, industry, communication and so on), storage of surface and ground water and living (housing).

With the advancement of civilization and technology, man has been harming the land by clearing forests and depleting many resources.

The soil profile of land determines its ability to serve socio-economic needs. Statistics depict that more than 5,000 million tonnes of top soil is eroded annually along with 5 million tonnes of nutrients. About one-third of it is lost to the sea, while the rest turns into silt on river beds leading to floods. In India, 38 per cent of the land area suffers from water-based erosion which needs immediate watershed development. In addition to this, 4 per cent of arid area also suffers from soil erosion. Thus, 42 per cent of the total land resources require immediate conservation management.

CASE STUDIES

I. 18 watt modern compact fluorescent lamp can replace a 40 watt tube light or a 100 watt incandescent lamp. It is an energy saver in the house.

II. Today, refrigerators are made with new and energy-efficient technologies. A refrigerator with a 200 litre capacity consumes 90 kWh now, instead of the 350 kWh consumed previously.

III. Instead of steel, light materials such as aluminium and fibre are used for making car bodies to reduce weight by up to 15 per cent, thereby increasing fuel economy by up to 6 to 8 per cent. Further, using bicycles is an excellent energy saving transportation method.

IV. Industries need to be developed using advanced energy efficient technologies, an area in which India lacks. For example, the steel industry in India consumes 9.5 million kilo calories per tonne of steel against 4.3 and 4.1 million kilo calories in Italy and Japan, respectively. Similarly, the cement industry in India consumes 2 million kilo calories to produce one tonne of cement compared to 0.82 and 0.92 million kilo calories in Germany and USA, respectively.

V. Fuel wood and *chulhas* are used by 50 per cent of the world population. This is one of the reasons for the depletion of forest cover.

VI. The first hydroelectric power dam in the world was built in Appleton, Wisconsin in 1982. In India, Jamshedji Tata built the first hydroelectric power dam in the Western Ghats of Maharashtra in the early 1900s to supply clean power to Bombay's Cotton and Textile Mills. He took the British Government's permission to build dams, namely the Andhra, Sirowata, Valvan and Mulshi hydel dams in the Western Ghats to generate electricity using high rainfalls in the hills as storage areas.

Land Degradation

Land degradation or the decline in land quality or land pollution is the result of both natural processes and anthropogenic activities. Human activities like deforestation, mining, overgrazing, intensive cultivation, dam and other developmental activities and solid waste disposal result in the exposure of soil and thus accelerate the process of land pollution. Land degradation in turn results in consequent surface run-off, floods, desertification, loss of soil productivity and also reduction of productive land for sustainable agriculture.

As per the Human Development Report 1998, about one sixth of the world's productive land, that is two million hectares has been degraded in the 53 years since 1945. Every year about 30 million acres of productive land gets eroded in India. This results in loss of food, cultivable land and thus food security. Land degradation is more severe in developing countries like India. Nearly half of the world's degraded land is in Asia and one-fourths of the world's moderate to strongly degraded land is in Africa. Two-thirds of the world's population lives in these degraded lands of Asia and Africa. The food shortage due to less cultivable land caused by land degradation is, therefore, a great concern to the world and necessitates urgent attention.

Landslides

A sudden, uncontrolled descent of a mass of earth and rock down a slope (mountain slope) under the force of gravity is called a landslide. As already mentioned, the mountainous regions of northern and north-eastern parts of India are vulnerable to landslides. Table 2.4 lists the landslide prone areas of India.

Table 2.4 Landslide Prone Areas of India

Regions	Intensity
Himalayas	High to very high
North-eastern hills	High
Western Ghats	Moderate to high
Eastern Ghats	Low
Vindhyachals	Low

Although landslides are a natural process, anthropogenic causes like deforestation, industrialization, construction of dams, hill roads, railway tracks and making place for human habitation accelerate the process. Thus, for the implementation of developmental projects proper precautionary measures should be taken.

Landslides not only destroy human life but also destroy plantation and wildlife. They also block roads causing immense casualties to human life and economic loss to the country. In remote areas, landslides disrupt emergency services such as water, power and communication. Further details are available in Unit-V of this book.

Soil Erosion

Soil erosion is one of the outcomes of land degradation caused both by natural processes as well as human activities. The continuous eroding away of the earth's crust by the forces of weather is known as soil erosion. Normal geological erosion is very slow and it tends to bring the earth's surface to a uniform level. When this soil erosion is accelerated due to different human activities it is called artificial or accelerated erosion. Deforestation and overgrazing are the most prominent causes of soil erosion. Depending upon the factors causing it, erosion is classified as wind erosion, gully erosion, water erosion, coastal erosion and so on.

Soil erosion is the washing or wearing away of the topmost layer of soil. By this process not only is the productivity of the eroded land destroyed but when the top soil run-off lands on top of other productive soil, it destroys the productive soil as well.

However, soil erosion can be prevented by using remedial measures. Wind erosion can be prevented by planting trees to break the wind force. Trees can also check soil erosion by holding together the soil particles. Water erosion can be checked by terracing, contour cultivation or through strip cropping.

Overgrazing

The increase in livestock population with simultaneous decrease in grazing land has resulted in the overexploitation of natural resources. India supports about 85 per cent of the world's total livestock such as buffaloes, cattle, goats and sheep, with only 1/40th of the total land area of the world at its disposal. The result is overgrazing on fallow, uncultivated forestlands. This uncontrolled grazing removes the ground vegetation, causing soil erosion and gradual depletion of soil organisms which turn the land to wasteland.

Soil conservation through a planned growth of fodder and grasslands on the outskirts of dense forests and wastelands can check soil erosion and simultaneously solve the problem of overgrazing.

Mining Activities

With the advancement of science and technology, our underground resources are now excavated through mining activities. For this usually, the two principles of opencast mining and underground mining are followed. In opencast mining, the underground resources are excavated directly by removing the topsoil and other strata. As a result, the total area gets destroyed and loses its productivity. In underground mining, the minerals are mined through tunnels. When the mine is abandoned after mining, the whole area collapses into a big depression and becomes a wasteland.

However, in both types of mining the mined materials are dumped near the mine. This not only causes unwanted occupation of land but the rainwater washes away the mining waste (rich in sulphuric acid) to the nearby agricultural fields and pastures. The debris blocks the drainage system leading to water-logging as well. This type of land pollution can be stopped only by large-scale planned landfilling followed by plantation.

Desertification

There are two main causes because of which a fertile piece of land or a forestland gets converted into a desert.

(i) Continuous climate change and

(ii) Abusive and overuse (exploitation) of land

The continuous pressure of the rise in population is squeezing the earth's forests to create more land for agriculture pastures and for the implementation of the results of technological developments. This is resulting in the denudation and degradation of the forest areas. This removal of vegetation cover of the land changes the microclimate of the area. The changes in rainfall, temperature, wind velocity and so on disturb the local ecological equilibrium. This in turn leads to soil erosion, loss of the nutrient cycle and finally to the onset of desert-like conditions. Thus, it is the improper land use practices and removal of vegetative cover of the land that is responsible for the climate change of the area.

The onset of desertification is marked first by the destruction of productive land in patches. Loss of forest cover decreases humidity or increases the temperature of the area. Thus, productive land gets slowly converted into a barren, sandy desert. At the time of Independence, India had 22 per cent land area under forest cover which has diminished to only 10 per cent within 50 years.

According to a report of the Government of Australia, in 1985, about two-thirds of the vegetative cover of Australia had been destroyed. Almost half the agricultural land had serious soil pollution and needed treatment to stop soil degradation. The situation is worse in Third World countries, where desertification remains unchecked. For example, it is reported that a new desert at a rate of 170 hectares per hour is being created around the area of the Sahara desert in Africa.

Role of an Individual in the Conservation of Natural Resources

The role of the individual can be best understood by the following words of Mahatma Gandhi, 'The earth provides enough to satisfy every person's need but not every person's greed.' The global ecosystem has a limited capacity for replenishing natural resources. The increasing overuse and misuse of natural resources will destabilize the natural balance. Non-renewable natural resources if not checked will get exhausted soon. Even renewable resources, namely those from oceans, forests, grasslands, wetlands and so on are also going to get exhausted or degraded in the future. Population growth and increasing consumerism of the affluent members of society put pressure on natural resources.

As an individual, we must decide what we want to leave for our children and at what cost do we get our resources. Our small actions can help a lot in conserving natural resources. The following are some illustrations:

(i) Switch off lights, fans, air conditioners and other electrical gadgets when they are not required.

(ii) Close the tap; mend the leakages to avoid wastages and to save water.

(iii) Do not use plastic products; instead use recyclable paper products.

(iv) Use energy-efficient lights and electrical gadgets to save electric energy.

(v) Use a pressure cooker. Keep the vessels covered with a lid. Keep cooled foodstuff out of the refrigerator to bring it to room temperature before cooking.

Let us play a positive role in conserving natural resources instead of waiting for the society and the government to do everything.

Equitable Use of Resources for Sustainable Lifestyles

The world is sharply divided between developed and underdeveloped, the haves and the have-nots. Ideally, the fast depleting natural resources should be evenly distributed among countries as well as among the communities within the countries. This will bring down disparities and will reduce overexploitation of natural resources. It will also lead to a sustainable lifestyle which primarily depends on the earth. Instead of overexploiting natural resources we may be able to lead a better lifestyle through the equitable use of natural resources such as sharing of community halls, ponds, grounds, forest products, staying in flats instead of houses and using public transport or car pools instead of using one vehicle each. Such a lifestyle will help sustain nature's wealth. A dense forest, green grassland, the clean water of a lake, the ocean sheltering all types of aquatic plants, animals and resources all these can be sustained in our lifestyle if we avoid wastefulness and overutilization of the fast-depleting natural resources and by equitable use and sharing.

SUMMARY

Depending on their origin, natural resources may either be biotic or abiotic. Based on their availability and utility, natural resources may also be classified into renewable resources and non-renewable resources.

Renewable Resources

Natural resources, which can be renewed as they are used, are known as renewable resources. These include resources such as forest cover, solar energy and hydroelectricity.

Non-Renewable Resources

There are some resources in nature which are formed over thousands of years and under favourable climatic conditions. These cannot be replaced easily. Such resources are called exhaustible or non-renewable resources.

Forest Resources

Forests cover a major part of the lithosphere of the earth and are a very important renewable natural resource. Tribal people in and around forests directly and the entire human population indirectly depend on forests for fuel, as a source of clean air and water and also forest products for food, clothing, housing and other necessary products.

Deforestation

Deforestation is the wanton cutting of trees including lopping, felling, removal of forest litter, browsing and grazing. Some of the causes of deforestation are urbanization, industrialization, developmental projects such as dams, loss of forest lands due to human activities such as agriculture, overgrazing and collection of fuel wood and other forest products.

Water resources

The hydrosphere includes all sources of water such as oceans, lakes, rivers, underground water etc. Water is the most important factor of life. Human civilization has flourished with the availability of water.

Mineral resources

A mineral is a naturally occurring substance in the earth's crust formed over a period of millions of years. India has a large number of economically useful minerals. About 25 per cent of the world's known mineral reserves are found in India. These underground minerals are extracted through mining operations. Since minerals are non-renewable natural resources their unplanned extraction and exploitation pose a threat to raw material resources for future generations and have caused irrecoverable environmental damage.

Food resources

Food is essential for the survival of all living beings on the earth. The capability of all people to access enough food at all times for an active and healthy life is known as food security. Food security depends on:

(i) Availability of enough food for all.

(ii) Accessibility of food for all.

(iii) Adequacy of food for all.

Energy resources

Energy can be obtained from various sources. With the exception of solar energy, energy can also be obtained from different resources such as fossil fuels, biomass, biogas, nuclear power, hydroelectricity, wind energy, tidal energy and geothermal energy. Depending on the source, energy can be broadly classified as renewable and non-renewable energy.

Non-renewable resources of energy include coal, oil, natural gas, forests etc, while renewable resources of energy include nuclear energy, hydropower, wind energy, ocean (tidal) energy, geothermal energy, dendrothermal energy, energy from urban waste etc.

Growing Energy Needs: The economic growth and development of a nation, despite increasing energy-efficiency due to technological developments, demand more energy to be generated and consumed. Demand and consumption of energy is, therefore, an index of a nation's development.

Land resources

Land is a critical natural resource but it constitutes only about 29 per cent of the earth's total surface. With the advancement of civilization and technology, man has been affecting the land by clearing forests and depleting many resources.

Soil erosion is one of the phenomena of land degradation caused by both natural processes as well as human activities such as deforestation, mining, overgrazing, intensive cultivation, river dams and other developmental activities, solid waste disposal etc.

Landslides: A sudden, uncontrolled descent of a mass of earth and rock down a slope (mountain slope) under the force of gravity is termed as landslides. The mountainous regions of the northern and north-eastern parts of India are vulnerable to landslides.

ESSAY TYPE QUESTIONS

1. What is a natural resource? Give an account of different natural resources with special reference to India.
2. What is the impact of the present trend of population growth on the availability of natural resources?
3. What are the main benefits of forest resources?
4. What is deforestation? How does deforestation affect us? State a few possible steps to check deforestation.
5. Explain the causes of timber extraction/logging in dense forests and also its impact on tribal people.
6. Explain briefly the necessity/benefit and the problems of building dams over rivers.
7. 'Construction of big dams is essential for India.' Explain your views against or in support of the statement.
8. What are the water resources on earth?
9. What is the future threat to the availability of fresh water?

10. Write a note on the conflicts over water around us.
11. Write briefly about the mineral resources of India.
12. Give a clear picture of the present-day world food problem.
13. Write a few lines on global food security in the near future.
14. Explain in detail the impact of overexploitation of productive land on the world food problem.
15. Describe the impact of modern agricultural practice on the environment.
16. Write a few lines on the growing energy needs of the present day urban people.
17. What are the main causes of soil erosion and how can it be prevented?
18. How does land degradation occur? Explain.
19. What is conservation of natural resources? Explain the role of an individual for the same.
20. Write a note on equitable use of resources required for a sustainable lifestyle.
21. Write a paragraph on the interaction between women and the forest.
22. Write short notes on the following:
    (i) Tidal energy.
    (ii) Geothermal energy.
    (iii) Dendrothermal energy.
    (iv) Wind energy.
    (v) Biomass energy.
23. Write short notes on:
    (i) Agro Forestry.
    (ii) Desertification.
    (iii) Formation of Bougainville Island State.

SHORT-ANSWER TYPE QUESTIONS

1. What is the importance of forest resources?
2. What is the importance of land resources on earth?
3. What type of environmental degradation occurs due to extensive use of water resources?
4. What is the future threat to the availability of fresh water on earth?
5. What are the major control measures of soil erosion?
6. Write short notes on:
    (i) *Chipko* Movement.
    (ii) *Narmada Bachao Andolan*.
    (iii) Cauvery Water Dispute.
    (iv) Krishna–Godavari Water Dispute.
7. What is food security?
8. Write a short note on the effects of modern agriculture on the world food problem.
9. Distinguish between renewable and non-renewable sources of energy with suitable examples.

10. Distinguish between nuclear fission and nuclear fusion.
11. 'Large hydel projects are ecologically disruptive.' Discuss the statement.
12. Write a short note on the classification of energy resources.
13. What is fossil fuel and what are the harmful effects of the combustion of fossil fuel.
14. Write a short note on Ocean Thermal Energy Conversion technology (OTEC).
15. Write a few lines describing the generation of energy from urban waste or municipal waste.

MULTIPLE CHOICE QUESTIONS

1. Which of the following is not a renewable resource?
   (a) Coal
   (b) Wind power
   (c) Geothermal Energy
   (d) Dendrothermal Energy
2. The main causes of deforestation are
   (a) overgrazing and agriculture.
   (b) industry and other developmental projects.
   (c) timber extraction.
   (d) All of the above.
3. Which of the following methods of timber extraction cause minimum environmental damage?
   (a) Clear felling.
   (b) Handlogging.
   (c) Selective logging.
   (d) Reduced-impact logging.
4. Forest is responsible for
   (a) watershed protection.
   (b) land erosion control.
   (c) providing economic and environmental benefits.
   (d) All of the above.
5. The main purpose of dam construction is
   (a) irrigation.
   (b) flood control.
   (c) hydroelectricity.
   (d) provide water to industry.
6. A sudden uncontrolled descent of a mass of earth under the force of gravity is called
   (a) soil erosion.
   (b) mining.

(c) landslide.
(d) earthquake.

7. The heat of the interior of the earth present at the volcanic regions, geysers or hot springs is called
   (a) geothermal energy.
   (b) dendrothermal energy.
   (c) nuclear energy.
   (d) wind energy.
8. ________ is the perpetual source of energy.
   (a) Nuclear reactors
   (b) Hydropower
   (c) Solar energy
   d) None of the above
9. The least quality of coal with 95 per cent carbon content and only 5 per cent volatile matter is
   (a) lignite coal.
   (b) anthracite coal.
   (c) bituminous coal.
   (d) peat coal.
10. All sources of water such as the oceans, lakes, rivers and underground water together constitute:
    (a) Hydrosphere.
    (b) Atmosphere.
    (c) Lithosphere.
    (d) Biosphere.
11. Famine is mainly the result of
    (a) deforestation.
    (b) overuse of surface water.
    (c) a prolonged drought.
    (d) All of the above.
12. Prospecting, exploration, development and exploitation are the four phases of
    (a) mining.
    (b) urbanization.
    (c) deforestation.
    (d) industrialization.
13. Burning of biomass produces
    (a) thermal energy.
    (b) bioenergy.
    (c) wind energy.
    (d) hydropower.

14. Water-logging and soil salinity are the outcomes of
    (a) overirrigation.
    (b) mining.
    (c) soil erosion.
    (d) acid rain.
15. The equitable use of resources is necessary for
    (a) sustainable development.
    (b) better lifestyle for all.
    (c) to sustain natural wealth.
    (d) All of the above.
16. India has second largest reserve of _______ after Brazil.
    (a) iron ore
    (b) bauxite
    (c) coal
    (d) copper
17. Hygroscopic and combined water
    (a) are used by plants during transpiration.
    (b) compensates for loss due to evapotranspiration by plants.
    (c) stored in the xylem tissues of plant.
    (d) are of no use to plants.
18. Any material that can be transformed into more valuable and useful product or service is called
    (a) resource.
    (b) mineral.
    (c) product.
    (d) None of the above.
19. The constituents of producer gas are
    (a) 25 per cent water vapour, 75 per cent methane.
    (b) 42 per cent methane, 8 per cent propane, rest carbon dioxide.
    (c) 55 per cent methane, 45 per cent carbon dioxide.
    (d) 45 per cent methane, 20 per cent carbon dioxide, 35 per cent propane.
20. The single largest river development scheme in India is
    (a) the Narmada Valley Development Project.
    (b) the Damodar Valley Development Project.
    (c) the Ganga Valley Development Project.
    (d) the Tapti Valley Development Project.

ANSWERS

1 (a)	2 (d)	3 (d)	4 (d)	5 (c)	6 (c)	7 (a)	8 (c)	9 (b)	10 (a)
11 (d)	12 (a)	13 (b)	14 (a)	15 (a)	16 (a)	17 (d)	18 (a)	19 (c)	20 (a)

UNIT III

Ecosystem

LEARNING OBJECTIVES

After studying Unit III, students will be able to:

- Define the concept of ecosystems.
- Describe the structure and function of an ecosystem.
- Identify producers, consumers and decomposers.
- Define Energy Flow in the ecosystem as well as terms such as Ecological Succession, Food Chain, Food Web and Ecological Pyramids.
- Describe the types of ecosystems and their characteristics.
- Identify the structure and function of the forest ecosystem, the grassland ecosystem, the desert ecosystem, and the aquatic ecosystems such as ponds, streams, lakes, oceans and estuaries.

3.1 Concept of Ecosystem

Life is found in the biosphere on this planet and the species interact with their biotic and non-biotic components. The study of the relationship between organisms and between the organism and the environment is known as ecology. The structural and functional unit of ecology is known as the ecosystem. The lifecycle of a tree or a fish in a pond indicates the interdependence of the organisms among themselves as well as with the environment. There is a continuous production and exchange of materials between the living and the non-living components of the ecosystem.

The part of the earth where these ecosystems operate is called the biosphere. It is the place where the atmosphere, hydrosphere and lithosphere meet thereby making life possible through their interactions. In some natural ecological groupings, plants and animals of a regional climate and soil-type interact to produce a characteristic land community known as biome.

In biomes, the emphasis is on biotic community whereas in the ecosystem it is on the interaction between plants, animals, and microbes. Biomes can be deserts, tundra, forest type, grassland or the savannahs. In each biome, the kind of climax vegetation (dominant vegetation) is uniform such as grasses, conifers and so on but the particular species of a plant varies. The climax vegetation of a biome depends on the physical environment and the two together determine the kind of animals that will be present there.

3.2 Structure and Function of an Ecosystem

The functional unit of the ecosystem consists of two distinct structural components namely biotic and abiotic components. The ecosystem may be small or a microecosystem such as a drop of pond water or large or a macroecosystem such as an ocean. Depending on how they are generated, ecosystems may be natural for instance a lake or man-made such as an aquarium and may either be permanent or temporary.

3.2.1 Biotic Component

The biotic component of an ecosystem consists of all living components of the environment, such as plants, animals, human beings and microbes. A detailed study of the biotic components reveals that all living organisms of our environment can be further categorized depending on their self-food producing capability. They can be categorized as autotrophic component or producers and heterotrophic component or consumers. Biotic components are depicted in Fig. 3.1.

Autotrophic Component or Producer

This category includes all those organisms, green plants, bacteria and algae which contain chlorophyll and are capable of converting solar energy into chemical energy and storing foodstuff in the presence of carbon dioxide and water. Autotrophic components produce carbohydrate type of food through photosynthesis and hence the name

$$6CO_2 + 6H_2O \xrightarrow[\text{Photosynthesis}]{\text{Solar Energy}} C_6H_{12}O_6 + 6O_2 \uparrow$$

Heterotrophic Component or Consumers

All other organisms, which cannot convert solar energy into food and depend on autotrophs to obtain their energy for survival, are called consumers or heterotrophs. Heterotrophs utilize, rearrange and decompose the complex material produced by autotrophs. Depending upon their feeding habits, the heterotrophs are classified as follows:

(a) **Primary Consumers:** Organisms or animals, which feed on green plants (autotrophs) to obtain energy for survival, are called primary consumers or herbivores. Cows, goats, horses, rabbits, insects and grasshoppers are examples of herbivores.

(b) **Secondary Consumers:** Animals which feed on herbivores are known as secondary consumers. Secondary consumers eat the flesh of primary consumers (that eat autotrophs) and are hence known as carnivores. For example, frogs and lizards are secondary consumers as they eat grasshoppers and other insects.

(c) **Tertiary Consumers:** Tertiary consumers are those that eat the flesh of secondary consumers, as is the case with a tiger or a lion, which eats a fox which in turn feeds on herbivores. Tertiary consumers are animals, such as lions, tigers and vultures. Since they are not killed and eaten by other animals, they are known as top carnivores.

Plant → **Insect** → **Frog** → **Snake** → **Eagle**

Autotroph (Producer) | Herbivore (Plants) | Primary Consumer | Secondary Consumer | Top Carnivore

Figure 3.1 Trophic Level in a Grassland Ecosystem

Saprophytes or Decomposers

The dead bodies of the producers and consumers (organic matter) are eaten and broken down into simple inorganic substances by certain microbes (bacteria and fungi). These simple substances are utilized again by the producers (plants) to prepare food. Hence, the decomposers play the vital role of releasing essential materials from the dead organic matter or the plant, thereby maintaining a continuous cycle of materials. Certain decomposers are also called scavengers. Water, carbon dioxide, phosphates, nitrogen, sulphates and a number of organic compounds are by-products of activity of organisms on dead organic matter.

3.2.2 Abiotic Components

All the non-living components of the environment such as light, temperature, humidity, moisture, solar radiation and salinity of soil as well as inorganic and organic compounds constitute the abiotic components of the ecosystem. The quantity of abiotic materials present in an ecosystem is called the 'standing stage'. All biotic organisms (such as plants, animals and microbes) interact only with the abiotic components of the environment.

3.3 Types of Ecosystem

The biosphere includes four major types of habitats namely marine, freshwater, terrestrial and man-engineered (agriculture).

Marine ecosystem: All the saline water reservoirs on earth such as the seas, oceans and estuaries with their characteristic biome form the marine ecosystem.

Freshwater ecosystem: Ponds, streams, rivers and lakes along with their flora and fauna form the freshwater ecosystem.

Terrestrial ecosystem: It includes desert ecosystem, grassland ecosystem, tree ecosystem, crop fields, forest ecosystem and so on.

Man-engineered ecosystem: Agriculture and aquaculture systems are man-engineered ecosystems. These ecosystems have all the essential components such as producers, consumers (both herbivores and carnivores), decomposers and abiotic materials. The main purpose of these ecosystems is to produce more yields in terms of grains, milk, fish or meat. This can be done by increasing the supply of energy in the form of labour, extra nutrients fossil fuels, fertilizers, and pesticides. This is a monoculture system as a single crop is grown whereas natural ecosystems are polyculture systems. Hence, natural ecosystems are more stable compared to man-made ecosystems which are fragile and highly productive and generate lots of pollutants.

3.4 Functional Components of an Ecosystem

The functional components of an ecosystem are:

(i) Biodiversity;
(ii) Productivity (primary and secondary);
(iii) Food chains and food webs;
(iv) Material cycling and energy flow;
(v) Balance of nature and
(vi) Succession and evolution of ecosystems.

3.4.1 Biodiversity

The diversity of organisms in an ecosystem is a very important feature. The diversity of species is the least at the poles and increases progressively towards the tropics. Diversity decreases due to stress, both natural and man-made. The greater the diversity, the better is the ecosystem. There are about 2, 89,100 species of plants and fungi and 10, 53,800 species of animals presently known to scientists. The following table gives an idea of the different types of living species in the biosphere which is an indicator of biodiversity:

Table 3.1 Total Number of Living Species Known to Scientists

Organisms	No. of Species known
Algae	26,900
Bacteria	3,060
Similar form of viruses	1,000
Fungi	28,983
Protozoa	30,800
Higher plants	2,48,400
Insects	7,51,000
Other animals	28,1000

Biodiversity in its broadest sense includes the following diversities:

Taxonomical: The existence of a variety of species, ecological, inter- and intra-specific relationships and niches of ecosystems;

Adaptational: Species adapting to the environmental changes for survival, reproduction and continence;

Genetic: Intra- and inter-specific genetic variations between the species that makes a differentiation between them and

Biochemical: Metabolic and chemical diversities synthesized by organisms.

Countries where there is an abundant diversity of plants and animals are called megadiversity countries. Brazil tops the list of being a megadiversity country with the highest number of flowering plant species. India with the prevalence of a wide range of soils, habitats and climatic conditions is also a country of megadiversity. However, the rate of deforestation in India (2.7 per cent), which is the highest rate among the megadiversity countries poses a great threat to the country's rich biodiversity.

The number of species in a community or its 'specie richness' by itself, is an inadequate indicator of the diversity of species. The relative abundance of each species is also important. Hence, ecologists consider both the richness and relative abundance of the species to measure its diversity. When an ecosystem is adversely affected by man-made or natural causes, some sensitive species might get eliminated. In other words, the richness of the species decreases but this decrease may favour the relative abundance of some others. For example, water hyacinths grow abundantly in a polluted pond at the cost of some sensitive species of organisms and fishes.

Ecologists express diversity in species by two common indices namely The Simpson Index and the Shannon–Wiener Index.

$$\overline{\text{H}} = \sum\left(\frac{n_1}{N}\right)\log_e\left(\frac{n_1}{N}\right)$$

The Shannon and Wiener Index is expressed as follows.

Where H = Shannon and Weiner index,
n_1 = Number of biomass or energy flow for each species or component in a community,
N = total number of biomass or energy flow in the community.

The higher the number of species the higher is the value of index for the community. Hence, the index can be used:

(i) To study the effect of pollution on species;

(ii) For comparison of communities in the same climatic area and

(iii) To compare aggregation of some species in different climate zones.

3.4.2 Productivity

Primary Productivity

The rate of food production in unit time on a unit area is known as productivity. Green plants or autotrophs contain chlorophyll with the help of which they convert solar energy into chemical energy. The total amount of solar energy converted by autotrophs into chemical energy (carbohydrates) is the Gross Primary Production (GPP). Plants use some amount of this GPP for respiration and maintenance of their body. The remaining amount of energy, known as Net Primary Production (NPP), is stored in the plant body and is used by all heterotrophs of the biosphere to support life. The rate of primary production (amount of energy fixed by autotrophs) depends upon:

(i) Availability of solar energy: This energy is more for the leaves facing the sun than for those which do not face the sun (leaves in shade).

(ii) Amount of energy expenditure by respiration and other maintenance processes.

It is observed that in the community of C4 type of plants such as sugarcane and some grasses where photorespiration is absent, the GPP rate is higher than crops in the field. Although the GPP is high in the trees of climax forests, the NPP is the least as most of it is used in respiration. On the contrary, in the agricultural system, the loss due to respiration is the least, hence NPP is the most.

Secondary Productivity

Secondary Productivity is the net quantity of energy transferred and stored in the reproductive and non-reproductive tissues of heterotrophs over a fixed period of time. Heterotrophs include all herbivores, carnivores and decomposers. Since heterotrophs depend upon autotrophs for food, the secondary productivity directly depends upon primary productivity. Secondary productivity also depends upon temperature, ecological and physiological factors and also on the body size of the heterotroph.

The tissues of green plants (autotrophs) mainly consist of carbohydrates while the tissues of animals (heterotrophs) contain fat and protein. Hence, secondary productivity gives more proteins and energy. However, on an average only about 15 to 20 per cent of primary productivity is converted to secondary productivity. For this reason, meat and milk are costlier than food grains and other plant products.

It has also been observed that among heterotrophs, the warm-blooded animals (mammals and birds) assimilate 70 to 80 per cent of the consumed food but spend 90 to 95 per cent of it on body metabolism. On the other hand, fish, prawn and other cold-blooded animals spend only 30 to 50 per cent of the assimilated food and they store about 50 to 70 per cent of the assimilated food. Hence, cultivation of fish and prawns is more profitable than mammalian meat.

3.4.3 Food Chains and Food Webs

As already discussed, in the ecosystem only green plants contain chlorophyll with the help of which they can convert solar energy to food which is being taken by different heterotrophs. Heterotrophs cannot produce food for themselves. Herbivores, carnivores, and decomposers are collectively known as heterotrophs. The food which is produced by an autotroph is eaten by herbivores (cows, goats, horses, rabbits and so on) which are eaten by small carnivores which in turn become food for bigger carnivores and the process continues.

Ultimately after the life cycle, the dead organism is decomposed by different decomposers (bacteria, fungi, microbes etc.) as shown in Fig. 3.2.

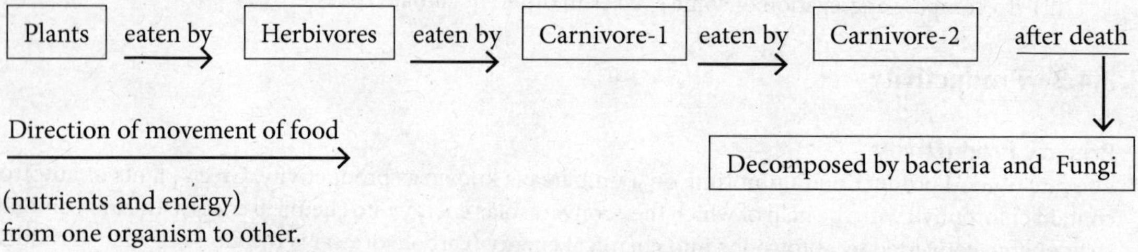

Figure 3.2 Flow Chart Showing the Movement of Food and Energy in a Food Chain

This sequence of organisms which feed on one another for their survival is known as the food chain. In a food chain, the links are known as trophic levels (food level). The plants are producers and form trophic level-1, the herbivores are primary consumers and form trophic level-2, the small carnivores are secondary consumers and form trophic level–3 and the big carnivores are tertiary consumers and form trophic level-4. They are also called top carnivores.

The food chain in the ecosystem helps to maintain:

(i) The biodiversity of nature.

(ii) The feeding relationship of nature.

(iii) Flow of energy of the ecosystem.

(iv) Passage of nutrients from one organism to another.

The only demerit of a food chain is that along with nutrients it also transports toxic substances from one organism to another which finally results in biomagnification. In the ecosystem, a number of food chains operate simultaneously. These food chains are inter-linked with one another to form a food web. For example, a plant may be food for any herbivore or carnivore such as human beings. A herbivore again becomes food for carnivore-1 or is directly eaten by the top carnivore. For example, a mouse feeding on food grains becomes food for a snake which is eaten by a hawk. The mouse can be directly eaten by a hawk. In this way, the inter-related complex food chain forms a food web. A food web constitutes a number of alternative paths for energy flow and provides greater stability to the ecosystem. A food web of a grassland ecosystem is shown in Fig. 3.3.

Ecological Pyramids

In 1927, scientist Charles Elton observed that the number of animals present at the top of the trophic level is much less compared to the number of animals present at the base of the food chain. He also plotted his findings on a graph to get a pyramid-like structure. He called this pyramid the Eltonian Pyramid after his name. It is also known as the ecological pyramid. Thus, an ecological pyramid is the graphical representation of the trophic structure (the position of organisms in the food chain) and function at successive trophic levels. The base of the pyramid consists of the food producer level. The successive levels make the tiers, with the top carnivores forming the apex. These ecological pyramids are of three types:

(i) Pyramid of numbers.

(ii) Pyramid of biomass.

(iii) Pyramid of energy.

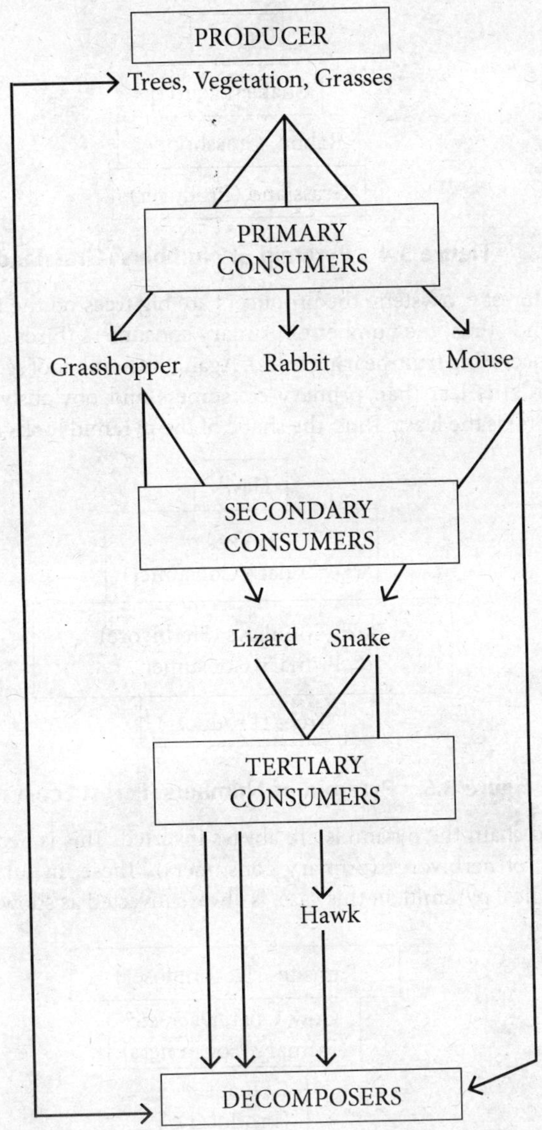

Figure 3.3 Simple Food Web of a Grassland Ecosystem

Pyramid of Numbers: This pyramid deals with the relationship between the number of primary producers and consumers (herbivore and carnivore) of different orders. Depending on the nature of the food chain, in the present ecosystem the pyramid of numbers may be upright or inverted. For example, in a grassland ecosystem, the number of grasses (producer) is always high followed by primary consumers (herbivores like rabbits and grasshoppers) that are less, the secondary consumers (carnivores like snakes and lizards) that are lesser and finally the top carnivore, in this case hawks, which are the least in number. So the pyramid is upright in this case as shown in Fig. 3.4.

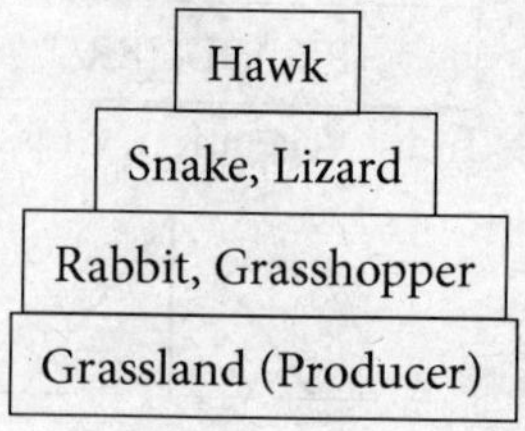

Figure 3.4 Pyramid of Numbers (Grassland)

On the other hand, in a forest ecosystem, the producers are big trees on the fruits of which birds and other primary herbivores depend. Thus, the number of primary consumers (birds and monkeys) is always greater than the number of producers (big fruit-bearing trees). Again the number of secondary consumers (carnivores such as snakes and lizards) is less than primary consumers and obviously the number of top (tertiary) carnivores (lions and tigers) is the least. Thus, the shape of the pyramid looks as shown in Fig. 3.5.

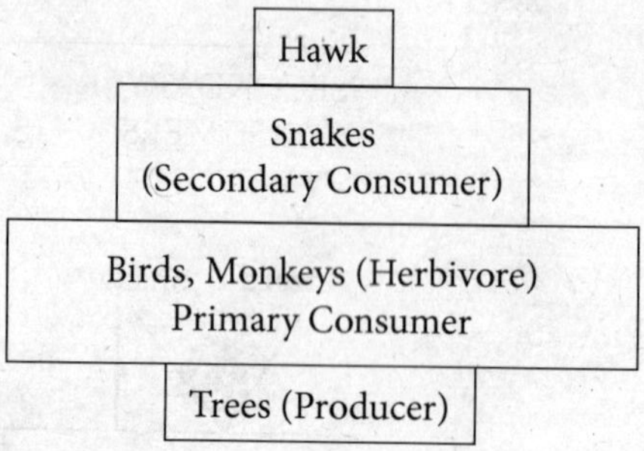

Figure 3.5 Pyramid of Numbers (Forest Ecosystem)

In case of a parasitic food chain the pyramids are always inverted. This is because a single plant (producer) supports a large number of herbivores (primary consumers). These, in turn, support a large number of parasites. Thus, the ecological pyramid, in this case, is always inverted as shown in Figure 3.6.

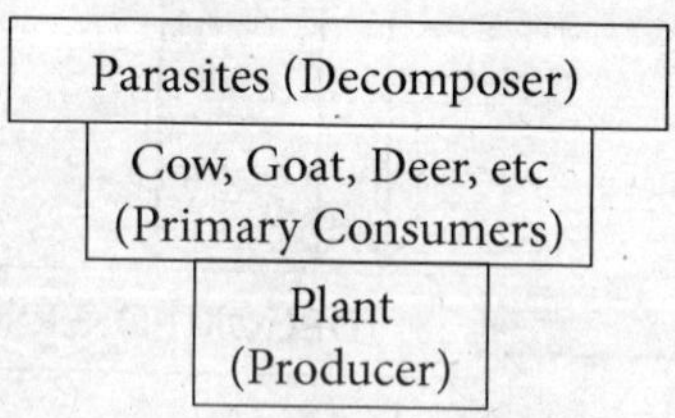

Figure 3.6 Pyramid of Numbers (Parasitic food chain)

As it is very difficult to count the exact number of all organisms, the pyramid of numbers cannot give a true picture of the trophic structure for an ecosystem. They generally vary with different communities having different types of food chains in the same ecosystem.

Pyramid of Biomass: In this concept, the individual in each trophic level is weighed instead of being counted. Thus, in a pyramid of biomass the total weight of each trophic level is represented. For most of the ecosystems on land (for example, forest and grassland), the biomass of producers is large (the base of the pyramid) and it gradually decreases with each successive layer resulting in an upright pyramid as shown in Fig. 3.7.

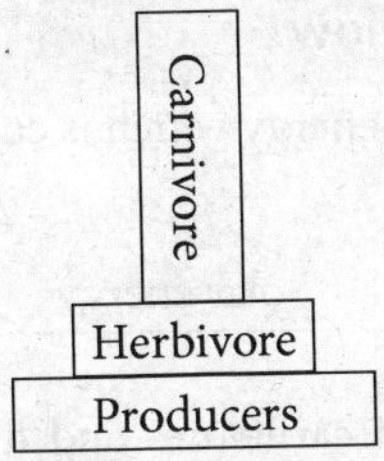

Figure 3.7 Pyramid of Biomass (g.dry – wt/Unit area) on Land (Forest, Grassland)

In the case of a pond ecosystem, the producers are tiny phytoplanktons which grow and reproduce rapidly. These phytoplanktons are consumed as fast as they reproduce (only survivors). In this case, the biomass of the consumer at any instant is more than the producer biomass. Thus, in the case of a pond or any aquatic ecosystem, the pyramid of biomass has an inverted shape as shown in Fig. 3.8.

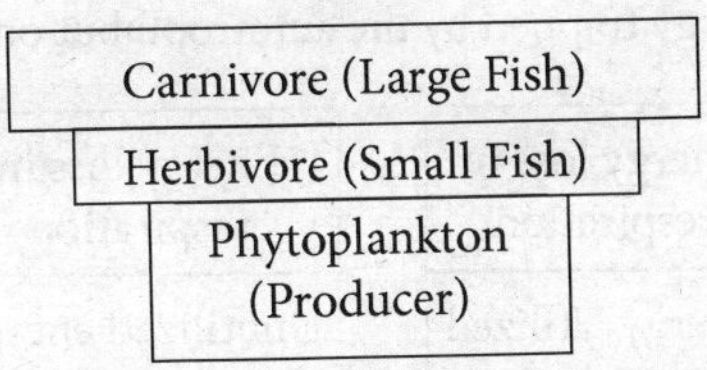

Figure 3.8 Pyramid of Biomass of an Aquatic Ecosystem

Pyramid of Energy: The pyramid of energy actually depicts the rate at which the food mass is passed through the food chain. It is based on the actual amount of energy that individuals take in, how much is burnt up in the metabolism, how much remains in the waste products and how much they store in the body tissue; this is a reflection of the laws of thermodynamics. Thus, the energy pyramid gives the best picture of the overall nature of the ecosystem. The actual amount of energy content in successive trophic levels from the producer to various consumers decreases. Hence, the shape of energy pyramid is always upright.

For example, a forest ecosystem receives 1,000 calories of sunlight in a day of which about 100 calories are stored in the plant. When any herbivore (primary consumer such as a deer or a goat) eats the plant, it receives 100 calories. But after its expenditure, its own metabolism lets it store 10 calories of energy. Thus, any carnivore (secondary consumer) eating the above herbivore will receive 10 calories. It can store one calorie of energy after its expenditure for metabolism. Thus, the energy pyramid looks as shown in Fig. 3.9.

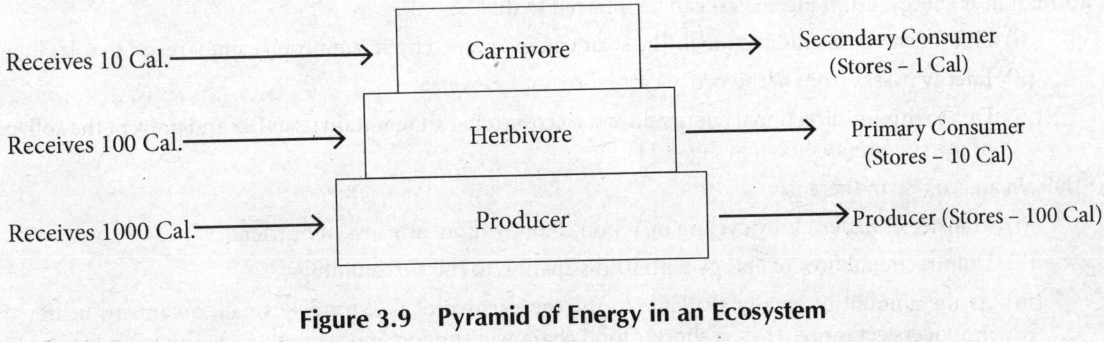

Figure 3.9 Pyramid of Energy in an Ecosystem

3.4.4 Material Cycling and Energy Flow

It is clear from the food chains that the solar energy which is converted by the autotrophs to chemical energy as carbohydrates, fats and proteins,

$$6CO_2 + 6H_2O \xrightarrow{\text{Solar energy}} C_6H_{12}O_6 + 6O_2$$

is transferred to herbivores then to different carnivores and finally to the decomposer level. Decomposers (bacteria, fungi, and microbes) break it into a simpler form which is used as nutrients by the autotrophs. Thereby a complete cycle of the essential nutrients takes place. But the energy does not cycle in the ecosystem rather its flow is unidirectional. According to the first law of thermodynamics, 'The total energy of the universe is constant although it can be transferred from one form to another.' That means energy can neither be created nor destroyed only its inter-conversion is possible. The solar energy trapped by plants is transmitted to herbivores grazing on them followed by the carnivores eating the herbivores and finally to the micro-consumers (bacteria, fungi and microbes). In each trophic level there is a loss of energy during transfer. So the top-level consumer does not get the total amount of energy trapped by the autotroph but only 10 per cent of it.

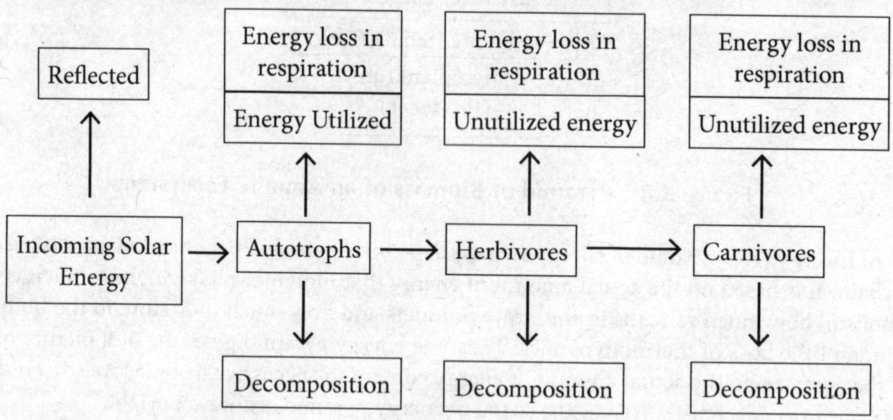

Figure 3.10 Flow of Energy in an Ecosystem

The dissipation of energy during its transmission from one trophic level to another is in agreement with the second law of thermodynamics which states that 'Processes occur spontaneously only if the sum total of the entropy of the system and its surrounding increases.' In other words, processes involving energy transformation will occur spontaneously if energy degradation takes place from a non-random to a random form of heat energy. From Fig. 3.10 it can be inferred that:

(i) Flow of energy is unidirectional. The solar energy trapped by an autotroph cannot revert to solar input.

(ii) Energy passes from herbivore to carnivore not vice versa.

(iii) Due to this unidirectional energy flow, the ecosystem can maintain its entity and prevent the collapse of the system (as shown in Fig. 3.11).

Thus, in an ecosystem there is:

(i) Transfer of materials by cycling in a food chain without any loss of nutrients.

(ii) Unidirectional flow of energy with its dissipation to the surroundings.

(iii) As the amount of energy available to the top carnivores is extremely small, organisms nearer the producers get more. Thus, a shorter food chain will support more numbers as shown in Fig. 3.12.

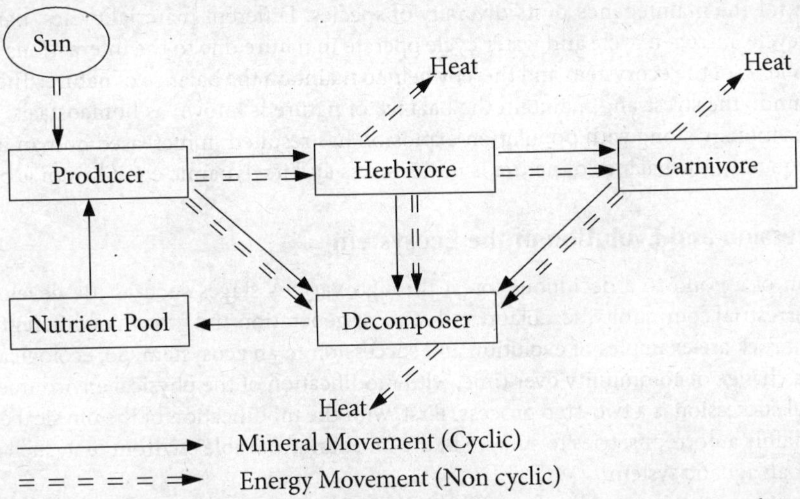

Figure 3.11 Flow Chart of Energy Mineral Movement in the Ecosystem o' Organisms

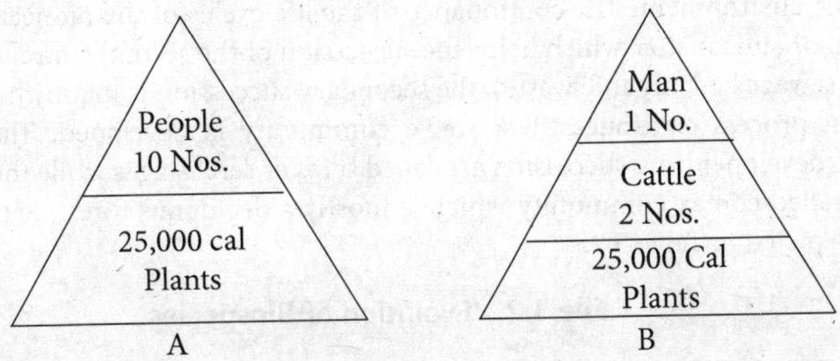

Figure 3.12 Relative Efficiency of Vegetarian and Non-Vegetarian Diets

In Fig. 3.12 A, a vegetarian diet of 25,000 calories supports 10 people, while in Figure 3.12 B, the same amount of plant food supports only one man who eats meat.

3.4.5 Balance of Nature

In an ecosystem the component parts like food chains, material cycling and energy flow are closely inter-related to different living organisms thereby maintaining a dynamic equilibrium amongst them. A fluctuation in any of its parameters will disturb the equilibrium and try to affect the total balance of nature. The science of systems of control in an ecosystem is known as cybernetics.

For example, in a forest, the population of carnivorous animals (lions and tigers), herbivorous animals (deer and rabbits) and autotrophs (food for herbivores) are interdependent on each other and maintain a dynamic equilibrium. An increase in the population of lions or tigers (carnivorous) will decrease the population of the herbivores such as, deer, rabbits, buffaloes and so on. As a result, the lion or tiger community will suffer from shortage of food and would either die or migrate to other areas. Ultimately the decrease in herbivore population will decrease the number of carnivores of that area. Similarly, if in a forest the population of herbivores increases then they will eat away more of the grasslands (autotrophs) making the forest an open land. Consequently, the herbivores will face food shortage and will migrate away. This in turn will affect the carnivore community as well. Hence, nature always tries to maintain the balance of the ecosystem for the maintenance of its diversity of species. Different material cycles, namely the oxygen cycle, carbon cycle, nitrogen cycle and water cycle operate in nature due to the interactions of the biotic and abiotic components of the ecosystem and thereby help to maintain the balance of nature. The tendency of the ecosystem to undo the stress and maintain the balance of nature is known as homeostasis. Development of science and technology along with population explosion has resulted in increased human interference with nature. Consequently, the balance of nature is under stress and its dynamic equilibrium is being disturbed.

3.4.6 Succession and Evolution of the Ecosystem

The conversion of a pond to a deciduous forest through various stages such as the development of mesic community, terrestrial community, scrubland and so on or generation and growth of different biospecies in an exposed parent rock are examples of evolution and succession in an ecosystem. So, ecological succession can be defined as a change of community over time, with modification of the physical environment.

Ecological succession is a two-step process. First, with the modification of the physical environment one community, mainly autotrophs, tries to survive in the most unfavourable environment, such as an open rock, sand dunes, or an aquatic system.

Evolution of a biospecies for the first time in a sterile area is known as primary succession. The first community is always autotrophic and is usually known as the pioneer community as its evolution occurs in a purely inorganic environment. The continuance of the life cycles of the pioneer community makes the environment rich in organic matter which helps the succession of the second community. As a lot of organic matter is available (sewage bed, organic waste), the secondary succession is mainly heterotrophic.

The succession process continues till a stable community is developed. The transitional series of communities which develop during succession are called seres or seral stages while the stable community that develops finally is called climax community which is mostly a deciduous forest. Series of different physical environments are depicted in Table 3.2.

Table 3.2 Evolution of Biospecies

Physical Environment	Biospecies
Aquatic system	Hydrach succession or Hydrosere
Exposed parent rock	Xosere
Dry sand (Desert)	Psammosere
Saline land	Halosere

The major terrestrial communities are different types of forests namely, moist tropical forest, montane subtropical forest, montane temperate forest, rain forest, deciduous forest, coniferous forest and so on. The major aquatic communities are inland water bodies, oceans, estuaries and so on.

Both terrestrial and aquatic communities show abundant biodiversity, depending on the variation of the environment. In an aquatic system, biodiversity depends on the variation of salinity of waters.

Biomagnification (Biological Magnification)

Biomagnification is an anthropogenic (man-made) process in the ecosystem through which accumulation of non-degradable pollutants (toxicants) increases in the food chain to a harmful concentration level. The transfer of food energy from primary producers to top carnivores through a series of organisms including repeated eating and being eaten is known as the food chain. The food chain helps in the dispersal of food energy. But sometimes, the concentration of a particular chemical increases manifold inside this food chain as the weaker organisms are eaten by the stronger ones. Thus, increasing the the concentration of the non-degradable substances in the food chain, which cannot be metabolized by living organisms. Herbicides, rodenticides and insecticides belong to this category. For example, pesticides such as Dichloro-Diphenyl Trichloroethane (DDT)–a broad-spectrum pesticide, are used in the fields at a very low concentration of 0.02 parts per million (ppm) to control harmful insects. At this concentration, its toxicity for human beings is very low. Microscopic plants take up the contaminated water and small fishes eat these plants. It also enters the body of the fish through its skin. In this way, it enters the food chain and travels through it. Being a fat-soluble chemical, DDT sticks to the fat tissue of the animal body and its concentration inside the organism's body increases.

The fishes are eaten by birds and with this step the concentration of DDT in the fat tissue of the body reaches 2,000 ppm after getting magnified in the food chain and becomes lethal for the host body. A similar effect is observed with heavy metals (mercury poisoning and arsenic poisoning) and radio nucleotides.

3.5 Different Ecosystems

Ecosystems can be defined as the structural and functional unit of ecology. Our earth is a giant ecosystem where abiotic and biotic components are constantly interacting with each other bringing structural and functional changes in it. Due to the vastness, it is subdivided into units of smaller ecosystems like:

(i) Terrestrial ecosystems such as forest, grassland, desert.

(ii) Man-Engineered ecosystems such as cropland.

(iii) Aquatic ecosystems such as freshwater, marine water.

These unit ecosystems are open systems with no constraints of boundaries and with constant interaction between biotic and abiotic components.

3.5.1 Forest Ecosystem

The forest as an ecosystem contains interacting biological communities (vegetation) and faunal communities which together interact with the physical environment resulting in an integrated structure. It is always the vegetation which forms the base of the food chain.

According to the forest survey report of 1993, about 19.5 per cent of the total geographical area of India is under forest cover. Depending on factors such as annual rainfall, its distribution over the year, mean monthly temperature, total annual dry and wet period and relative humidity, the Indian forest can be classified into 16 different types. However, the different components of all forest ecosystems are the same.

(i) **Abiotic Component:** All inorganic, organic (litters, debris) substances present in the environment and minerals present in the forest constitute the abiotic components. Mainly, the amount of sunlight depends on the stratification condition of the trees.

(ii) **Biotic Component:** The biotic component consists of all living components of the environment which constitute producers, consumers and decomposers.

Producers: The vegetation of the forest is the producer. The term vegetation includes big trees, medium-sized bush and small herbaceous plants. All vegetation contains chlorophyll and performs photosynthesis. The herbaceous vegetation contains maximum green photosynthetic tissues, produces the maximum and being annual, it also decomposes very fast. Thus, herbaceous vegetation contributes to the nutrient cycling and boosts the production in forests.

Consumers:

Primary Consumer: Insects like ants, beetles, flies, spiders, birds and other herbivores such as deer, squirrels, shrews, mongooses and elephants graze over the primary producer and convert it into secondary production. Thus, herbivores are the links between the primary producer and the carnivores.

Secondary Consumer: Carnivorous animals such as snakes, birds, fox and jackal which are the predators of herbivores come under this category. These animals regulate the population size of herbivores and thereby their grazing activity. In this way, the base of the food chain is maintained.

Tertiary Consumers: These are top carnivores, such as lions, tigers and hawks that feed on secondary consumers. Thus, there exists a complete balance between different groups of animals and plants and the forest ecosystem is naturally conserved. When there is a loss of this balance (biodiversity), the destruction of the forest begins.

Decomposers: These organisms remain confined to the soil of the forest floor and have the capacity to degrade all dead plants, herbivore and carnivore tissue to release nutrients into the soil. These nutrients are again used by the producer. A wide variety of micro-organisms such as fungi, bacteria, mites, nematodes, protozoa and earthworms are present in the forest soil to perform the role of decomposers. The rate of decomposition is more rapid in tropical and subtropical forests than in a temperate one. Thus, decomposers act as scavengers of the forest, have links with all groups of plants and animals and help in recycling the nutrients.

3.5.2 Grassland Ecosystem

Continental interiors, especially in temperate regions with low rainfall are dominated by grasses. Grasslands come under the terrestrial ecosystem and occupy about 19 per cent of the earth's total surface. The abiotic and biotic components of a grassland ecosystem are described as follows:

(i) **Abiotic Component:** The nutrients of the environment, such as C, H, O, N, P, S and so on are supplied by carbon dioxide, water, nitrates, phosphates and sulphates present in the air and soil of the area.

(ii) **Biotic Component:** The biotic component consists of the following:

Producers:

These are mainly grasses, a few forbs (herbaceous flowering plants) and shrubs that contribute to primary production.

Consumers:

Primary Consumers: Herbivores such as bison, antelope, cattle and rodents feed on grasses and insects such as leptocorisa, dysdercus, cicineella, some termites and milllipedes feed on leaves of grasses.

Secondary Consumers: These are carnivores such as foxes, jackals, snakes, frogs, lizards and prairie dogs that feed on herbivores, the primary consumers.

Tertiary Consumer: Birds such as hawks feed on secondary consumers in a grassland ecosystem.

Decomposers: Microbes, some bacteria, actinomycetes and fungi such as mucor, aspergillus, penicillium and rhizopus become active in the decay of dead organic matter and bring back the minerals/nutrients to the soil. This soil is rich in mineral content and is useful for farming.

3.5.3 Desert Ecosystem

Continental interiors with very low sporadic rainfall and with low humidity are converted to deserts. The sun's rays easily penetrate the atmosphere making the ground temperature very high during the day. The nights are very cold. The species composition is quite varied because of the extreme climatic condition.

Biotic Components

The biotic component consists of the following:

Producers: Drought-resistant vegetation like euphorbias, sage bush and cacti are common here. Lower plants such as lichens and xerophytic mosses may also thrive in an oasis area.

Consumers: A large number of nocturnal animals, mainly various reptiles and insects are seen here. Some birds and camels that feed on shoots of plants are also present here.

Decomposers: A few fungi and thermophillic bacteria are present in this ecosystem. As the vegetation is less, the decay is also less and is managed by these decomposers.

3.5.4 Aquatic Ecosystem

About 70 per cent of the earth's total surface is under the aquatic ecosystem. An aquatic ecosystem can be freshwater, marine or estuarine ecosystem. Thus, a wide variety of species are found in the aquatic ecosystem.

Freshwater ecosystem: A freshwater ecosystem is of two types:

Lotic: Having flowing water. For example, freshwater streams, springs, rivulets, brooks, rivers and so on.

Lentic: Having stagnant or still water, for example, ponds, swamps, bogs and lakes besides others. Lotic water bodies usually start from narrow springs and waterfalls and tend to broaden and become deep and slow-moving. So, in the upstream only those organisms are available which can maintain their position in the fast flowing water and can adhere to an exposed surface. Adhering organisms associated with large aquatic plants are known as periphyton. The main producers of the lotic system are algae and the organic matter brought in from the surrounding terrestrial ecosystem. Consumers are fishes like trout and salmon found upstream and carp and catfish found downstream. Thus, the nutrient level is always higher downstream.

Lentic water bodies include ponds, swamps and lakes. Being a still water system they vary widely in chemical, physical and biological characteristics. Lentic water bodies are considered to have three zones namely littoral, limnetic and profoundal.

The littoral zone extends from the shore line to the innermost rooted plants and is dominated by floating vegetation rooted at the bottom (for example, reed, cattail and water lilies). Frogs, snails, snakes and a large number of adult insects and their larvae are found here.

The limnetic zone is the zone of open water down to the depth where light penetrates. Phytoplanktons such as diatoms, green, blue-green algae, zooplanktons from protozoan to micro-arthropods, fish, amphibians, the nekton and larger insects are available in this region.

The profundal zone is present below the limnetic zone. This zone gets food from the limnetic zone and consists mainly of decomposers. The nekton in this zone varies with temperature and nutrient condition. The components of a still water body like a pond (freshwater ecosystem) are self-regulating and self-sufficient. The components of an aquatic system may be classified into two types, biotic and abiotic.

(i) **Abiotic Component**: Abiotic components in this case are carbon dioxide, oxygen, calcium, nitrogen, phosphorous, amino acids and water.

(ii) **Biotic Component**: The biotic component consists of the following:

Producers: The autotrophic green plants and some photosynthetic bacteria fix the radiant energy with the help of nutrients obtained from the mud of the pond.

Consumers:

Primary Consumers: Herbivores like zoo plankton and small invertebrates such as copepods feed on the producers.

Secondary Consumers: Small carnivores like small fishes feed on primary consumers.

Tertiary Consumers: Large fishes that feed on the small fishes are the tertiary consumers.

Decomposers: Finally, bacteria and fungi present at the base of the pond decompose the organisms and help in the release and recycling of nutrients.

Marine (Ocean) Ecosystem: Three-fourths of the earth's surface is covered by oceans with an average salinity of 35 parts per thousand (ppt). The concentration of nutrients in the marine (ocean) ecosystem is low. However, a marine ecosystem can be divided into littoral, neritic, pelagic and benthic zones. The shoreline between the land and the open sea is termed as littoral zone. Waves and tides have the maximum effect in this zone. In the rocky shore region, organisms such as barnacles, starfish, and algae are found, whereas in the sandy shore area, organisms burrowing and adhering to sand are dominant. In bays, algae are found on the surface of mudflats. Often photosynthetic bacteria are present below the algae. Coral reefs are also formed by colonial coelenterates.

Just above the continental shelf lies the neritic zone. This zone is rich in nutrients (washed from land) and hence rich in species too. Sunlight also penetrates the neritic zone. Hence, the productivity of this zone is high. Phytoplankton and zooplankton are abundant here and support fishing grounds. However, pollution affects this zone first.

The open sea constituting 90 per cent of the total ocean surface forms the pelage zone. Photosynthesis is carried out by phytoplanktons present in this zone. Zooplanktons, shrimps and jelly fish are also found here. However, though this zone is very large in area, it is low in nutrients and productivity. Fin and blue whale are found here. Organisms of the pelagic zone are present below the light penetration zone and totally depend on the rain of detritus of upper regions for their nutrition. In deeper water, many animals have poor vision, fishes are biome-nascent. Some deep-water fishes have a light-producing organ.

The floor of the ocean constitutes the benthic zone. It stretches from the edge of the continental shelf to the deepest ocean trenches. Organisms present here are heterotrophic. Rooted animals such as sponges, sea lilies, sea fans and so on are present here. While snails and clams remain embedded in the mud, starfish, sea cucumbers and sea urchins move on the surface.

Estuarine ecosystem: Coastal bays, river mouths and tidal marshes form estuaries. Here, freshwater from the rivers mixes with the ocean water. The degree of salinity depends upon the amount of freshwater flow and tidal inflow. Estuaries are more productive than adjacent rivers or oceans due to the high concentration of the nutrients received from land as well as the sea. Rooted plants are supported in shallow water of lesser salinity than the sea.

Organisms present in estuaries are those which are capable of tolerating fluctuation in the salinity of water. Some oysters, crabs and sea shrimps are found here. Estuaries contain producers, such as seaweed, marsh gases, benthic algae and phytoplanktons. They are also used as nurseries by deep-water fishes to bring up their younger ones.

3.6 Biogeochemical Cycles

The bodies of living organisms of both plants and animals are made up of carbon, hydrogen, nitrogen, oxygen and phosphorous and minute quantities of calcium, iron, copper and molybdenum. These elements are present in the environment and are taken in by the plants, (autotrophs) they reach all the heterotrophs through the food chain and are finally returned to the environment through the decomposers. This cycling of matter through the atmosphere, hydrosphere, biosphere and lithosphere is known as biogeochemical cycle. Biogeochemical cycles are classified as follows:

(i) Gaseous cycle, represented by oxygen, nitrogen and carbon cycle.

(ii) Sedimentary cycle, represented by the phosphorous cycle. Sulphur is found both in the atmosphere as well as in gases such as hydrogen sulphide and sulphur dioxide and also in the earth's crust. So, sulphur represents a combination of gaseous and sedimentary cycles.

Some important cycles are discussed here.

Oxygen Cycle: Oxygen constitutes nearly 21 per cent of the atmosphere by volume. Oxygen is also present in water and rocks in a combined state. The oxygen cycle is based on the exchange of oxygen among the various environmental segments namely atmosphere, lithosphere, hydrosphere and biosphere.

Oxygen is a vital component for the survival of the biosphere. It is required by most of the plants and animals for enzymatic oxidation of organic food that sustains growth and general metabolism. Hence, it is absorbed from the environment during aerobic respiration. Green plants release oxygen in the environment during photosynthesis.

There is also an exchange of oxygen between the atmosphere and all water surfaces on the earth. In the atmosphere, at the troposphere level there is constant conversion of oxygen to ozone in the presence of high-energy radiation. The ozone layer shields the harmful ultraviolet radiation from entering the earth's atmosphere. As a result of all these processes the total amount of oxygen in the biosphere is constant and the oxygen cycle becomes stable.

$$O_2 + hv \longrightarrow O + O_2$$

$$O_2 + hv \longrightarrow O + O$$

$$O_2 + O \underset{}{\overset{hv=220\text{–}330nm}{\rightleftharpoons}} O_3$$

Nitrogen Cycle: The nitrogen cycle consists of various steps namely fixation, ammonification, nitrification and denitrification. Nitrogen and its compounds are essential constituents of protein which is the building block of all living organisms. There is a continuous exchange of nitrogen within the ecosystem thus maintaining the nitrogen cycle as shown in Fig. 3.13.

Although there is an inexhaustible supply of nitrogen in the atmosphere, most living organisms cannot use it directly. Thus, for the use of living organisms the elementary nitrogen has to be converted to its compound form such as ammonia, nitrates or nitrites. This conversion of nitrogen to its compounds is called 'fixation of nitrogen'. This can be achieved:

(i) by atmospheric thunder and lightning (to a limited extent);

(ii) by certain free-living bacteria and blue green algae and

(iii) by industrial processes.

Atmospheric thunderstorm and lightning convert gaseous nitrogen to nitrates and it reaches the soil through precipitation and is used by plants.

Leguminous plants such as pea, beans and so on contain nodules in their roots where symbiotic bacteria (such as rhizobium) live. These bacteria can fix atmospheric nitrogen to nitrates used by the plants. The blue-green algae can also fix atmospheric nitrogen as ammonium ion, which is further oxidized by certain bacteria to form nitrates and nitrites.

Certain plants can directly absorb ammonia as a source of nitrogen. Industrially prepared nitrogen compounds are also added to the soil to replace the nitrogen that has been removed during harvesting.

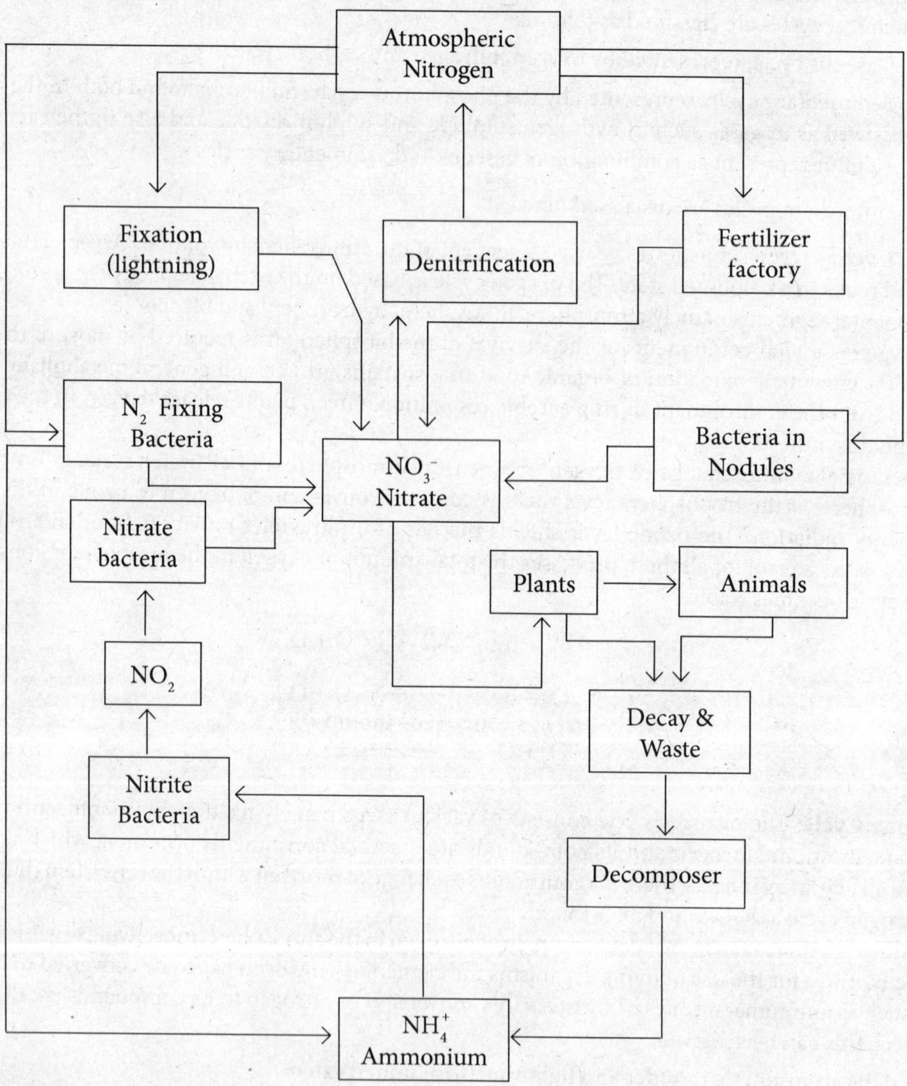

Figure 3.13 Nitrogen Cycle

The death and decay of plants, animals and also the excreta of animals are major sources of nitrogenous compounds (proteins) to the soil. Some soil micro-organisms break down these nitrogenous compounds to N_2 in the soil (denitrification).

The nitrates synthesized by the bacteria in the soil are taken up by plants and they go through higher trophic levels of the ecosystem. The death and excretion of living organisms return nitrogen to the soil as ammonia. A certain amount of soluble nitrates is lost to the system with run-off along with surface water or groundwater. In the ocean and soil, the denitrifying bacteria convert these nitrates and nitrites to elemental nitrogen which escapes to the atmosphere and thus completes the cycle and maintains the atmospheric balance of nitrogen concentration. The application of fertilizers only alters and upsets the huge stock of atmospheric oxygen. The addition of nitrogenous compounds to lakes enhances the growth of phytoplankton and causes the decay of the lake.

Carbon Cycle: Carbon is a vital requirement for the production of carbohydrates by plants during photosynthesis. Plants absorb carbon from the atmosphere as carbon dioxide. From plants, this carbon goes to animals as carbohydrates through food and from animals to different micro-organisms. These micro-organisms or decomposers decompose the dead organic matter while carbon returns to the atmosphere as carbon dioxide (CO_2).

Some dead plants and animals get buried in sediments before they decompose for hundreds of millions of years; they are gradually transformed into coal, oil or natural gas. On burning these fossil fuels, the stored carbon in them escapes to the atmosphere as CO_2.

Oceans are the richest source of carbon where it is stored as carbonate and bicarbonate ions and an exchange with the atmosphere occurs continuously. The oceans contain 50 times more dissolved CO_2 than the atmosphere. A major portion is present below the thermocline where the temperature of the water prevents rapid exchange with the atmosphere. The ocean regulates the atmospheric CO_2 level to 0.032 per cent despite the photosynthesis uptake. A small amount of carbon dioxide is also returned to the atmosphere through respiration and different metabolic activities of the living organisms of the biosphere. The carbon cycle is shown in Fig. 3.14.

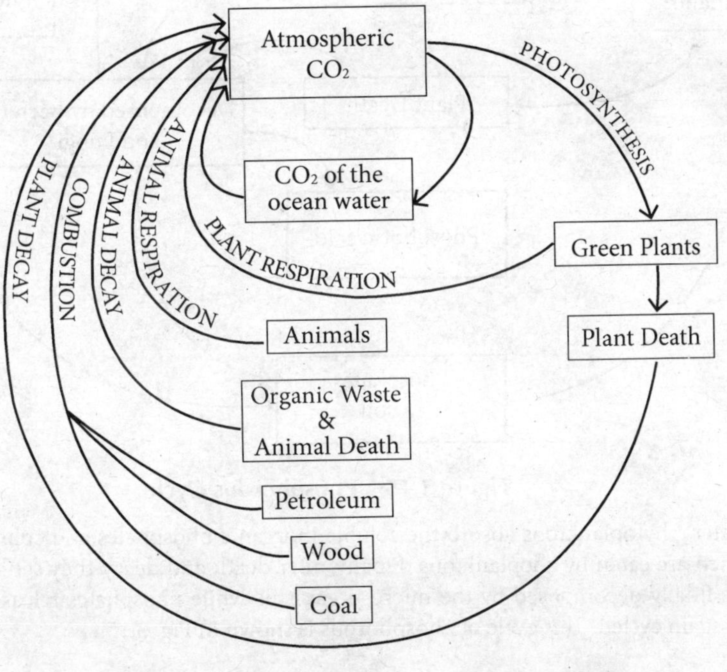

Figure 3.14 Carbon Cycle

Although carbon is the basic requirement for the succession of the ecosystem, its percentage in the atmosphere is only 0.032 per cent compared to oxygen which is 20.95 per cent and nitrogen which is 78.804 per cent. But this minor carbon content is always maintained by the exchange of carbon dioxide between the atmosphere and organisms, and atmosphere and the ocean. With the onset of industrialization and urbanization, the natural balance of carbon in the environment appears to be disturbed in spite of the carbon cycle.

Phosphorous Cycle: The phosphorous cycle is a sedimentary one and comes mainly from phosphatic rocks. Phosphorous in terms of phosphates (PO_4^{-3}) is required for the growth and maintenance of animal bones and teeth. Organic phosphates are required for cell division and production of DNA and RNA. It is also present in the ATP (Adenosine Triphosphate) of living organisms.

Phosphorous is present as inorganic soluble and insoluble phosphates in soil and rocks. Terrestrial plants absorb the inorganic phosphates from the soil and convert them into organic phosphates. Heterotrophs get the phosphates from the autotrophs and finally on the death and decay of living organisms these are decomposed by the micro-organisms to the soil. A part of these are absorbed in the soil while a major part leeches out to rivers and then to the oceans.

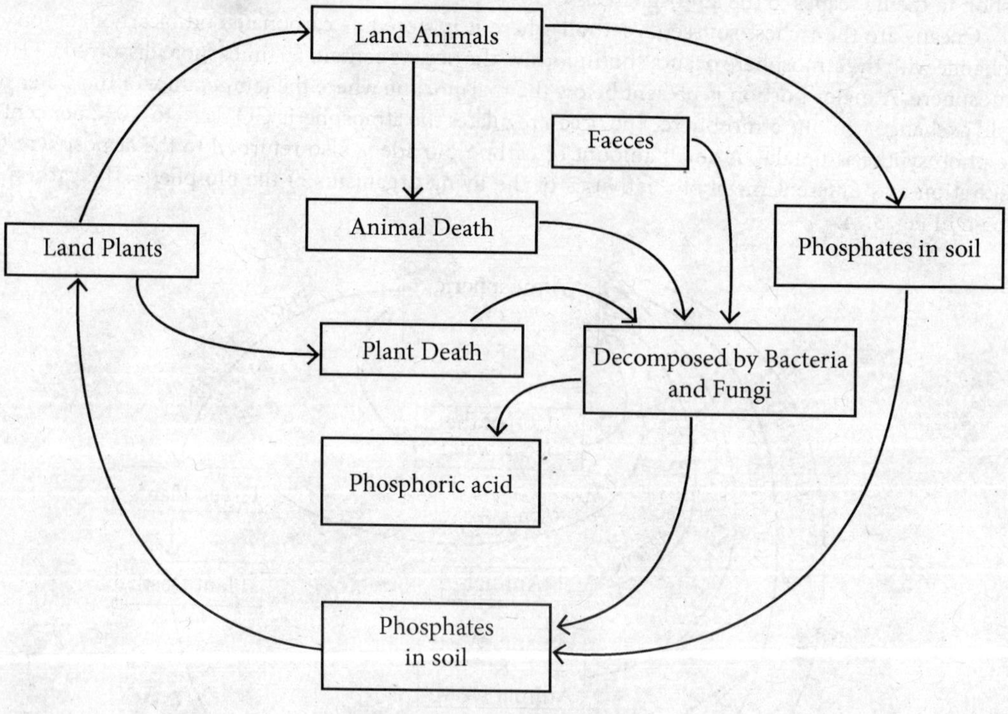

Figure 3.15 Phosphorous Cycle

In freshwater, phytoplanktons absorb the soluble inorganic phosphates and convert them into organic phosphates which are eaten by zooplanktons. Finally, after death and decay they settle at the bottom of the water. They are finally decomposed by the micro-organisms while phosphates released in water as soluble phosphates are again cycled. This cycle of phosphorous is shown in Fig. 3.15.

The rate of recyclization of phosphorous back to the soil is much lower than the rate of loss of phosphorous from the soil. For this, super phosphates and trisuper phosphates are added to agricultural fields as fertilizers. This leads to phosphorous pollution of rivers and lakes, causing eutrophication.

Sulphur Cycle: Sulphur is an essential constituent of certain amino acids and vitamins of the B-complex group. So plants and animals depend on a continuous supply of sulphur. It is present in the atmosphere as hydrogen sulphide (H_2S) and sulphur dioxide (SO_2) gas. In nature, these gases are emitted from volcanoes only. At present, during the burning of fossil fuels these gases are emitted at an alarming rate. In soil, sulphur is present as sulphides, sulphates and organic sulphur. It is taken by the plants and through the food chain it returns to the soil. Fungi like Aspergillus and Neurospora decompose the sulphur from proteins as sulphates by aerobic decomposition. These sulphates undergo further cyclization. In an aerobic condition, H_2S is produced from the decomposition of proteins by bacteria like Escherichia.
These days' wide industrialization and rapid urban development has increased the concentration of sulphur in the atmosphere where it reacts with moisture and causes acid rain. As a result, soil and aquatic environment becomes more acidic and detrimental to organisms. The sulphur cycle is shown in Fig. 3.16.

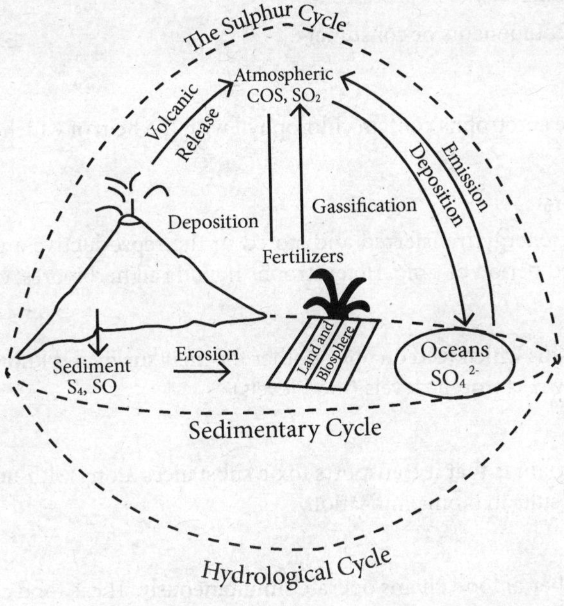

COS = Carbonyl Sulphide
SO_2 = Sulphur Dioxide

Figure 3.16 Sulphur Cycle

Hydrological (Water) Cycle: The hydrological cycle helps in the exchange of water between the atmosphere, lithosphere, hydrosphere and biosphere. About one-third of solar radiation absorbed by the earth is used for deriving the hydrological cycle which includes evaporation of water from all water surfaces, cloud formation and precipitation. The water cycle has been discussed in detail in Unit-V on hydrosphere and water pollution.

SUMMARY

The study of the relationship of organisms among themselves and between the organisms and the environment is known as ecology. The structural and functional unit of ecology is known as ecosystem. The part of the earth where these ecosystems operate is called biosphere. It is the place where the atmosphere, hydrosphere and lithosphere meet and make life possible through their interactions. The functional unit of an ecosystem consists of two distinct structural components namely biotic components and abiotic components.

Abiotic Components

All the non-living components of the environment such as light, temperature, humidity moisture, solar radiation and salinity of soil as well as inorganic and organic compounds constitute the abiotic components of an ecosystem.

Biotic Components

Depending on their self-food producing capability, biotic components are further categorized as:

(i) Autotrophic components or producers.

(ii) Hetrotrophic components or consumers.

Productivity

Green plants, that is, the autotrophs contain chlorophyll with the help of which they convert solar energy into chemical energy.

Secondary Productivity

It is the net quantity of energy transferred and stored in the reproductive and non-reproductive tissues of heterotrophs over a fixed period of time. Heterotrophs include all herbivores, carnivores and decomposers.

Food Chains

The sequence of organisms which feed on one another for their survival is known as the food chain. In a food chain, the links are known as trophic levels (food levels).

Biomagnification

The demerit of a food chain is that it transports toxic substances along with nutrients from one organism to another which finally results in biomagnification.

Food Web

In an ecosystem, a number of food chains operate simultaneously. These food chains are interlinked with one another to form a closed, web-like structure called a food web.

Ecological pyramid

The ecological pyramid is the graphical representation of the trophic structure (an organism's position in the food chain) and functions at successive trophic levels. The base of the pyramid consists of the food producer level and the successive levels make the tiers with the top carnivore forming the apex. These ecological pyramids are of three types namely:

(i) Pyramid of Numbers.

(ii) Pyramid of Biomass.

(iii) Pyramid of Energy.

Biogeochemical Cycles

Both plants and animals are made up of carbon, hydrogen, nitrogen, oxygen and phosphorous along with minute quantities of calcium, iron, copper and molybdenum etc. These elements are present in the environment, which are taken in by the plants (autotrophs) and through the food chain they reach all the heterotrophs and are finally returned to the environment through the decomposers. This cyclic movement of matter through the atmosphere, hydrosphere, biosphere and lithosphere is known as biogeochemical cycle. It is classified as follows:

(i) Gaseous cycle represented by oxygen, nitrogen and carbon cycles.

(ii) Sedimentary cycle represented by phosphorous cycle.

ESSAY TYPE QUESTIONS

1. Write a short note on the functional components of an ecosystem.
2. Describe the fundamental role of producers, consumers and decomposers in an ecosystem.
3. Write a note on material cycling and energy flow in an ecosystem
4. What is an ecological pyramid? Differentiate between pyramids of number, biomass and energy.
5. Describe your concept of ecological succession.
6. Discuss the importance of biogeochemical cycles for the sustenance of an ecosystem.
7. Describe forest ecosystem.
8. What is a food chain? Describe its role in the ecosystem.
9. Write a note on grassland ecosystem.
10. Describe aquatic ecosystem giving special emphasis on ocean ecosystem.

SHORT-ANSWER TYPE QUESTIONS

1. Define the concepts of ecosystem and ecology.
2. What are the biotic and abiotic components of an ecosystem?
3. Define the term biosphere.
4. Write a short note on man-engineered ecosystem.
5. 'Natural ecosystem is more stable than man-engineered ecosystem.' Explain the statement.
6. What is a food chain and a food web? Explain the difference between the two.
7. What is an ecological pyramid?
8. Make a comparison of pyramid of numbers, biomass and energy.
9. Define biomagnification.
10. What are biogeochemical cycles?
11. What do you understand by fixation of nitrogen? How can it be achieved?
12. Write a short note on the phosphorous cycle.

13. Write a note on lentic and lotic water systems.
14. Give examples of different biotic components of a pond ecosystem.
15. Who are the producers, consumers and decomposers in a forest ecosystem?

MULTIPLE CHOICE QUESTIONS

1. The study of interaction between the living species and the environment is called
   (a) biology.
   (b) antology.
   (c) ecology.
   (d) zoology.

2. The interdependence of the living organisms among themselves and with the environment is called
   (a) ecology.
   (b) ecosystem.
   (c) biology.
   (d) antology.

3. Ecosystem consists of
   (a) biotic components only.
   (b) only abiotic component.
   (c) both biotic and abiotic components.
   (d) None of the above.

4. The quantity of the abiotic material present in an ecosystem is known as
   (a) concentration.
   (b) standing stage.
   (c) non-living stage.
   (d) None of the above.

5. The names of major habitats present in the biosphere are
   (a) freshwater and terrestrial ecosystem.
   (b) marine and man-engineered ecosystem.
   (c) All of the above.
   (d) None of the above.

6. Man engineered ecosystem consists of
   (a) desert and forest ecosystem.
   (b) grassland and tree.
   (c) agriculture and aquaculture ecosystem.
   (d) All of the above.

7. The sequence of organisms which feed on one another for their survival is known as
   (a) passage of nutrients from one organism to other.
   (b) food chain.
   (c) trophic level.
   (d) biodiversity.

8. The food chain in an ecosystem helps to maintain
   (a) the feeding relationship in nature, thus biodiversity.
   (b) flow of energy in the ecosystem.
   (c) passage of nutrients in the ecosystem.
   (d) All of the above.

9. The phenomenon of accumulation of non-biodegradable contaminants in higher trophic level is known as
   (a) bioprospecting.
   (b) biomagnification.
   (c) biopiracy.
   (d) bioremediation.

10. The dissipation of energy during its transmission from one trophic level to another is in agreement with
    (a) First law of thermodynamics.
    (b) Second law of thermodynamics.
    (c) Third law of thermodynamics.
    (d) None of the above.

11. Which of the following statements is not correct?
    (a) Flow of energy in an ecosystem is unidirectional, from herbivores to carnivores.
    (b) In a food chain, transfer of materials takes place through cycling.
    (c) Top carnivores always get more energy than organisms present nearer to the producer.
    (d) Unidirectional flow of energy prevents the ecosystem from collapsing.

12. The science of systems of control in an ecosystem is known as
    (a) syncology.
    (b) cybernetics.
    (c) biocoenoses.
    (d) None of the above.

13. The importance of ecosystem lies in
    (a) transfer of food.
    (b) flow of energy.
    (c) cycling of materials.
    (d) Both b & c.

14. The graphical representation of an organism position as well as function at successive trophic levels is called
    (a) food chain.
    (b) food web.
    (c) ecological pyramid.
    (d) biogeochemical cycle.

15. The cyclization of matter through the atmosphere, hydrosphere, biosphere and lithosphere is known as
    (a) food web.
    (b) material cycling.
    (c) biogeochemical cycle.
    (d) biomagnification.

16. In case of a parasitic food chain, the shape of the pyramid of number is always
    (a) upright.
    (b) linear.
    (c) inverted.
    (d) Not certain.

17. The shape of the pyramid of biomass for a pond or any aquatic ecosystem is
    (a) inverted.
    (b) upright.
    (c) linear.
    (d) Not certain.

18. The graphical representation of the interrelation of producer and consumer in an ecosystem is termed as
    (a) ecological niche.
    (b) ecological pyramid.
    (c) trophic levels.
    (d) food web.

19. The main source of energy in an ecosystem is
    (a) sugar stored in plants.
    (b) solar energy.
    (c) heat released during transpiration.
    (d) heat released during fermentation.

20. A food chain consists of
    (a) producers, consumers, decomposers.
    (b) producers, carnivores, decomposers.
    (c) primary producer, herbivores, carnivores.
    (d) producers, primary consumers, carnivores.

ANSWERS

1 (c) 2 (b) 3 (c) 4 (b) 5 (c) 6 (c) 7 (b) 8 (d) 9 (c) 10 (b)
11 (c) 12 (b) 13 (d) 14 (c) 15 (c) 16 (c) 17 (a) 18 (b) 19 (b) 20 (a)

UNIT IV
Biodiversity and its Conservation

LEARNING OBJECTIVES

After studying Unit IV, students will be able to:

- Define biodiversity and explain the need for its conservation.
- Identify and define genetic, species and ecosystem diversity.
- Explain the biogeographical classification of India.
- List the values of diversity.
- Describe consumptive use and productive use.
- Identify the social, ethical, aesthetic and optional values of biodiversity.
- Describe biodiversity at the global, national and local levels.
- Define hotspots in diversity and explain why India is a megadiversity nation.
- Identify and describe threats to biodiversity, such as habitat loss, poaching of wildlife, man–wildlife conflicts and list some endangered and endemic species of India.
- Explain conservation of biodiversity in terms of in-situ and ex-situ conservation of biodiversity.

4.1 Biodiversity and its Conservation

4.1.1 Introduction

Biodiversity is the variety and the number of living organisms (both plants and animals) present in the ecosystem. At the 1992 Convention on Biodiversity held in Rio de Janeiro, it was defined as 'the variability among living organisms from all sources including inter alia, terrestrial, marine and other aquatic ecosystems and the ecological complexes of which they are part. This includes diversity within species, between species and of the ecosystem.'

Diversity is a rule of nature and the policy of the natural habitat. So, there is variability (difference) of genes within and between the species and also diversity of ecosystems. It is impossible to have a species with zero diversity.

Food chains that form a food web, link the species to one another. This keeps the energy flowing continuously inside the ecosystem. Any loss in species (biodiversity) means the breaking of a link in the food chain which in turn affects all those who benefit from the chain. Human beings, who are at the bottom of the chain are the ultimate sufferers. Hence, every effort has to be made to conserve biodiversity.

Being a combination of genes, species and the ecosystem itself, biodiversity represents the quality and characteristic features of life in an ecosystem. This diversity can be divided into:

(i) Genetic diversity.

(ii) Species diversity.

(iii) Ecosystem diversity.

Genetic Diversity

When there is a variation of genes within the same species (single population) and also among geographically separated populations it is called genetic variation. It is responsible not only for the difference in characteristics but also for the adaptation of organisms to a particular habitat or environment. Within the same species, while some individuals are taller than others, some have brown or blue eyes.

A change in external as well as internal factors is responsible for genetic variations. The species that is spread out over a large area interbreeds thereby spreading its genes but the species that is confined to a small area, has a low, very localized gene flow.

According to an estimate, there are 10,000,000,000 different genes distributed all over the biosphere though all of them do not make similar contributions towards genetic diversity.

Species Diversity

This diversity provides a quantitative idea of the number of species and the variety of species present in a particular ecosystem. Table 4.1 gives us the number of various life forms (species) described so far on earth.

Table 4.1 Species Diversity in the Biosphere

Sl. No.	Life Forms (Species)	No. Described
1	Virus	1,000
2	Bacteria	3,060
3	Cyanobacteria	1,700
4	Fungi	28,983
5	Algae	26,900
6	Lichens	18,000
7	Bryophytes	16,000
8	Pteridophytes	11,299
9	Gymnosperms	929
10	Dicotyledons	1,70, 000
11	Monocotyledons	50,000
12	Invertebrates	9,89,761
13	Mammals	4,000
14	Other vertebrates	48,853
15	Other forms	27,400
	Grand Total	13,92,485

So far there are approximately 13.92 million species on earth and the expected number is about 25–30 million, with tropical and sub-tropical parts contributing around 70 per cent of the global biodiversity. Presently, about 3,000 scientists are engaged in exploring and identifying these life forms.

Species diversity is the most basic way to keep an account of biodiversity as it includes all forms of life from micro-organisms such as viruses and bacteria to multi-cellular kingdoms of plants, animals, and fungi.

Ecosystem Diversity

A broader scale of biodiversity depicts the differences between different habitats, ecological processes, and the ecosystem in which the species exist. Based on the physical structure and species composition, the ecosystem can be divided into:

(i) Terrestrial Ecosystems such as forests, grasslands, deserts and so on.

(ii) Aquatic Ecosystems: These are of two types:

(a) Freshwater, consisting of lotic and lentic.

(b) Marine, consisting of oceans and estuaries.

(iii) Artificial or Man-made Ecosystem consisting of lakes, croplands and so on.

4.1.2 Biogeographical Classification

The scientific study of the geographic distribution of plants and animals is called biogeography. This science of biodiversity patterns, distribution and abundance of species, population, health and so on, is of utmost importance to us as we depend on plants and animals for food, clothes, shelter, medicines, oxygen, water and energy. For example, people living on islands can understand the importance of the species of a particular community. As per the island biogeography, the number of species on islands of similar habitat and latitudes depends on the size and location of the islands.

The entire biosphere based on oceans, mountains, deserts and so on, can be split into eight distinct biogeographic regions based on the combination of plants, animals, and climate:

(i) Nearctic region of North America,

(ii) Neotropical region of South America,

(iii) Palearctic region of Eurasia and North America,

(iv) Afrotropical region of India and South East Asia,

(v) Oriental region of India and South East Asia,

(vi) Australasian region,

(vii) Oceanic region and

(viii) Antarctic region.

In the absence of any precise boundary to demarcate two sub-regions of specific climates and relative isolation, these regions too are further divided into biogeographical sub-regions. For instance, Philippines is a part of the Indo-Malay region which stretches from humid tropical India through the Malay Archipelago to New Guinea.

These sub-divisions are further divided into biogeo zones or biogeographic zones or biomes. Biomes are spatial delineation of the country into contiguous and more or less homogeneous zones where plants and animals live together in related diversity.

With about 8 per cent of biodiversity concentrated in about 2 per cent of the earth's space, India is one of the 12 megadiversity countries in the world. To ensure proper planning and conservation of the environment at the national, state, and local levels, India needs to be further classified into different biogeographic zones and then into biotic provinces or land regions or ultimately into biomes. The total area of 32, 87, 263 sq km in India has been classified into 10 biogeographical zones listed in Table 4.2.

Table 4.2 Biogeographic Zones of India

Sl. No.	Biogeographic zones	No. of biotic provinces in the zone	Area of India (in per cent)	Speciality
1	Trans–Himalaya	2	5.6	
2	Himalaya	4	6.4	

Sl. No.	Biogeographic zones	No. of biotic provinces in the zone	Area of India (in per cent)	Speciality
3	Desert	2	6.6	
4	Semi-arid	2	16.6	
5	Western Ghats	2	4.0	Contains 60 per cent of the biodiversity of India. One of the hotspots of diversity in India
6	Deccan Peninsula	5	42.0	
7	Gangetic Plain	2	10.8	
8	Coasts	2	2.5	
9	North East	2	5.2	
10	Islands	2	3.0	

4.1.3 Values of Diversity

Nature gives us sustenance and biological resources form the basis of all life on earth. The fundamental, social, ethical, cultural and economic value of these resources has been recognized in religion, art and literature from the earliest days of recorded history. Thus, conservation of these resources is very essential and has now become a global issue. For conservation of biological resources an economic evaluation of biodiversity is essential. Depending on the economic evaluation different funding agencies such as banks, governments, and so on provide monetary assistance for the conservation of biodiversity. But the hidden cost of loss of biodiversity is very difficult to determine. However, economists and biologists have categorized the values of biodiversity by assessing how people benefit from them. These values are of:

(i) market-place resources,
(ii) unharvested resources and
(iii) future resources.

The economic values of biodiversity can be categorized as:

(i) **Direct Use Values:** These are values directly used by people.
(ii) **Indirect Use Values:** They provide benefits even if not used by people.
(iii) **Social Values:** These provide some usable or non-usable benefits in the future.
(iv) **Existence (Ethical) Values:** These provide benefits in the present and may not benefit our future generations.

Direct Use Values

These are values directly consumed by the people and include food, products for developing pharmaceuticals, for developing and maintaining the genetic basis for agriculture, supporting industry through timber extraction, fisheries, poultry, dairy farming, and so on. Direct values may be of both consumptive and productive use.

Consumptive Use Values: These include food products and fossil fuel (wood) which does not figure in the national or international market but are consumed locally. About 3,000 plant species, 200 of which have been domesticated are used as a food source. Presently, 20 per cent of these plants provide more than 80 per cent of our food. To meet the huge demand breeders opt for the hybrid variety. This variety has better resistance to drought and disease and has wide genetic diversity. Compared to plants, a smaller variety of animals are used by humans for food. However, trees provide 3.8 million cubic meters of food annually by way of fuel, timber or pulp.

Productive Use Values: These are values of products that are harvested from the wild and sold in national and international consumer markets. For instance, many of the raw materials used for drug manufacturing are either plant-based or animal-based. Globally, 3.5 billion people consume herbal medicines. Raw materials for industries such as fibres, resins, dyes, waxes, lubricants, perfumes, pesticides, timber and so on are either plant or animal-based and have a high monetary value in the market. In fact, the cost of the productive use value of biodiversity is constantly rising.

Indirect Use Values

These are as important for the well-being of human beings as are direct values. Indirect Values are obtained without using up the resources and include soil formation and protection, watershed protection, waste disposal, pollination, nutrient cycle, oxygen production, carbon sequestration, control of floods, climate regulations, recreation and eco-tourism, educational and scientific value and environmental monitoring.

Social Values

Aesthetic, recreational, cultural and spiritual, social values are ideals and beliefs which people preach and uphold in order to structure the traditions, institutions and laws of the society they live in. The relationship between human beings and nature and the one between society and nature is very important. These values evolve and undergo changes with changes in circumstances and relationships. There can be a marked difference in landscape and biodiversity preferences according to age, socio-economic factors and cultural and religious influences. The following six values provide a definition of social values:

(i) Human conquest of nature carries a moral responsibility for the perpetuation of other life forms. (Hornday, 1914)

(ii) Wanton consumption and merciless slaughter of wildlife is uncivilized. (Hornday, 1914)

(iii) Aesthetic and intellectual contemplation of nature is integral to the biological and cultural inheritance of many people. Monuments of nature, great works of art and architecture, should be guarded from ruin. (Conwentz, 1909)

(iv) Healthy ecosystems are necessary to safeguard economic growth, quality livelihoods and social stability. (Ehrlich & Eherlich, 1992; Daily, 1997; Carson, 1999)

(v) It is prudent to maintain the earth's genetic library from which society has derived the basis of its agriculture and medicine. (Myers, 1979)

(vi) Society has a moral duty to permit traditional people inhabiting natural landscapes to choose their own destiny in time frames appropriate to their history and culture. (WCED, 1987)

Social movements along with the efforts of non-governmental organizations at the national and international levels, play an important role in the conservation of nature. Today, there is a marked shift towards development. In that context if nature is divided into discrete units and assigned a monetary value it can be treated as a commodity and conservation can be treated as a free market delivery of economic goods and services. The social values in conserving nature can then become more effective.

Aesthetic Values: The use of biodiversity for recreation, scientific investigation, and eco-tourism is also growing rapidly. The fast growing 'leisure industry' has now begun to value nature for its aesthetics, and treats it as part of the cultural heritage. People too prefer to visit a reserve forest and enjoy nature-based activities such as hiking, trekking, bird and wildlife watching, fishing and photography. The aesthetic value of our ecosystem contributes to the emotional and spiritual well-being of a highly urbanized population. Partly due to the government's contribution, eco-tourism contributes financially to conservation. Thus, nature provides opportunities for outdoor recreation to millions of nature lovers if it is conserved. In fact, the travel and tourism economy in India accounts for 5.6 per cent of the total Gross Domestic Product.

Option Values: The potential of a species to provide economic benefits to human beings in the future is called option value. The biotechnologists working towards generating new species to fight and cure deadly diseases such as cancer and AIDS are a fine example of option value.

Existence (Ethical) Values

Unlike the other values, ethical values are an intrinsic part of nature. It is very difficult to quantify their economic value. But, these play a major role in the protection and conservation of biodiversity along with all the other values. All religions, cultures and philosophies, stress on being ethical. The ethical obligations for protecting biodiversity are:

(i) Protect other species from extinction.
(ii) Do not waste resources.
(iii) Remember that all species are interdependent and have a right to exist.
(iv) People must take responsibility for their actions.
(v) People must feel responsible towards their future generations.
(vi) Remember that nature has spiritual and aesthetic values that can be transformed into economic values.
(vii) Keep in mind that nature matters to us and so our actions must not harm it.

The ethical values of biodiversity teach us:

(i) How to improve quality of life.
(ii) How to conserve natural resources.
(iii) How to enrich environmental quality, culture, religion and aesthetics of society.

4.1.4 Hotspots in Diversity

According to British ecologist Norman Myers, certain ecosystems despite their small size account for a high percentage of global biodiversity. Many of these areas also suffer from logging, overexploitation of land due to excessive agriculture, hunting and climatic changes. Myers was the first to devise the concept of biodiversity hotspots so as to identify these areas and preserve the endemic species there.

Biodiversity hotspots are environmental emergency rooms (store houses) of the earth. They are biologically rich areas with a large percentage of endemic species. For example, a terrestrial biodiversity hotspot is based on plant diversity that has:

(i) At least 0.5 per cent or 1,500 of the world's 3, 00,000 species of green plants
(ii) Has lost 70 per cent of its primary vegetation.

Coral reefs and multiple taxa (species of coral, snails, lobster and fishes) signify marine hotspots. Most hotspots are found in tropical and sub-tropical areas because warm, moist tropical environments are conducive to the growth and reproduction of the species present there. Species and ecosystem diversity varies with altitude and depth. For instance, the mountainous environment (orobiome) is vertically divided into montana, alpine and nival ecosystems and diversity in the aquatic ecosystem (both marine and freshwater) species decreases as we go deeper. Biodiversity also tends to increase from the Poles to the Equator.

Keeping in mind Myers' definition of hotspots, biologists have identified areas of high endemism with species richness and labelled them as hotspots. Hotspots are defined as the localized concentration of biodiversity, and are in need of sincere conservation action. Conservation International has identified 25 terrestrial biodiversity hotspots around the world for conservation.

The identified hotspots around the world are:

(i) Tropical Andes,
(ii) Meso-American forests,
(iii) Caribbean,
(iv) Brazil's Atlantic forests,
(v) Choco Darien/Western Ecuador,
(vi) Brazil's Cerrado,
(vii) Central Chile,
(viii) California Floristic Province,
(ix) Madagascar,
(x) Eastern Arc and coastal forests of Tanzania/Kenya,
(xi) Western African forest,
(xii) Cape Floristic Province (South Africa),
(xiii) Succulent Karoo,
(xiv) Mediterranean Basin,
(xv) Caucasus,
(xvi) Sunderland,
(xvii) Wallace (Eastern Indonesia),
(xviii) Philippines,
(xix) Indo-Burma (Eastern Himalayas),
(xx) South-Central China,
(xxi) Western Ghats of India and the Island of Sri Lanka,
(xxii) South-West Australia,
(xxiii) New Caledonia,
(xxiv) New Zealand and
(xxv) Polynesia and Micronesia Island complex including Hawaii.

A recent global study conducted over four years, by nearly 400 scientists and other experts has identified nine new hotspots; bringing the total to 34. These new hotspots are home to 75 per cent of the world's most threatened mammals, birds and amphibians. Originally, these hotspots covered 16 per cent of the earth's surface which has now reduced to 2.3 per cent due to human encroachment and habitat destruction.

The nine new hotspots are:

(i) East Melanesian Island,
(ii) Madrean Pine-Oak Woodland on the US-Mexico border,
(iii) Japan,
(iv) Horn of Africa,

(v) Irano-Anatolian region of Iran and Turkey,

(vi) Mountains of Central Asia,

(vii) Maputaland-Pondoland-Albany in southern Africa (parts of Mozambique, South Africa and Swaziland),

(viii) Himalayan region and

(ix) Eastern Afro-Montana along the eastern edge of Africa from Saudi Arabia to Zimbabwe.

However, World Wildlife Fund (WWF) replaced the biodiversity concept that Myers had devised in 1977 with the 'Global 200' Strategy in 1998. Global 200 expands the conservation priorities to 233 eco-regions, comprising 19 terrestrial, freshwater and marine major habitats thereby covering major biodiversities of the planet.

4.2 India—A Megadiversity Nation

India has a rich heritage of forests, wetlands and marine areas, which range from the temperate forest to coastal land and tropical rain forest to the alpine region. This richness makes it one of the 12 megadiversity nations of the world.

As per the statistics of the Ministry of Environment and Forest, India accounts for 7.31 per cent of the total fauna, and 10.88 per cent of the total flora of the world. It has different biogeorgaphic zones and 25 biotic provinces and also hosts mega fauna such as rhinoceros, tigers, elephants and so on. Table 4.3 gives a clearer picture of the percentage and ranking of India's biodiversity.

Table 4.3 Comparison between the number of species in India and the World and Percentage and Ranking of India

GROUP	Number of species in		percentage of species of India with respect to the global total	India's ranking in the world
	INDIA	WORLD		
Mammals	350	4,629	7.6	8th
Birds	1,224	9,702	12.6	8th
Reptiles	408	6,550	6.2	5th
Amphibians	197	4,522	4.4	15th
Angiosperm	15,000	25,000	6.0	15th
Total Plant Species	45,000			
Butterflies & Moths	13,000			
Total Insects	50,000			
Fishes	2,546	21,730	11.7	

Of the 75.23 million hectares of forest in India, 40.61 million hectares are classified as reserved and 21.51 million hectares as protected area. This includes over 40 wildlife sanctuaries and 70 national parks spread across 1, 40,000 sq km. The remaining 13.11 million hectare forest area is maintained as unclassified. Marine protected area covers 2, 76,042 hectares, supporting economically valuable ecosystems such as mangroves, estuaries, lagoons and coral reefs.

Over 70 million years ago, India was formed when a giant continent split up, resulting in the formation of Gondwanaland and the southern land mass. India was attached to Africa, Australia and Antarctica. Subsequently, due to tectonic movements, India shifted northward to converge with the northern Eurasian continent across the Equator. When the intervening Tethys Sea started drying up, plants and animals that evolved in Europe and the Far East began migrating to India.

Subsequently, the Himalayas grew to form a natural barrier in the north, along with the three seas – Arabian Sea, Bay of Bengal and the Indian Ocean – in the south.

Some of the other prominent features of India as a megadiversity nation are its three important biomes: tropical humid forests, tropical dry deciduous forests and the arid or semi-arid deserts.

India has 25 hotspots mainly in the Western Ghats and the Eastern Himalayas. It ranks seventh in its contribution to world agriculture. India has more than 34,000 cereals and 22,000 pulses in its gene bank.

4.2.1 Endangered and Endemic Species of India

Natural and anthropogenic causes have always remained a great threat to biodiversity. Developmental works are only accelerating habitat loss and pushing wildlife (both fauna and flora) towards extinction (1,000 to 10,000 per year). Based on this, the International Union for Conservation of Nature and Natural Resources (IUCN) has categorized wild flora and fauna into eight categories. The list containing these categories is known as the Red List. These categories are: extinct, extinct in wild, critically endangered, endangered, vulnerable, lower-risk, data-deficient and not evaluated. This data, recorded in the Red Data Book is updated every four years. According to the Red Data Book, a threatened species is one whose natural habitat is disturbed. As a result, the species population decreases rapidly and there is a fear that the species may become extinct. As per Schedule I, of the Wildlife (Protection) Act, 1972 of India, a species is considered endangered when its number reduces to a critical level. The species is then provided with legal protection. So far, 38 species of birds, 18 of amphibians and reptiles and 81 species of mammals have been labelled endangered in India.

A species faces a very high risk of extinction in the wild when there is a suspected reduction of at least 50 per cent over the last 10 years or three generations, whichever is longer, based on the following:

(a) direct observation,
(b) an index of abundance appropriate for the species,
(c) a decline in area of occupancy,
(d) extent of occupancy and/or quantity of habitat,
(e) actual potential levels of exploitation and the effects of introduced species,
(f) hybridization,
(g) pathogens,
(h) pollutants,
(i) competition or parasites.

Endangered Species of India

Andaman Shrew (*Crocidura andamanensis*) (endemic to India)
Andaman Spiny Shrew (*Crocidura hispida*) (endemic to India)
Asian Elephant (*Elephas maximus*)
Banteng (*Bos javanicus*)
Blue Whale (*Balaenoptera musculus*)
Capped Leaf Monkey (*Trachypithecus pileatus*)
Chiru (Tibetan Antelope) (*Pantholops hodgsonii*)

Fin Whale (*Balaenoptera physalus*)
Ganges River Dolphin (*Platanista gangetica gangetica*)
Golden Leaf Monkey (*Trachypithecus geei*)
Hispid Hare (*Caprolagus hispidus*)
Hoolock Gibbon (*Bunopithecus hoolock, previously Hylobates hoolock*)
Indian Rhinoceros (*Rhinoceros unicornis*)
Indus River Dolphin (*Platanista minor*)
Kondana Soft-furred Rat (*Millardia Kondana*) (endemic to India)
Lion-tailed Macaque (*Macaca silenus*) (endemic to India)
Markhor (*Capra falconeri*)
Marsh Mongoose (*Herpestes palustris*) (endemic to India, it was previously considered to be a sub-species of *Herpestes Javanicus*)
Nicobar Shrew (*crocidura nicobarica*) (endemic to India)
Nicobar Tree Shrew (*Tupaia nicobarica*) (endemic to India)
Nilgiri Tahr (*Hemitragus hylocrius*) (endemic to India)
Parti-colored Flying Squirrel (*Hylopetes alboniger*)
Peter's Tube-nosed Bat (*Murina grisea*) (endemic to India)
Red Panda (Lesser Panda) (*Ailurus fulgens*)
Sei Whale (*Balaenoptera borealis*)
Servant Mouse (*Mus famulus*) (endemic to India)
Snow Leopard (*Uncia uncia*)
Tiger (*Panthera Tigris*)
Wild Water Buffalo (*Bubalus bubalis* previously *Bubalus arnee*)
Woolly Flying Squirrel (*Enpetaurus cinereus*)

Endemic Species

Species that have very restricted distribution and are found over relatively small ranges are called endemic species. Since their ecological requirements are met over a small area these species remain restricted to a particular area as rare or endemic species.

About 33 per cent plants of the world are endemic to India. North-east India, the Western Ghats, north-western and eastern Himalayas, a small pocket of the Eastern Ghats and of course Andaman & Nicobar Islands are rich in endemic species. In fact, as per the 1983 Botanical Survey of India, the Andaman & Nicobar Islands boast of at least 220 species of endemic flora in India. Agastyamalai Hills, Silent Valley, New Amarambalam Reserve, Periyar National Park, all the mountains of Western Ghats and eastern and western Himalayas are known for their conservation.

There are 44 endemic species of mammals, confined to a small range within the Indian Territory. The Western Ghats have been identified as the abode of four of these endemic species:

(i) Lion Tailed Macaque (Macaca Silenus)
(ii) Nilgiri Leaf Monkey (*Trachypithecus johnii* called Nilgiri *Langur* by the locals)
(iii) Brown Palm Civet (*Paradoxurus Jerdoni*)
(iv) Nilgiri Tahr (*Hemitragus Hylocrius*)

Endemic bird species are not a very common sight in India. About 55 of the endemic bird species can be spotted along the mountain ranges in eastern India. The other places where they can be found are south-west India (the Western Ghats) and the Andaman and Nicobar Islands.

As for the other animals, the number of endemic reptiles and amphibians in India is high. There are 187 reptiles and 110 amphibians endemic to India. In fact, India is the only country which is abode to eight amphibian genera. The most notable among them is the monotypic *Melanobatrachus*, which has only one species known only from a few specimens collected in the Annamalai Hills, in 1870.

4.2.2 Hotspots of India

The high endemism of Indian biodiversity is under constant threat. Among the 34 hotspots of the world, two are located in India and they extend to the neighbouring countries. The two hotspots are located in the eastern Himalayas, covering the Indo-Myanmar region and the Western Ghats, extending to Sri Lanka. About 30 per cent of the total flora recorded in the world are endemic to India and are concentrated in these two regions. About 62 per cent of the known amphibian species, reptiles, swallow-tailed butterflies and some mammals are endemic to the Western Ghats.

Eastern Himalayas

The eastern Himalayas include Bhutan, north-eastern India, southern, central and eastern Nepal and Yunan province in South West China. The species geographic distribution shows a distinct growth in the flora and fauna of these areas. The eastern Himalayas have a greater variety of oaks and rhododendrons than the Western Himalayas because of the higher rainfall and warmer conditions in the eastern Himalayas. Many deep, semi-isolated valleys are exceptionally rich in endemic flora. For example, of the 4,250 plant species in Sikkim in an area of 7,298 sq km, 2,550 or 60 per cent are endemic. India has 2,000 or 36 per cent of the endemic plant species out of 5,800, while Nepal has 7,000 of which many overlap with those of India, Bhutan and Yunan. Yunan has 500 endemic plants, nearly 8 per cent while Bhutan has 5,000, which is 15 per cent of the total plant species that are endemic to the eastern Himalayas.

The discovery of a new large mammal *Muntiacus gongshanensis* and four new genera of flowering plants in South East China have confirmed the findings of a study that said north-east India along with Yunan is an active centre of organic evolution. This hotspot is also home to 163 globally-threatened species, including three of Asia's largest herbivores – the Asian Elephant (*Elephas Maximus*), the great One-Horned Rhinoceros (*Rhinocero Unicorn*), Wild Water Buffalo (*Bubalus Bubalis*), its largest carnivore, the tiger (*Panthera Tigris*) and several large birds like vultures, adjutant storks and hornbills.

Earlier, clubbed with the Indo-Burma biodiversity hotspot, the eastern Himalayan region now stretches across the Indo-Burma hotspot and the Himalayan hotspot. The Himalayan hotspot was identified as a new hotspot in 2005. Table 4.4 lists the biodiversity and endemism of eastern Himalayas.

Table 4.4 Diversity and Endemism of Eastern Himalayas Hotspot

Taxonomic Group	Species	Endemic Species	Present Endemism
Plants	10,000	3,160	31.6
Mammals	300	12	4.0
Birds	977	15	1.5
Reptiles	176	48	27.3
Amphibians	105	42	40.0
Freshwater Fishes	269	33	12.3

Western Ghats (Southern India) and Sri Lanka

The tropical rain forests of Maharashtra, Karnataka, Tamil Nadu and Kerala in the west and south of India are very rich in biodiversity. In fact, the Western Ghats feature in the 34 hotspots of the world. The two main centres of diversity are Agasthymalai Hills and the Silent Valley/New Ambalam Reserve Basin. The Agasthyamalai Hills in the south harbour, houses the highest level of plant diversity. This hotspot is home to

11,000 animal species. Of the 140 mammal species present, 20 are endemic, the most prominent being the lion-tailed macaque and Asian elephants. Similarly, of the 450 bird species, as many as 35 are endemic. The hotspot also contains 6,000 vascular plant species, of which more than 3,000 (52 per cent) are endemic to the area. Table 4.5 gives a clear picture of the diversity and endemism of the Western Ghats.

Table 4.5 Biodiversity and Endemism of the Western Ghats Hotspot

Taxonomic Group	Species	Endemic Species	Present Endemism
Plants	5,916	3,049	51.5
Mammals	140	18	12.9
Birds	458	35	7.6
Reptiles	267	174	65.2
Amphibians	178	130	73.0
Freshwater Fishes	191	139	72.8

4.3 Threats to Biodiversity

The human drive for 'development', has led him to exploit more natural resources than are actually needed to improve the living conditions. This is responsible for endangering other species of the biosphere. These human actions are beginning to threaten biodiversity. The human urge to transform habitats and exterminate rivals and competitors has led to a lot of harm being caused to all ecosystems and species. Some of the major threats to biodiversity are:

(i) loss/degradation of habitat,
(ii) overexploitation of resources,
(iii) pollution,
(iv) extinction of species due to aggressive non-native species and
(v) global environmental changes.

4.3.1 Degradation of Habitat

A habitat is where every living being finds food, water and shelter to survive and a safe place to reproduce and bring up their offspring. So, loss of habitat is actually the greatest threat to the world. As per a global study by IUCN in 2000, 89 per cent of all threatened birds, 83 per cent of all threatened mammals and 91 per cent of all threatened plants have already been affected by loss or degradation of habitat. This can be caused by natural disasters like flood, fire, hurricanes and erosion. The human need for wood, minerals and water (dams) could also be responsible for this loss. Air and water pollution along with global climate changes also affects sensitive species.

Deforestation for agriculture (*jhum* cultivation), clearing of land for developmental work, overgrazing and so on, are also responsible for fragmenting habitats into small, isolated, scattered populations that are vulnerable to inbreeding, depression, high infant-mortality and susceptible to environmental stochasticity and possible extinction.

Changes in forest composition, quality and habitat-type, lead to a decline in primary food species for wildlife and eventually to loss of habitat. However, statistics show agricultural practices as one of the major causes of loss of habitat.

Table 4.6 Cropland in Million Hectare p.a.

Region	1900	1980	Percentage Change
Sub Saharan Africa	73	222	+204
Latin America	33	142	+330
North America	133	203	+53
South Asia	89	134	+51
South-East Asia	15	55	+267
China	89	134	+51
Europe	145	137	-5
Ex-USSR	147	233	+58

(Source: Pearce and Moran, 1994)

According to a recent estimate, at least 120 out of the 620 living primate species (apes, monkeys, lemurs and others) will become extinct in the wild, in the next 10 to 20 years. Large animals like tigers, mountain gorillas, pandas, Indian lions, tropical orchids and spotted owls often suffer more because they need larger areas for survival. The only species that benefit from human activity are rats, cockroaches, house flinches and so on.

4.3.2 Overexploitation of Resources

Unlimited extraction (through mining, fishing, logging, harvesting and poaching) and development work (human settlement, industry and associated infrastructure) are the major factors that contribute to the overexploitation of resources. As a result of this overexploitation, tigers, giant pandas, black rhinoceros, musk deer, cod and several whale species are on the verge of extinction.

4.3.3 Pollution

Loss of biodiversity due to pollution is very common these days. When we pollute nature with the waste generated by us, only the biodegradable waste gets broken down slowly and gets recycled. But the non-biodegradable or less biodegradable waste remains in the environment and enters our food chain. This waste travels through the food webs, gets biomagnified and reaches the tissues of all living species. These wastes are very toxic and sometimes their toxicity increases with time. A very common example of this is the organic pesticide DDT which affects all types of birds (peacocks, hawks, kites, and so on). Therefore, pollution in various forms is responsible for global climatic changes and for the extinction of most of the species till date.

4.3.4 Extinction of Species due to Aggressive Non-native Species

Despite its importance, this aspect is often overlooked particularly in island areas. When two or more species are inter-dependent or a particular species has strong links with another, the Domino Effect takes place causing extinction of the weaker species. It is the reported cause of extinction of almost 50 per cent species on islands all over the world since 1600 AD.

4.3.5 Global Environmental Change

Scientists feel that 35 per cent of the world's existing terrestrial habitat may face extinction due to global warming. Global warming is a result of the accumulation of Greenhouse gases. It causes the global environment to change and leads to the extinction of many species which fail to adapt and acclimatize to the changing environmental conditions.

However, poverty, macro-economic policies, international trade factors, policy failures, poor environmental laws or weak enforcement of the same, unsustainable developmental projects and a lack of local control over resources as well as population pressure are some of the underlying causes of biodiversity loss. Increase in the collection of fuel wood, fodder and grazing of animals belonging to local communities also take a toll on the forest and its biodiversity.

4.4 Threats to Indian Biodiversity

With 7.31 per cent species of fauna and 10.78 per cent floral species in the world, India is very rich in biodiversity. It has 89,451 animal species and several floral species, one-third of which are endemic to the country. These species are concentrated in the North East, Western Ghats, North West Himalayas, Lakshadweep and the Andaman and Nicobar Islands. But today, this rich biodiversity is under severe threat because of:

(i) loss/degradation of habitat due to agriculture, extraction,

(ii) fragmentation and overexploitation of resources,

(iii) poaching and international trade of wild species and products,

(iv) economic and social causes such as poverty, government policies, environmental laws and enforcement, population pressure and unsustainable development projects and

(v) deforestation due to the collection of fuel wood, fodder, overgrazing and agriculture.

Hunting and poaching alone are responsible for bringing to the verge of extinction as much as 37 per cent of the birds, 34 per cent of the mammals and 8 per cent of the plants, in addition to many reptiles and fishes. In fact, some animals such as tigers are more in demand than others which leads them to be poached more often.

Islands are particularly susceptible to invasion by alien species. This poses a serious threat to 30 per cent of the birds and 15 per cent of the plant species.

4.4.1 Combating the Problem

The fact that the world has become conscious of the value of, and a threat to, biodiversity was proven when at the International Convention on Biological Diversity at Rio de Janeiro in 1992, the focus was on the sustainable use of the components of biodiversity. It was decided that the strategies for sustainable conservation of biodiversity should be:

(i) Worldwide reduction of industrial and domestic pollution.

(ii) Controlling overexploitation of natural resources. Bodies such as the International Whaling Commission and Convention on International Trade on Endangered Species (CITES) are already active in this field.

(iii) Agricultural activities with conservation measures should be encouraged. Organic farming, which promotes habitat diversity, should be promoted.

(iv) The government should set up parks and reserves to protect and rehabilitate wildlife and natural vegetation.

(v) The government should formulate and strictly implement conservation legislations.

(vi) Progress in combating the alarming loss of biodiversity depends not just on the commitment and sacrifices of individuals but also on the actions of the government.

4.4.2 Poaching of Wildlife

Simply put, poaching is an illicit wildlife trade. It is the illegal killing of wildlife for sale in the international trade market. The animals are killed not only for their meat but also for their hides, and different parts of their body that are used as clothing, for food, to make folk medicine, or jewellery or as trophies. Some people poach just for the thrill of it, while others poach for trade. Poachers operate in groups throughout the year and are interested in any marketable animal that is available. Antlers are sold as trophies and are also used in folk medicine. The gall bladder of a bear can bring $18,000 in Asia. Its paws, claws and teeth are used in taxidermy and folk art trade. The horns and skin of a rhinoceros for traditional medicine, tusks of elephants for ivory, eggs from the paddle fish for caviar, musk of the musk deer for perfume, meat of pangolin, the skin, hide and bones of tigers, leopards, deer and bear are all much in demand.

It is the duty of every citizen to stop poaching and conserve wildlife by:

(i) Trying to identify poaching offences in your area.

(ii) Reporting poaching incidents to the local wildlife enforcement officer, local poaching hotline or to the state level officers.

(iii) Discussing the value of wildlife and threats posed by poaching with your near and dear ones.

(iv) Encouraging effective wildlife legislation.

(v) Encouraging the publication of articles against poaching, in local newspapers, journals, television and different mass media, distribution of pamphlets or arranging for lectures meant for a variety of audience.

(vi) Refusing to purchase products that you suspect have been illegally obtained from wild animals.

(vii) Improving wildlife law enforcement, including sufficient patrol officers with proper funding, effective penalties and supporting the judicial system.

4.4.3 Elephant Poaching and Ivory Trade

Despite the fact that CITES has banned international trade in ivory (elephant tusks) in 1990 and provided massive funds for the protection of elephants, poaching has continued albeit at a lower rate. As per statistics, between January 1, 2000 and May 21, 2002, more than 5.9 tonnes of ivory, 2,542 tusks and 14,648 pieces of ivory were seized worldwide; all these facts mean that more than 2,000 elephants have been killed.

Over the last 16 years, more than 289 adult elephants have been poached for ivory in Orissa in India. Illegal ivory trade is extremely lucrative. Ivory sells at Rs 12,000 to 15,000 per kg. If the trend continues, Orissa will soon lose its exalted status of being famous for its magnificent elephants since the time of Ashoka, the Great.

CASE STUDY

Tiger Poaching

In 1900, the population of Royal Bengal Tigers in India was 40,000. It came down to 1,800 in 1972 which prompted the Indian government to launch Project Tiger for the conservation of these big cats. As a result, the tiger population increased to 4,200 by the early 1990s. This came down again to 3,500 due to increased habitat loss and poaching besides other reasons. The population of the South Chinese or Indo-Chinese or Sumatran tiger has also reduced and they are close to extinction. The demand for tiger bones and other body parts for use in oriental medicine is also responsible for bringing the tiger to near extinction.

In 1994, trade in tiger parts was banned in China, Taiwan, Hong Kong, South Korea and most of South East Asia but it is still legal in Japan and North Korea. Medicines containing tiger parts are still used in Canada and the United States. In fact, as per statistics, one tiger per day is poached in India. If the trend persists, this large cat will become extinct in India in the next five to 10 years. In India, the well-organized poachers face little or no opposition at all from ill-equipped, unarmed wardens and rangers. Although a number of legislations have been enacted in India and a good number of tiger reserves have been created; the enactment of the law with such few wardens is very difficult. In India, tiger poaching is rampant in Madhya Pradesh, Uttar Pradesh, Orissa, West Bengal, Assam and Arunachal Pradesh. A tiger cell has been created in Madhya Pradesh with a view to protect the tigers and seize body parts. The largest seizure took place in 1993, in Delhi, when 400 kg of tiger bones, eight skulls, 58 leopard skins and the skin of a number of other animals was seized and taken into custody.

4.5 Man–Wildlife Conflict

When wild animals leave the protected areas (forests) to raid human settlements in search of food and water it gives rise to a conflict between man and wildlife.

The main reason for this conflict is the growing anthropogenic pressure on wildlife habitat which results in:

(i) Fragmentation and honeycombing of animal habitat.

(ii) Loss of corridors and migratory routes for long-range animals such as elephants, big cats (tigers, leopards, bears) besides others.

(iii) Loss of food and water in their habitat due to the shrinking of forest cover and loss of biodiversity.

When wild animals destroy crops causing economic and food losses to farmers, affect water supplies, kill or injure humans and cause havoc in the lives of human beings, they retaliate by killing the wild animal.

The conflict is fast becoming a critical threat to the survival of many globally endangered species such as the Sumatran tiger (*Panthera Tigra Sumatrae*), Asian Lion (*Panthera Leo Persica*), Snow Leopard (*Uncia Uncia*), Red Colobus Monkey (*Procolocus kirkii.*) and many more. It has also been observed that the more volatile species are more prone to extinction because of injury and death caused by humans, traffic (road, railway track) or other accidents (for example, falling into traps, wells, poisoning, electric fences and so on).

Considering that this conflict will always remain, strategies are being evolved by government wildlife managers, scientists and local communities, not just for the protection of humans but also for the conservation of biodiversity (wildlife).

This multi-faceted problem can be minimized with good management practices and approaches involving low-cost strategies such as electric fencing, community-based natural resource management schemes, incentives, and insurance programmes along with regulated harvesting and wildlife or human translocation.

Man–Tiger Conflict in Sumatra

A study by Nyphus and Tilson reveals that the man–tiger conflict is more common in intermediate disturbance zones than in high or low disturbance zones. Intermediate disturbance zones are isolated human settlements surrounded by extensive tiger habitats. There are less chances of conflict in logged, degraded and heavily used areas or in and around protected areas where human entry is prohibited by natural barriers or due to the presence of guards. But in Sumatra, tiger attacks have been recorded around different national parks due to the lack of spatial separation. Hence, for their conservation priority should be given to the security of large animals around reserve borders and in buffer zones.

Man–Monkey Conflict in Zanzibar

The farmers in Zanzibar Island consider the Red Colobus Monkey the third most serious vertebrate pest after the medium and large-sized animals, which threaten their crops. The Red Colobus is an endangered species. Only 1,500–2,000 individuals reside on Unguja Island. Although, the farmers feel that consumption of coconuts by the Red Colobus is a threat to the crop yield, the truth quite the contrary. The monkeys prune small and immature coconuts thereby increasing the yield. In fact, they account for a 2.8 per cent increase in potential harvest. Secondly, they are also a source of income through eco-tourism.

Man–Snow Leopard Conflict in India and Mongolia

Conflict between agro-pastoralists and wildlife is increasing day by day in Kibber Wildlife Sanctuary in Himachal Pradesh, India. Almost all the livestock deaths are caused by snow leopards. In 1995, snow leopards killed 18 per cent of the total livestock-holding. In retaliation, villagers captured and killed almost all the pups of the Tibetan Wolf in the 1980s. Similar incidents were also noted in Mongolia where economic losses were attributed to the snow leopards and the Tibetan Wolf and the pastoralists killed them in retaliation.

Man–Wildlife Conflict in India

In India, people living in and around the protected areas mainly depend on forest products, agriculture and agro-pasture. As a result, very often man–wildlife conflicts result in crop loss, injury or loss of human life and sometimes the death of wild animals.

About 1,07,770 people live in 117 villages in and around the Sariska Tiger Reserve Project in Rajasthan. Agriculture and rearing livestock are the main sources of livelihood for them. Many species of wild herbivores such as the *Nilgai* and wild boars are to be blamed for 50 per cent of the damage to their crops while *sambar*, *chital*, the common *langur*, rhesus monkeys and parakeets are blamed for the rest. Wild carnivores such as tigers and leopards are responsible for livestock loss. Tigers prey on big domestic animals like cattle and buffalos while leopards prey on goats, sheep and calves.

Man–Elephant Conflict

Once upon a time the forests of Orissa were home to thousands of elephants. But the establishment of Brutanga Irrigation Project in Nayagarh district, the coming up of a large number of steel and iron projects in Jaipur, Keonjhar and Sundergarh districts, the proposed Vedanta Alumina's refinery in Kalahandi district, the Hirakud Dam, the Rengali Irrigation Project and thermal power plants at Talcher in Orissa have caused a severe dent to the wildlife population. Owing to severe pressure on their habitat and food loss, elephants are in their worst-ever confrontation with people. Between 1995–1996 and 2003–2004, a total of 259 persons were killed by wild elephants in Orissa.

But in this case, the good news is that people have understood the problem and are cooperating in regenerating forests, especially in the Dhenkanal district of Orissa. Elephants have also begun to move towards the newly generated forests.

Man–Bear Conflict

In 1998–1999 and 2002–2003, bear attacks in the districts of Angul, Rairakhole, Nabarangpur and Baripada claimed 24 lives. In most cases, the bears attacked when people went inside the forest to pick *mohua* flowers, *kendu* fruits or honey.

Man–Leopard Conflict

From 1991-1992 to 2003, 78 instances of depredation were noted, most of which occurred in Sundergarh and Athmallik districts of Orissa. Man–leopard conflicts not only cause cattle loss but are also responsible for human injury.

Man–Crocodile Conflict

As per the 2004 census, Bhitarkanika National Park in Orissa is home to 1,358 estuarine crocodiles. Between 1998-1999 and 2002-2003, eight people were killed in crocodile attacks. Most of the killings occurred when victims went to collect prawn seeds from the sanctuary crossing the buffer area.

Measures Taken

To prevent man–wildlife conflicts, elephant-proof trenches are dug and rubble walls and energy fences are erected. Awareness is spread among people through newspapers, electronic media and by the distribution of pamphlets. Anti-depredation committees are formed to keep track of problem animals or groups and inform villagers and forest departments in the case of any approaching emergencies. High intensity focus lights, fire torches, drums and crackers are used to ward-off problem-causing animals from the site to the interiors of the forests. Apart from this, compassionate payments are also made to victims sustaining severe losses.

4.6 Conservation of Biodiversity

Conservation of biodiversity is aimed at the protection, preservation, management or restoration of natural resources such as the forests and their flora, fauna, and water. Thus, biodiversity conservation includes:

(i) Protection of all critically endangered, endangered, vulnerable, rare and other species of life present in the ecosystem.

(ii) Preservation of all varieties of old and new flora, fauna and microbes.

(iii) Protection and preservation of critical habitats, unique ecosystems.

(iv) Regulation of international trade in wildlife.

(v) Reduction of pollution.

(vi) Increase in public awareness.

Conserving biodiversity becomes a problem when there is lack of resources and a need to use the land for human activities. The term hotspot is used to define regions of high conservation priority with their biodiversity richness and high endemism and a high threat.

Conservation efforts are often focused on a single species. This is called 'keystone species' because the idea of conserving one species over others is more appealing. For example, conservation of tigers over say Zayante band-winged grasshoppers is not only more appealing and convincing, but it also attracts more resources, which can be used for the conservation of an endangered habitat.

However, the process of conservation can be broadly divided into two types:

(i) **In-situ Conservation:** In this type of conservation, the natural process and its interaction with the habitat as well as with all the elements of biodiversity are conserved. The establishment of protected areas such as national parks, sanctuaries and biosphere reserves is an example of in-situ conservation.

(ii) **Ex-situ Conservation:** In case of complete degradation of a habitat, in-situ conservation is not possible, as the endangered species need special care. In such cases, the endangered species is removed from the area and kept under total human supervision in places such as zoos, botanical gardens and seed banks. This is called ex-situ conservation.

4.6.1 In-situ Conservation

The basic principle of in-situ conservation is the protection and management of components of biological diversity through a network of protected areas in their natural habitat. In this method, the total ecosystem is protected by eliminating the factors that are harmful to the existence of the species concerned. Not only, do the endangered species benefit from this, but all the constituent species present in that ecosystem benefit as well. In-situ conservation is a cheap, convenient and natural way of conservation. The species are allowed to grow in their own natural habitat with the conservationists playing a supportive role. As the species grow in their natural habitat, they face natural calamities such as rain, floods, droughts and snow, and thereby evolve into better-adapted forms. For this reason, the wild species are more resistant to the prevailing environmental conditions than the domesticated or hybrid varieties.

However, the main disadvantage of in-situ conservation is that it requires a large area for the complete protection of biodiversity. This implies a restriction of human activity and a greater overlap or interaction of wildlife with local residents near a reserve forest. People living on the outskirts of a natural reserve depend on the forest for their livelihood. At present there are 7,000 protected areas, parks, sanctuaries and natural reserves in the world, covering more than 650 million hectares of the earth's surface, which is about 5 per cent of the total global land area.

National Parks and Sanctuaries

These are small reserves for the protection and conservation of a few species in their habitat. A national park has a well-defined boundary. Sanctuaries do not have a well-defined boundary and tourists are allowed inside a sanctuary.

Natural Reserve or Biosphere Reserve

These are large, protected areas where the entire biotic spectrum of the climatic zone is preserved. These have boundaries properly identified by legislation. Exploitive human activity or tourists are allowed only up to the outskirts of these reserves areas, which are also scientifically managed.

Project Tiger

The tiger is the finest symbol of earth's natural heritage but tiger sightings these days are very rare in India because of poaching. Tiger poaching is a recurrent problem in countries such as, India, Bangladesh, Bhutan, Cambodia, China, North Korea, Indonesia, Laos, Malaysia, Nepal, Myanmar and Thailand. Almost all the body parts of a tiger are traded for huge amounts of money. Many believe that tigers have healing powers. They believe that tiger bones cure rheumatism, muscular weakness, back pain and enhance longevity. Tiger skins can fetch \$1,50,000, the soup made from its penis is said to increase one's sexual prowess, the whiskers are potent poisons, tiger brain is used to treat acne, tiger tail mixed with soaps cures skin diseases and pills made from its eyes purportedly calms convulsions. Thus a tiger is considered equivalent to a big bag full of money.

Some species of this big cat are already extinct while others are endangered or close to extinction. According to the WWF, tigers are hunted primarily for the use of their body parts in Chinese medicine; these patented Chinese medicines have a huge demand in Asia. Tigers are also poached for souvenirs such as, their skin and mounted heads.

Efforts are being made to preserve this magnificent predator from extinction. Former Indian Prime Minister, the Late Indira Gandhi, launched Project Tiger in 1972, for the conservation and upliftment of the tiger population in India. At present, India has 27 tiger reserves, which extend from the high Himalayan region to the mangrove swamps of the Sundarbans and the thorny scrubs of Rajasthan. Of these 27 tiger reserves, Manas National Park of Assam has been declared a World Heritage Site by UNESCO. Table 4.7 gives a detailed list of the tiger reserves in India. However, more wildlife conservation laws and greater awareness among people are still required for the success of Project Tiger.

Table 4.7 Tiger Reserves of India

Sl. No.	Year of creation	Name of tiger reserve	State	Total area in sq. km.
1	1973-74	Bandipur	Karnataka	866
	1999-2000		Nagarhole (Extension)	642
2	1973-74	Corbett	Uttaranchal	1316
3	1973-74	Kanha	Madhya Pradesh	1945
4	1973-74	Manas	Assam	2840
5	1973-74	Melghat	Maharashtra	1677
6	1973-74	Palaman	Jharkhand	1026
7	1973-74	Ranthambore	Rajasthan	1334
8	1973-74	Simlipal	Orissa	2750
9	1973-74	Sundarbans	West Bengal	2585
10	1978-79	Periyar	Kerala	777
11	1978-79	Sariska	Rajasthan	866
12	1982-83	Buxa	West Bengal	759
13	1982-83	Indravati	Chattisgarh	2799
14	1982-83	Nagarjunasagar	Andhra Pradesh	3568
15	1982-83	Namdapha	Arunachal Pradesh	1985
16	1987-88	Dudhwa	Uttar Pradesh	811
	1999-2000		Katerniaghat (Extension)	511
17	1988-89	Kalakad-Mundanthurai	Tamilnadu	800
18	1989-90	Valmiki	Bihar	840
19	1992-93	Pench	Madhya Pradesh	758
20	1993-94	Tadoba-Andheri	Maharashtra	620
21	1993-94	Bandhavgarh	Madhya Pradesh	1162
22	1994-95	Panna	Madhya Pradesh	542
23	1994-95	Dampha	Mizoram	500
24	1998-99	Bhadra	Karnataka	492
25	1998-99	Pench	Maharashtra	257
26	1999-2000	Pakhui-Nameri	Arunachal Pradesh	1206
27	1999-2000	Bori, Satpura	Assam	1486

4.6.2 Ex-situ Conservation

Due to the degradation and fragmentation of habitat, a large number of species are on the verge of becoming extinct. Ex-situ conservation aims at protecting and preserving such endangered species in zoos, nurseries and laboratories. Breeding plants and animals under human care is the strategy employed by ex-situ conservation. Although, earlier it was not practiced for wildlife conservation today with the advancement of science and technology the practice has emerged as a well-defined technology for the purpose. The following are the two main steps for ex-situ conservation:

(i) Identification of the species to be conserved.
(ii) The selection of method to be followed for its ex-situ conservation.

(i) **Identification of the Species to be conserved:** Those species that are at the maximum risk of extinction are chosen for preservation. The life cycle of the species, its degree of specialization, rich location, dispersal ability, adult survival and atrophic status are studied for the final selection.

(ii) **Methods for Ex-situ Conservation:** From the study on the selected species, the method for its growth, reproduction and survival under ex-situ conservation is decided. The various methods adopted for ex-situ conservation of the critically endangered species are as follows:

(a) *Long-Term Captive Breeding*: If the species is being pushed into extinction due to habitat loss or by some adverse external conditions then they are removed from their natural habitat for long-term captive breeding. Captive breeding can increase their population and help the species to survive. Thereafter, as most of these species cannot survive in their wild habitat they are kept in zoos and botanical gardens under proper care.

(b) *Short-Term Propagation and Release*: If the population of a particular species decreases suddenly due to some temporary setback then it is removed from its natural habitat, maintained with ex-situ conservation methods, bred and later released into their natural habitat. Ex-situ crocodile conservation is an example of this method.

(c) *Animal Translocation*: If the population of a particular species decreases suddenly then some animals of the same species are brought from a similar habitat and released in the less populated area. For example, if the number of male tigers decreases in habitat number one, then male tigers of the same species are brought from some other area and released in this habitat so as to increase the tiger population here. However, the capture, transfer and release of wildlife from one area to another require maintenance of the species in captivity for a short period.

(d) *Animal Reintroduction*: When an animal becomes extinct from its natural habitat, attempts are made to reintroduce the species there. For this, newborn animals bred in captivity or animals caught in infancy and kept in captivity for some time then they are selected and released into the habitat from where the original population has disappeared. It is important to rehabilitate the reintroduced species or they too may suffer the same fate as the original species. For this purpose, proper maintenance of the natural habitat and constant observation of the reintroduced species is very important. These days radio collars are used for observation.

The capture, transfer and release of animals from one locality to another is difficult, so special drugs are administered to the target animal from a distance to immobilize the animal. Special emphasis is laid on the nutrition and health care of the animals by administering preventive medicines and systematic vaccinations to them.

Artificial insemination, embryo transfer and cryo-preservation of gametes and embryos are the techniques used to maintain the genetic diversity of nature.

Biosphere Reserves of India

Biosphere reserves are protected areas of representative ecosystems of terrestrial as well as coastal areas. They are internationally recognized under the Man and the Biosphere (MAB) Programme initiated by UNESCO in 1971. A biosphere reserve is aimed at:

(i) In-situ conservation of biodiversity of natural and semi-natural ecosystems and landscapes.

(ii) Contribution to sustainable economic development of the human population living within and around the biosphere reserve.

(iii) Providing facilities for long-term ecological studies, environmental education, training, and research and monitoring.

Thus, these reserves could serve as a referral system for monitoring and evaluating changes in the natural ecosystem. A biosphere reserve is classified into three zones:

Core Zone: This zone is meant for the conservation of biological diversity and is securely protected. Non-destructive research work and low-impact activities like education and ecotourism can be conducted here.

Buffer zone: This zone surrounds the core zone and is used for cooperative activities such as environmental education, recreation, basic and applied research and so on.

Transition area: It surrounds the buffer zone and may be used for agricultural activities, settlement of local communities, NGOs', cultural groups and by other stakeholders for economic interests and sustainable development of the area's resources.

Globally, 425 biosphere reserves have already been established in 95 different countries since 1979. The Government of India constituted a panel of experts in 1979, to identify potential areas of biosphere reserves under the MAB Programme of UNESCO. The experts identified 14 sites to be declared as biosphere reserves. Of them, 13 sites were declared biosphere reserves in 2005 and later Achanakmar-Amarkantak was declared the 14th biosphere reserve of India. Table 4.8 lists the biosphere reserves of India that have been declared till date.

Table 4.8 List of Biosphere Reserves of India

Sl. No.	Name	Date of establishment	Area (km²)	Location
1	Achankamar-Amarkantak	2005	3,835.55 (core 551.55, buffer 3283.96)	Covers part of Anupur and Dindori district of Madhya Pradesh and Bilaspur of Chhatisgarh state
2	Agasthyamalai	12 Nov. 2001	1,701	Neyyer, Peppara and Shendurney wildlife sanctuaries and their adjoining areas in Kerala
3	Dehang-Debang	2 Sept. 1998	5,111.50 (core 4,094.80, buffer 1, 016.70)	Part of Siang and Dibang valley in Arunachal Pradesh
4	Dibru-Saikhowa	28 July 1997	765 (core 340, buffer 425)	Parts of Dibrugarh and Tinsukhia district of Assam
5	Great Nicobar	6 Jan. 1989	885 (core 705, buffer 180)	Southernmost island of Andaman and Nicobar
6	Gulf of Mannar	18 Feb. 1989	10,500 (Total gulf area)	Indian part of Gulf of Mannar between India and Sri Lanka
7	Khangchenzonga	7 Feb. 2000	2,619.92 (core 1,819.34, buffer 800.58)	Part of Khangchenzonga hills and Sikkim
8	Manas	14 March 1989	2,837 (core 391, buffer 2446)	Parts of Kokrajhar, Bongaigaon, Barpeta, Nalbari, Kamrup and Darang districts of Assam
9	Nanda Devi	18 Jan. 1988	5,860.69 (core 712.12, buffer 5148.57, transition 546.34)	Part of Chamoli, Pithoragarh and Bageshwar districts (Uttaranchal)
10	Nilgiri	1 Sept. 1986	5,520 (core 1240, buffer 4,280)	Part of Wayanad, Nagarhole, Bandipur and Madhumalai, Nilambari, Silent Valley and Siruvani hills (Tamilnadu, Kerala and Karnataka)

Sl. No.	Name	Date of establishment	Area (km²)	Location
11	Nokrek	1 Sept. 1988	82 (core 47.48, buffer 34.52)	Part of Garo hills (Meghalaya)
12	Pachmarhi	3 March 1999	4,926	Parts of Betul, Hoshangabad, Chindwara district of Madhya Pradesh
13	Simlipal	21 June 1994	4,374 (core 845, buffer 2,129, transition 1,400)	Part of Mayurbhanj district of Orissa
14	Sunderbans	29 March 1989	9,630 (core 1,700, buffer 7,930)	Part of delta of Ganges and Brahmaputra river system (West Bengal)

World Heritage Sites

The World Heritage Site list was established in November 1972 at the 17th General Conference of UNESCO, under the terms of the convention concerning the protection of world culture and natural heritage. The main responsibility of the World heritage committee was to provide technical cooperation under the World Heritage Fund to safeguard these sites. Table 4.9 lists the world heritage sites of India.

Table 4.9 The World Heritage Sites of India (Natural)

Sl. No.	Name of the site	Location
1	Kaziranga National Park	Assam
2	Manas Wildlife Sanctuary	Assam
3	Keoladeo National Park	Rajasthan
4	Sundarbans National Park	West Bengal
5	Nanda Devi National Park	Uttar Pradesh

Ramsar Sites in India

A close observation of the current list of Ramsar sites in India (Table 4.10) represents only a fraction of the diversity of wetland habitats existing in the country.

Table 4.10 List of Ramsar Sites in India

Sl. No.	Name
1	Ashtamudi Lake
2	Bhitarkanika Mangroves
3	Bhoj Wetlands
4	Chilika Lakes
5	Deepor Beel
6	East Kolkata Wetlands
7	Harike Wetlands
8	Keoladeo National Park
9	Kolleru Lake
10	Loktak Lake
11	Point Climere Wildlife and Bird Sanctuary
12	Pong Dam Lake

Sl. No.	Name
13	Sambar Lake
14	Sasthamkota Lake
15	Tsomoriri
16	Vembanad-kol Wetland
17	Wular Lake
18	Kanjli Lake

4.7 Bioprospecting and Biopiracy

Bioprospecting is the collecting, cribbling of biological samples (of plants, animals and micro-organisms) and gathering indigenous knowledge to help in discovering genetic or biochemical resources. The main objective of bioprospecting is the development of new life-saving drugs, crops that provide better economical benefits or industrial products.

Prior to 1992, biological resources were considered the common heritage of mankind. Anybody could collect and carry samples away from the place without permission.

In 1992, the Convention on Biological Diversity (CBD) established the Sovereign National Rights over biological resources. According to this, the biodiversity-rich countries are committed to:

(i) Conserve biodiversity.

(ii) Develop it for sustainable use.

(iii) Share the benefits resulting from their use.

Thus, the biodiversity rich countries have to not only allow bioprospecting but also be vigilant so that equal benefit is available to the communities that traditionally use the resources, to corporations (usually in developed countries) and to universities collecting the bioresource. Bioprospecting must also obey the national laws and respect the rules of international treaties like:

(i) Informed consent: The source country must know, what will be done, which benefits will be shared and must give permission for collection.

(ii) Fair agreement on benefits of sharing: Benefits may include support for conservation, research, equipment, technologies, knowledge, transfer, development and royalties.

CASE STUDY

Jeevani Drug Issue

The local Kani tribals in the Thiruvanthapuram district of Kerala, claim that they can live for days together without food and still be able to perform rigorous physical work just by eating a few fruits of a plant named *Aarogyapaccha* everyday. Scientists from the Tropical Botanic Garden Research Institute (TBGRI) learnt about the use of this plant and carried out a detailed investigation. The findings revealed that the plant is really a source of health and vitality. The leaves of the plant had anti-stress, anti-hepatoxic and immunodulatory/immunorestorative properties. Eventually, the TBGRI scientists prepared the drug *Jeevani* from *Aarogyapaccha* and three other medicinal plants. Later, in 1995, a license to manufacture Jeevani was given to Arya Vaidya Pharmacy, Coimbatore for a period of seven years, for a fee of Rs 10 lakh. The Kani tribals were to get 50 per cent of this amount along with 50 per cent royalty obtained by the TBGRI on the sale of the drug. This is bioprospecting in which all the rules set for the purpose are followed.

When bioprospecting ignores all the principles described above it becomes biopiracy. Translational companies race against one another to manufacture pharmaceuticals and agricultural products from the genetic material of medicinal plants and food crops. They also collect micro-organisms, animals and even genes of indigenous people, for their research activities. Thereafter, these companies rush to apply for a patent of the so-called 'new' product or technology, so that they can get larger profit by stopping other companies from using the raw material or by selling the technology.

Farmers, indigenous communities and citizen groups protest against these companies being given patent rights because it is the local communities who are responsible for identifying and evolving the use of these biological species (plants, animals, genes and so on). These companies claim exclusive rights to produce and sell the products made by them. Third World communities are afraid that in future they would have to pay a higher price for the material which they had identified and developed. This injustice (biopiracy) is now being fought by farmers, people and public interest groups. Many NGOs such as Rural Agricultural Foundation International (RAFI) and Genetic Resources Action International (GRAIN) have been raising their voice against these unscrupulous companies.

In India, M. D. Nanjundaswamy of the Karnataka Farmers Union is leading farmers against the patenting of seeds, plants and the operations of foreign grain companies in the country. In 1993, half a million farmers rallied in Bangalore to protest against the Uruguay Round Treaty which opened the door to the patenting of genetic materials, seeds and plants. Such agitations have been held in South America, Asia and the Pacific and also at the Beijing UN Women's Conference where 118 indigenous groups from 27 countries signed a declaration demanding 'a stop to the patenting of all life forms' which is 'the ultimate commodification of life which we hold sacred.' They demanded that the Human Genome Diversity Project be stopped and patent applications for human genetic materials be rejected.

Patenting of Life

A bacterium that can digest oil has been derived through genetic engineering. Cells from a human spleen have been transplanted in mice that are genetically predisposed to get cancer. These are derived from living organisms and all these have been patented as human inventions. As these inventions affect the society and the life of the owner, the question is how can one get the right to privatize ownership of life? Campaigns have been launched by religious heads and NGOs against the patenting of life.

Biopiracy—the Turmeric Patent

In 1995, two US-based Indians were granted the US patent 5,401,504 on the 'use of turmeric in wound healing' both externally and orally. This patent, which was assigned to the University of Mississippi Medical Centre, USA also granted them exclusive right to sell and distribute turmeric.

Turmeric is a tropical herb of East India. Its rhizomes in the powdered form are used as a dye, a cooking ingredient and in a number of medicinal uses, such as a blood purifier, in treating the common cold and an anti-parasite for skin infections. Turmeric has been used in India for thousands of years. Hence, concern grew about the economically and socially damaging impact of this legal biopiracy. Two years later, in 1997, the Council for Scientific and Industrial Research (CSIR) filed a complaint challenging the novelty of the University's discovery. The US patent office investigated the validity of the patent and revoked it in 1997.

Some other examples of US patents related to Asian materials widely known for their medicinal applications are bitter melon from China (Pollack, 1999); products obtained from the neem tree (*azadirachta indica*) of India, some varieties of chick peas and basmati rice (by Rice tec) some hybrid varieties of Bolivian quinua by the University of Colorado Scientist (Gari, 2000). Similar cases in Amazonian, Andean countries, Meso-America, Africa and Asia have created widespread awareness about the value of genetic resources and the new modes of biopiracy.

SUMMARY

Causes of Biodiversity Loss

Today, the world has become conscious of the value of biodiversity and the threats facing it. The International Convention on Biological Diversity at Rio de Janeiro, in 1992 focused attention on the sustainable use of the components of biodiversity.

Threats to Indian Biodiversity

India is a rich storehouse of biodiversity, with 7.31 per cent of the world's species of fauna and 10.78 per cent of the species of flora. The country has a total of 89,451 numbers of animal species and several floral species, one-thirds of which are endemic to the country.

Loss and fragmentation of habitat has generated small, scattered and isolated populations. The strategies for sustainable conservation of biodiversity should follow agricultural activities that are coupled with conservation measures.

Poaching of wildlife

Poaching can be simply defined as illicit wildlife trade. It is the illegal killing or taking of any wildlife for sale, in the international trade market.

Man–Wildlife Conflict

The conflict between man and wildlife occurs when wild animals leave the protected areas (forests) to raid human settlements in search of food, water and to migrate long distances or to raid crops. The main cause of man–wildlife conflict is the growing anthropogenic pressure on wildlife habitat. This results in:

(i) Fragmentation and honeycombing of animal habitats.

(ii) Loss of corridors and migratory routes for long-range animals such as elephants and big cats (tiger, leopard, bear etc.).

(iii) Loss of food and water in their habitat due to shrinkage of forest cover and loss of biodiversity.

Conservation of biodiversity includes measures like protection, preservation, management or restoration of natural resources including all the critically endangered, endangered, vulnerable, rare and other species of life present in the ecosystem. Conservation efforts are often focused on a single species called 'keystone species' because some species act as the key to the functioning of a habitat and their loss could lead to greater average change in the entire ecosystem. The process of conservation can broadly be divided into two types:

(i) **In-situ Conservation**: In this method, the species are allowed to grow in their own natural habitat. In-situ conservation protects the entire ecosystem; hence a large number of species, along with the flag-ship species, are conserved and the food chain and food web remains intact.

(ii) **Ex-situ Conservation**: Ex-situ conservation aims at the protection and preservation of endangered species away from their natural habitat under human care in zoos, nurseries and laboratories.

National Parks and Sanctuaries

National parks and sanctuaries are small reserves for the protection and conservation of one, two or very few species in their habitat. A national park has a well-defined boundary. Sanctuaries do not have a well-defined boundary and tourists are allowed inside a sanctuary.

Natural Reserve or Biosphere Reserve

These are large protected areas where the entire biotic spectrum of the climatic zone is preserved. These have boundaries properly identified by legislation. Exploitive human activity or tourists are allowed only at the outskirts of these reserve areas which are also scientifically managed. Biosphere reserves are internationally recognized through the Man and Biosphere (MAB) Programme of UNESCO initiated in 1971.

Project Tiger

Former prime minister of India, Late Indira Gandhi, launched the program in 1972 for conservation and upliftment of the tiger population in India.

World Heritage Sites

The World Heritage Sites list was established under the terms of the Convention concerning the protection of World Culture and Natural Heritage in November 1972 at the 17th General Conference of UNESCO. The main responsibility of the committee is to provide technical cooperation under the World Heritage Fund to safeguard these sites.

Ramsar Sites in India

Ramsar Sites in India represent only a fraction of the diversity of wetland habitats existing in the country.

Bioprospecting and Biopiracy

Bioprospecting is the collecting, cribbling (sieving) of biological samples (of plants, animals, micro-organisms) and gathering indigenous knowledge to help in discovering genetic or biochemical resources.

ESSAY TYPE QUESTIONS

1. Define biodiversity. How is it related to the availability of genes, species and ecosystem of a region?
2. 'Biodiversity of a region is the totality of genes, species and ecosystem of the region.' Explain.
3. Write a short note on biogeographic classification of India.
4. Justify the status of India as a megadiversity nation.
5. What are the values of biodiversity? Differentiate between direct use values and indirect values.
6. What are endemic species? Discuss the status of India as the abode of endemic flora and fauna.
7. What are the major threats to biodiversity? Discuss.
8. Write a note on the cause, effect and combating of the problem of man–wildlife conflict.
9. Write a note on efforts taken for biodiversity conservation.
10. Differentiate between in-situ and ex-situ conservation principles.

SHORT-ANSWER TYPE QUESTIONS

1. What do you mean by the word biodiversity? Why is it necessary to conserve biodiversity?
2. Explain the role of biodiversity in genetic variation of species.
3. Write a short note on the necessity of biogeographic classification of the biosphere of the earth.
4. Name the biogeographical regions of the earth.
5. What are values of biodiversity? Write a short note on economic values of biodiversity.
6. What is consumptive use value?
7. Write the salient points of social use values.

8. What are biodiversity hotspots? How many global hotspots are recorded till date?
9. Name two hotspots of India. Justify the objective of identifying these areas as biodiversity hotspots.
10. What are endemic species? Name some endemic species of India.
11. What are the major threats to the biodiversity of India?
12. How does one combat the threat to biodiversity?
13. What is poaching of wildlife? How can it be controlled?
14. Differentiate between bioprospecting and biopiracy?
15. Name the biosphere reserves of India. How does it help in conservation of biodiversity?

MULTIPLE CHOICE QUESTIONS

1. The scientific study of the geographic distribution of plants and animals is called
   (a) biodiversity.
   (b) biogeography.
   (c) ecology.
   (d) biology.

2. The entire biosphere is distributed into following number of biogeographic regions:
   (a) Six.
   (b) Eight.
   (c) Nine.
   (d) Twelve.

3. The total area of India is classified into following number of biogeographical zones:
   (a) Six.
   (b) Eight.
   (c) Nine.
   (d) Ten.

4. Biodiversity hotspots are also known as
   (a) evergreen forests of tropic region.
   (b) biologically rich areas with large percentage in endemic species.
   (c) desert areas.
   (d) All of the above.

5. Species with very restricted distribution over relatively small ranges is called
   (a) endangered species.
   (b) extinct species.
   (c) endemic species.
   (d) None of the above.

6. The major threats to biodiversity is due to
   (a) habitat loss/degradation.
   (b) pollution and global climatic changes.
   (c) extinction of species by aggressive non-native species.
   (d) All of the above.

7. Protection and preservation of endangered species away from their natural habitat under human care in zoos, nurseries and laboratories is known as
   (a) in-situ conservation.
   (b) ex-situ conservation.
   (c) biodiversity conservation.
   (d) None of the above.

8. Protection of endangered species by preserving the entire ecosystem is known as
   (a) in-situ conservation.
   (b) ex-situ conservation.
   (c) biodiversity conservation.
   (d) None of the above.

9. The concept of biodiversity hotspots is given by
   (a) F.P. Odum.
   (b) Norman Myers.
   (c) James Lovelock.
   (d) Rachel Carson.

10. Which of the following is an endemic species found in Western Ghats, India?
    (a) Marsh Mongoose.
    (b) Indian Rhinoceros.
    (c) Brown Palmcivet.
    (d) Flying Squirrel.

11. Which of the following is not a world heritage site of India?
    (a) Sundarbans National Park.
    (b) Manas Wildlife Sanctuary.
    (c) Simlipal.
    (d) Kaziranga National Park.

12. Which of the following is a Ramsar site in India?
    (a) Sambar Lake.
    (b) Dal Lake.
    (c) Ansupa Lake.
    (d) Dimna Lake.

13. Which of the following is an in-situ tiger reserves in India?
    (a) Dudhwa.
    (b) Gulf of Myanmar.
    (c) Western Ghats.
    (d) Agasthyamalai.

14. Which of the following is not a biosphere reserve of India?
    (a) Sundarbans.
    (b) Great Nicobar.
    (c) Periyar National Park.
    (d) Khangchenzonga.

15. Which of the following is a biodiversity hotspot in India?
    (a) Succulent Karoo.
    (b) Mediterranean Basin.
    (c) Sundland.
    (d) Eastern Himalayas.

16. Which of the following animals is endemic to India?
    (a) Snow Leopard.
    (b) Blue Whale.
    (c) Asian Elephant.
    (d) Red Colobus Monkey.

17. The variety and the numbers of living organisms present in an ecosystem is called
    (a) biodiversity.
    (b) biopiracy.
    (c) biogeography.
    (d) bioprospecting.

18. Variation of genes within the same species is
    (a) genetic diversity.
    (b) species diversity.
    (c) biodiversity.
    (d) ecosystem.

19. Extinction of a weaker species by an aggressive alien species is the result of
    (a) endemism of weaker species.
    (b) habitat loss.
    (c) the Domino Effect.
    (d) All of the above.

20. Which is not a zone of the biosphere?
    (a) Core Zone.
    (b) Buffer Zone.
    (c) Transition Zone.
    (d) None of the above.

ANSWERS

1 (b)	2 (b)	3 (d)	4 (b)	5 (c)	6 (d)	7 (b)	8 (a)	9 (b)	10 (c)
11 (c)	12 (a)	13 (a)	14 (c)	15 (d)	16 (b)	17 (a)	18 (a)	19 (c)	20 (d)

UNIT V
Environmental Pollution

LEARNING OBJECTIVE

After reading this chapter, students will be able to:

- Define and identify the causes, effects and control measures of air pollution, water pollution, soil pollution, marine pollution, noise pollution, thermal pollution, and nuclear hazards.
- Explain solid waste management.
- Describe the causes, effects and control measures of urban and industrial wastes.
- Explain the role of individuals in the prevention of pollution.
- Describe case studies.
- Outline the fundamentals of disaster management in the event of flood, earthquakes, cyclones, and landslides.

5.1 Introduction

Environmental pollution dates back to the time when man discovered the use of fire. The burning of fossil fuels (wood, coal, and oil gas) releases a number of poisonous gases into the atmosphere. Environmental pollution includes air, water, noise, and soil pollution as well as pollution caused by radiation.

Pollution is a broad term that includes the accumulation of matter due to unknown activities in the environment; rates of flow which exceed the capacity of the ecosystem to either neutralize or disperse them below harmful levels. An undesirable change in the physical, chemical and biological characteristics of air, land and water that is likely to affect human life or the lives of desirable species adversely is also termed as pollution. Industrial processes, living conditions and cultural assets that are likely to cause wastage or deterioration of raw material resources can also be termed as pollution.

At the beginning of the human civilization, our environment was pure, virgin, uncontaminated and undisturbed; it was most supportive and hospitable to living organisms. The advancement of science and technology led to the exploitation of all types of natural resources. Progress in agriculture, followed by rapid industrialization has left us with barren lands, contaminated soil, depleted wildlife, polluted rivers and also exhausted our natural resources.

Today, environmental pollution is a serious problem. Air, water, and soil are essential for survival of life on earth but unfortunately pollution is causing them irreparable harm. Beyond certain limits air pollution can cause illness and even death. Polluted water causes fish and other aquatic life to perish and is also dangerous to human health.

Soil pollution reduces the amount of land available for growing crops, fruits, and vegetables. All these factors have led the people to become conscious about and aware of the adverse effects of pollution.

Since pollution is mostly caused by things that are of use to mankind, the problem becomes complicated to deal with. For example, automobile exhaust is a major cause of air pollution, but it provides transport for millions of people all over the world. Similarly, chemicals contained in fertilizers and pesticides are essential for agriculture but their overuse can ruin the soil. Scientists and engineers, all over the world are working on ways to reduce pollution and governments are also actively making and enforcing laws for the same purpose. Thus, it is our responsibility to step forward and do what we can to reduce pollution.

5.2 Air Pollution

Air pollution is one of the most dangerous forms of environmental pollution in current times. As a result of natural and man-made activities, gases such as CO, SO_2, H_2S and oxides of nitrogen, mist particulates and aerosol are continuously released into the atmosphere and get dispersed in the air through air currents. These gases exist upto 2,000 feet above ground level. This disturbs the dynamic equilibrium of the atmosphere and causes health hazards to all organisms. The average composition of clean air near the sea level is shown in Table 5.1.

Table 5.1 Average Composition of Clean Dry Air Near Sea Level (ppm by volume)

Components	Concentration in average concentration ppm	Volume percentage
Major		
Nitrogen	7,80,900	7.809×10^1
Oxygen	2,09,400	2.094×10^1
Minor		
Argon	9,300	9.3×10^{-1}
Carbon dioxide	318	3.18×10^2
Trace		
Neon	18	1.8×10^{-3}
Helium	5.2	5.2×10^{-4}
Krypton	1.0	1.0×10^{-4}
Methane	1.3	2.0×10^{-4}
Hydrogen	0.5	1.0×10^{-7}
Carbon monoxide	0.1	1.0×10^8
Ozone	0.02	
Nitrogen dioxide	0.001	
Sulphur dioxide	0.0002	

Except argon and other inert gases almost all the components of the earth's atmosphere are primarily produced and removed by biological activities and are greatly affected by biota. Although there is a natural alternation of the atmosphere due to biotechnology, the relative proportions of the mixture of gases are constant. It has been observed that the mass of the atmosphere is so large that the total mass of even a trace gas is quite large.

5.2.1 Sources of Air Pollution

The heavily polluted air is a result of natural and man-made pollution. Air pollutants are present in the atmosphere in concentrations that disturb the dynamic equilibrium of the atmosphere and thereby affect human beings and the environment. The pollutants are neither formed nor distributed locally or regionally. Hence, the sources of pollutants are classified as point source or non-point source.

Pollution caused by automobile discharge can be defined as non-point source, while the emission of gases from an industry through a chimney can be described as point source. In both the cases the pollutants are carried to far away places through air currents. For example, sulphur dioxide emissions in the United Kingdom causing acid rain in Scandinavia or the dust from the Sahara Desert reaching West Germany through air currents is well known. However, sources of air pollution can be broadly classified into (i) man-made or anthropogenic sources and (ii) natural sources.

Anthropogenic or Man-made Sources

Man-made sources include population explosion, burning of fossil fuels, vehicular discharges, rapid industrialization, agricultural activities and modern warfare. All these lead to global warming.

(i) **Population explosion**: The fast growth rate of population will create several serious problems in a chain reaction. Increase in population will create global warming and the emission of Greenhouse gases. These in turn will result in a rise in sea levels. Population explosion will also cause an increase in demand for food followed by the need for more land for cultivation thus causing destruction and loss of forest cover and wildlife.

(ii) **Burning of fossil fuels and fires**: Burning of conventional fossil fuels such as coal, lignite, petroleum and natural gas produces gaseous by-products which are poisonous, these include CO_2, CO, CH_4, SO_2 and oxides of nitrogen. These gases pollute the air and make it unfit for breathing. However the quality and concentration of these pollutants depends on the type of fuel used. The smoke from the chimneys of a factory and the dust from anthropogenic activities contain a large amount of sulphur and nitrogen, respectively. Sulphur and nitrogen burn in the atmosphere to produce their oxides. Both are highly soluble in water and form sulphuric, sulphurous and nitric acid, respectively. These acids come down to the earth along with rain in the form of acid rain or acid precipitation. They are responsible for the destruction of the ecosystem and corrosion in factories.

(iii) **Vehicular discharges**: In urban areas, about 75 per cent of the air pollution is caused by automobile emissions. Automobiles run mainly on petrol or diesel. They pollute the air not only with exhaust gases but also with tiny bits of lead from tetraethyl lead that is contained in gasoline so as to prevent engine knock. A mixture of carbon monoxide (about 77 per cent), oxides of nitrogen (about 8.4 per cent), hydrocarbons (about 14 per cent) and leaded gas along with some particulate of lead is emitted due to incomplete combustion of hydrocarbons (petrol and diesel) in internal combustion engines. These discharged gases react with oxides of nitrogen in the presence of sunlight to produce highly toxic photochemical smog. Number of automobiles is continuously on the rise; their growth is shown in Fig. 5.1.

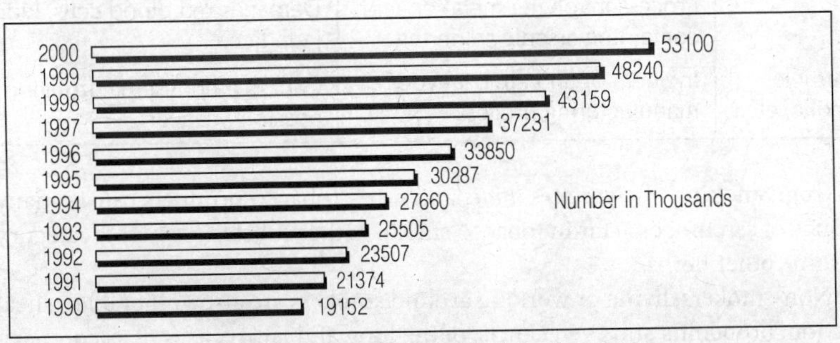

Source: Transport Commissioner Office – Maharashtra State

Figure 5.1 Growth of Automobiles in India

(iv) **Rapid industrialization:** The advancement in science has led to the establishment of several industries such as the chemical industry, paper and pulp mills, cotton mills, metallurgical plants smelters, petroleum refineries, mining, synthetic rubber industries and tanneries. These industries are responsible for 20 per cent of the air pollution as they discharge pollutants of the gaseous, liquid and particulate type. The most common pollutants are CO, CO_2, NO, NO_2, SO_2 and H_2S.

(v) **Agricultural activities:** Different types of insecticides, pesticides and herbicides used in agriculture pollute the air through air currents thereby end up making the air hazardous for both human and animal health.

(vi) **Modern warfare:** To a large extent, radioactive rays from nuclear reactors, nuclear explosions and modern war explosives cause extensive air pollution and suffering to mankind.

Table 5.2 Common Air Pollutants, their Sources and Effects on Human Beings

Pollutant	Source	Pathological effect on human beings
Sulphur dioxide	Combustion of coal and oil	Causes chest constriction, headache, vomiting and, ultimately, death due to respiratory ailments.
Nitrogen oxide	Soft coal, automobile exhaust	Inhibits cilia action so that soot and dust penetrates far into the lungs.
Hydrogen sulphide	Refineries, chemical industries and bituminous fuels	Causes nausea, irritation of eyes and throat.
Carbon monoxide	Burning of coal, gasoline motor exhaust	Reduces oxygen carrying capacity of blood.
Hydrogen cyanides	Blast furnace, fumigation, chemical plants	Interferes with nerve cells, produces dry throat, indistinct vision, headache.
Ammonia	Explosives, dye-making, fertilizer plants and lacquers	Inflames upper respiratory passage.
Phosgene	Chemical and dye-making industry	Induces coughing, irritation and fatal pulmonary oedema.
Aldehydes	Thermal decompositions of oils, fats and glycerols	Irritates nasal and respiratory tracts.
Arsenic	Process involving metal or acid containing arsenic soldering	Damages red blood cells, kidneys and causes jaundice.
Suspended particles (ash, soot, smoke, etc.)	Incinerator and almost every manufacturing process	Causes emphysema, irritation in the eyes and possibly, cancer.

(vii) **Smoking:** Smoking of cigarettes, *bidis* and other tobacco products causes many diseases due to the presence of carcinogenic tar in tobacco smoke. Other substances in this category are opium, *dhatura* and some other herbs.

Non-smokers, living or working around smokers are also vulnerable to health problems such as irritation, bronchitis and even cancer of the lung and larynx as well as coronary diseases. In spite of common knowledge and the statutory warning that states, 'Smoking is injurious to health', smoking is very common and continues to be a major social problem.

5.2.2 Types of Air Pollutants

Depending upon the generation of different air pollutants, they are grouped as (a) Primary pollutants and (b) Secondary pollutants. Common air pollutants are listed with their sources and effects in Table 5.2.

Primary Pollutants

A primary pollutant can be defined as a harmful chemical that directly enters the air as a result of either natural or man-made activities. The following five types of primary pollutants account for more than 90 per cent of the nationwide air pollution.

(i) Oxides of Carbon
(ii) Nitrogen oxides (NO)
(iii) Hydrocarbons (HC)
(iv) Sulphur oxides (SO)
(v) Particulate matter

(i) **Oxides of Carbon:** Oxides of carbon, mainly carbon monoxide (CO) and carbon dioxide (CO_2) are produced from burning of fossil fuels, automobile discharge and from various other natural and anthropogenic sources.

Carbon Monoxide: About 80 per cent carbon monoxide is produced by automobile exhaust. CO is also produced by oxidation of methane by anaerobic decomposition of organic matters in swamps and humid tropics. A small amount is emitted from volcanoes, forest fires and by incomplete combustion of fuels. On inhalation, CO passes into the blood stream through the lungs. Here, carbon monoxide displaces oxygen from haemoglobin to produce carboxyl haemoglobin.

$$O_2Hb + CO \longrightarrow COHb + O_2$$

As a result it reduces the oxygen carrying capacity of blood. The level of COHb in the blood is directly related to the concentration of inhaled air. It can be fatal at a concentration exceeding 1,000 ppm.

Poisoning by way of CO occurs due to lack of áwareness, which is a probable cause of many accidents. Progressively higher exposure to CO causes impairment of the central nervous system functions, changes in cardiac and pulmonary functions, drowsiness, coma, respiratory failure and finally death.

Carbon Dioxide: It is the basic end-product obtained on burning fossil fuels, paper, leaves and other carbon-containing material. Carbon dioxide is used by plants for photosynthesis. Although carbon dioxide has no direct effect on health, but with higher concentrations (above 10 per cent) **it causes global warming, acid rain and Greenhouse Effect.**

(ii) **Nitrogen Oxides:** About 90 per cent of oxides of nitrogen are emitted from diesel engines. Apart from automobile discharge, it is also produced from electric power plants, industries, commercial and residential units.

At normal temperature, oxides of nitrogen are relatively harmless but at high temperature, atmospheric nitrogen combines with oxygen to produce nitric oxide (NO) which causes asphyxiation at high concentration. Besides, this high concentration of nitric oxide is the root cause of lung cancer, pneumonia, inflammation and internal bleeding.

(iii) **Hydrocarbons:** In India, automobiles are the chief source of hydrocarbon pollutants in the atmosphere. Besides this, hydrocarbons are also emitted in huge quantities from plants, particularly of Coniferae and Mysta Ceace family. For example, eucalyptus, cottonwood, oak, sweet gram and spruce trees emit hemiterpene that is isoprene. They are also produced by anthropogenic activities such as the production of coke and smouldering of refuse piles near coal mines.

Hydrocarbons at high concentrations (500–1,000 ppm) are carcinogenic and therefore harmful for the lungs. Aromatic hydrocarbons like benzene and toluene are more dangerous, inhalation of their vapour causes irritation of the mucous membrane. Increased concentration of hydrocarbon vapours increases mucus secretion leading to blockage of the respiratory tract, as a result of which a person coughs continuously. This creates pressure on the trachea of the lungs and finally the alveoli membrane bursts leaving very little area for the exchange of oxygen and carbon dioxide.

Hydrocarbons react with nitrogen oxides and produce photochemical smog that causes irritation to the eyes, nose and throat, as well as respiratory distress. Plants exposed to high levels of hydrocarbons display yellowing of the leaves.

(iv) **Sulphur Oxides:** Besides natural sources, coal-fired electrical power plants and industries are the main sources of sulphur oxide (SO_2) pollutants.

Sulphur dioxide reacts with ozone, hydrogen peroxide or atmospheric water vapour to produce a toxic acid called sulphuric acid (H_2SO_4). It comes down on the earth with rain or snow as acid rain. It corrodes limestone, metals and even clothes. It is also very injurious to the respiratory system, mainly to the lungs.

(v) **Particulate Matter:** Air-borne, small, solid particles or liquid droplets are collectively known as particulate matter or aerosols. The size of this particulate matter varies from a diameter of 2×10^{-8} cm to 5×10^{-2} cm. Particles in cigarette smoke have diameters as small as 10^{-6} cm, whereas those produced from a cement factory have a diameter of about 10^{-2} cm.

These air pollutants or aerosols remain in the atmosphere under gravitational pull. Depending on the size, their lifetime (settling time) varies from a few seconds to several months. Aerosols like fog, plant spores, bacteria and pollen are of natural origin and do not cause any atmospheric pollution. On the other hand, smoke; cement dust, fuel dust, quartz, asbestos, powder, oil smoke, tobacco smoke, fly ash from power plants, mining operation, radioactive aerosols, volcanic eruptions, spraying of salt and other solid particles from the sea and ocean are air pollutants. The second type of aerosols remain suspended in the air and get absorbed by substances such as lead, hydrocarbons, sulphur and nitrogen oxides thus producing smog. The concentration of particulate matter in air varies from several 100 per cm^3 in clean air to 105 per cm^3 in highly polluted air. In urban areas, it is 6×10^{-5} gm to 2×10^{3} gm per cm^3. However, the chemical nature of the particulates as well as, the size is more important than their concentration in a particular area. This particulate matter may be organic or inorganic in nature.

Organic particulate matter: The natural particulate matter present in aerosol is known as organic particulate matter. For example, particulates that originate from volcanoes, dust storms, forest and grassland fires and living vegetation

Inorganic Particulate Matter: This mainly comprises, metal oxides formed during the burning of fossil fuels, industrial processes, vehicular exhaust and acid rain. During combustion of pyrites, coal iron oxide is formed and from its ash, calcium oxide is released into the atmosphere.

$$3FeS_2 + 8O_2 \longrightarrow Fe_3O_4 + 6SO_2$$

From ash $$CaCO_3 \xrightarrow{\text{heat}} CaO + CO_2$$

Effect of Particulates on Human Beings: Suspended particulate matter of smaller sizes can penetrate deep into the lungs. Thus, suspended particulate matter of less than 10 micron size (abbreviated as PM10) is called Respirable Suspended Particulate Matter (RSPM). The Environmental Protection Agency (EPA) has further divided these tiny particles which make up about 55 per cent of the total mass of suspended particles in air, into fine particles of 2.5 micron or less and coarse particles (PM10). These are filtered out by the nose while PM2.5 particles can penetrate deep into the alveoli of lungs and damage lung tissue. The further a particle penetrates into the lungs, the higher is its concentration and the more is the damage it causes to the lungs. This may lead to severe breathing trouble and diseases like Black lungs disease (coal miners), silicosis (from inhalation of free silica) and bronchitis.

RSPM can also create mutagenic and carcinogenic effects by carrying toxic chemicals such as lead compounds, metals like Cd, Be, Hg, As and hazardous organics absorbed in the particulates such as polycyclic aromatic hydrocarbon. These can penetrate deeper into the respiratory tract and act more efficiently in delivering toxic carcinogen pollutants into the human respiratory system.

The RSPM levels in India, are found to exceed the National Ambient Air Quality Standard (NAAQS) (annual average) in residential areas of Hyderabad, Vishakapatnam, Delhi, Ahmedabad, Parwanoo, Bangalore, Cochin, Dehradun, Thiruvananthapuram, Mumbai, Nagpur, Pune, Sholapur, Angul, Rourkela, Jaipur, Chennai, Kanpur, Lucknow and Kolkata.

According to the 2002-2003 annual report of the Central Pollution Control Board Of India, the RSPM levels in the industrial areas of Ahmedabad, Dehradun, Sholapur, Thiruvananthapuram, Jaipur, Kanpur and Kolkata are also found to exceed the NAAQS as are (annual average) residential areas of Kozhikode and Shillong and industrial areas of Hyderabad, Vishakapatnam, Bangalore, Mysore, Cochin, Kottayam, Kozhikode, Palakkad, Mumbai, Nagpur, Pune, Rourkela and Chennai.

Effect of Particulates on Plants: The presence of suspended particulate matter adversely affects the plant kingdom. The deposition of toxic substances makes the soil unsuitable for plant growth. These particles also get deposited on plant leaves and block the stomata of the plant, thereby decreasing the rate of transpiration, respiration and photosynthesis causing the vegetation to perish.

Effect of Particulates on Materials: The suspended particulate matter accelerates the corrosion of metals. It is more common in urban and industrial areas than in villages. They also corrode buildings and sculptures.

Effect of Particulate Matter on Climate: Particulates reduce the visibility by adsorbing and effectively scattering solar radiation. They influence the climate through formation of clouds, rain and snow and acts as nuclei on which water condenses. Hence, it also chemically affects the nature of precipitation (and alkalinity). Thus, separation and removal of these particulates from the gas stream is very essential to control air pollution. Various methods are available for the purpose which is discussed in the following section.

Secondary Pollutants

Secondary pollutants are formed by primary pollutants through their reaction with normal atmospheric compounds.

Ozone is a secondary pollutant formed by photochemical reaction between primary pollutants and natural atmospheric gas. Ozone affects the respiratory and nervous system. It also damages rubber products and textiles. PAN, a secondary pollutant is formed when hydrocarbon radicals (primary pollutant) react with nitrogen dioxide.

Peroxy Acetyl Nitrate causes photochemical smog and also causes irritation of the eyes, nose and throat, and respiratory distress.

Disposition of the toxic secondary pollutants, either in gaseous form or as particulates on soil make the soil unsuitable for plant growth. Acid rain for a longer period decreases the pH of the soil and makes it acidic. The effect is more pronounced in industrial areas. Particulates on the other hand, deposit on the leaves of plants and block their stomata. As a result the rate of photosynthesis and transpiration decreases and leads to decreased crop yield and retarded growth. In the Sacramento Valley of California, the presence of SO2 in the air and metallic pollutants in soil destroyed the vegetation over an area of 260 sq km and affected the growth on a further 320 sq km of land.

The presence of ozone and Peroxy Acetyl Nitrate (PAN) in the atmosphere attacks plants and oxidizes their SH groups of proteins to disulphides, as a result of which individual enzyme activity is inhibited. O_3 and PAN also affect the photosynthesis process in plants. Even SO_2 has been found to have adverse biochemical effects on plant metabolism.

Cattle are also found to be affected by air pollution. In 1957, the Smithfield cattle show was held under heavy smog. But after the show the cattle developed breathing problem and many died. Ozone and PAN irritate the eyes, impair vision and create breathing problems leading to asthma in human beings.

5.2.3 Air Pollution and Biosphere

Biosphere denotes the domain of living organisms and their interaction with the atmosphere, lithosphere and hydrosphere collectively known as environment.

Atmospheric pollutants are mainly present to a great extent in the troposphere and the lower stratosphere. Depending on the extent of air pollution the atmosphere can be divided into three layers. The first layer up to 100 m from the ground is highly polluted in general and in urban and industrial areas in particular. A part of the pollutants present in this layer is absorbed by vegetation, water surface and buildings. The next layer, 100–2,000 m high, contains lesser amounts of pollutants such as turbulent air currents, drizzle, rain and fog. They divert and dilute the concentration of pollutants there. The third layer containing a large amount of water vapour and clouds, extends up to the troposphere. Here, cloud water droplets dissolve some pollutants and bring them back to the earth as rain. The pollutants from volcanic eruptions, nuclear explosions and products of photochemical reactions stay in the stratosphere for a long time due to lesser movement of air currents in the stratosphere.

A large quantity of particulates and smoke from fossil fuels, particularly over urban and industrial areas, increases atmospheric turbidity and reduces visibility. These particulates absorb and reflect solar radiation which causes a 20 per cent decrease in solar flux in urban areas as compared to the rural areas. Particulates also participate in cloud formation resulting in the formation of smog over urban areas. This also leads to erosion and corrosion of metals and materials and even damages plants.

Effect on Biosphere

The secondary pollutants formed in the atmosphere by the reaction of primary pollutants, water, oxygen and ultraviolet rays of the sun already present in the atmosphere affect the soil, vegetation, crops, animals and human beings adversely.

Table 5.3 represents the formation of some toxic secondary pollutants.

Table 5.3 Formation of Secondary Pollutants

Primary pollutant	Reactions	Secondary pollutant
CO_2	$\xrightarrow{H_2O}$	H_2CO_3
H_2S	$\xrightarrow{2O_2}$	H_2SO_4
SO_3	$\xrightarrow{H_2O}$	H_2SO_4
NO_2	(O_2) $\xrightarrow{H_2O}$	HNO_3, HNO_2, N_2O
$HF_{(g)}$	$\xrightarrow{H_2O}$	$HF_{(l)}$
SiF_4	$\xrightarrow{H_2O}$	H_2SiF_6

5.2.4 Meteorology and Air Pollution

Meteorology can be defined as the study of the behaviour of atmospheric variation that determines the extent of air pollution. When factors like wind, whose speed and direction of flow directly determines the extent of dispersion and dilution of air pollutants are called primary meteorological factor, while precipitation, humidity and solar radiation that control the dispersion of the pollutants indirectly are called secondary meteorological parameters.

Geography

Solar energy is received in larger amounts at the equator than at the poles due to the shape of the earth and its orientation. On the other hand the transmitted radiation is uniformly distributed at the latitudes. As a result the equator becomes warmer than the poles. As a counter action, heat energy is transferred from the equator to the poles as wind. Pollutants in the air also move with it and get diffused and diluted. During the day, land and sea breeze and mountain valley winds develop due to this very phenomenon.

During daytime, land warms up faster than water. The air over the land gets warm, becomes lighter and rises. As a result, cool heavy air over the ocean flows towards land creating a sea breeze. This sea breeze carries the pollutants generated near the sea to inland areas. Secondly, in the evening, the land and the air over it cools faster than over the oceans and air flows from the land to water carrying the pollutants out to sea, but the morning breeze may bring the pollution back to the land.

As for mountain valley winds, the temperature difference is created as a result of the difference in height. The air above the ground warms up faster and moves up during the day and the reverse occurs in the evening. Mountain valley winds diffuse pollutants in unpredictable ways.

Climate

The climate of a region is the average pattern of variation of weather over a period of time. Hence, the climate of a region depends upon availability and also evaporation of water mass, reflection of solar radiation, heat storing capacity, topography and texture of the region.

Due to technological advancement, more air pollutants and heat are released in urban areas than rural areas. These form a 'micro climate' over the urban areas due to population explosion, excess heat release, automobile exhaust, industry emissions and concrete jungles. Hence, human beings exert a great influence on meteorological changes.

Weather

The average condition of temperature, wind, humidity, cloudiness and precipitation over time periods ranging from a few hours to a few days at a particular place is known as the weather of that area. So, weather can change within hours, days or even weeks.

The phenomenon of weather takes place in the lower atmosphere or troposphere. During the day, earth and the air near it is heated up by the sun. This warm air, carrying pollutants rises upward and cooler air comes down resulting in a vertical mixing of the atmosphere. Thus, under normal conditions the temperature of air near the earth's surface decreases with the increase of altitude. After the sun sets, the ground air cools faster than the air above creating an inversion.

This situation with warm light air above and cold heavy air below suppresses the vertical mixing over altitude range where the temperature has increased. Inversion also occurs when the temperature decreases, when the altitude is not sufficient to overcome the stratifying effect of density variation, during which inversion lasts for days causing serious air pollution commonly known as smog. Rain, snow, fog and hailstorms bring down pollutants suspended in the atmosphere to the earth.

Wind

Atmospheric motion or wind is the result of difference in temperature between the equator and the poles. Wind helps greatly in dispersion and dilution of gases and particulate matter.

Gas and particulate matter are dispersed mainly from point sources (chimneys). They depend on wind and atmospheric temperature gradient of the place. The stronger the wind and higher the temperature, the greater is the dispersion of gases and particulate matter. The behaviour of a plume of smoke from a chimney changes with the course of the day and depends upon the degree of clouds present, humidity and temperature of the area.

A great tragedy took place in Bhopal, Madhya Pradesh in December, 1984 when 45 tonnes of Methyl Isocyanate (MIC) gas escaped accidentally from the Union Carbide factory. The poisonous cloud spread with the cool night time north wind nearly all over Bhopal, killing 3,000 people and affecting about 60,000 people. But the concentration of the killer MIC gas was the densest in the 1.5 km radius of the factory where casualties were the highest. The Bhopal tragedy could have been worse if the wind was stronger and the killer gas had spread over a bigger area around the factory.

5.2.5 Air Quality

According to the World Health Organization (WHO), air pollution is a situation where the outdoors or the atmosphere contains concentrated materials that are harmful to the biosphere. The quality of air or the extent of pollution can be determined by measurements made either at the source (chimney) or the air around it (ambient). Now, remote measurements can be made at large heights above the ground with sophisticated instruments.

Two methods are used for ambient air quality measurements:

(i) **Stock emissions:** Where a hole is made in the stock air and samples are analysed on the spot.

(ii) **Ambient air quality:** Although the principle is the same, a large volume of sample is required, as the pollutants are dispersed over a large volume. Obviously, this method depends on climatic conditions such as velocity and direction of the wind. The Central Pollution Prevention and Control Board, New Delhi, has fixed the standard for ambient air quality in India under the Act of 1981. If ambient air does not meet this standard it can be considered polluted in a legal sense.

Table 5.4 Ambient Air Quality Standards in India (concentration mgm)

Area category	Suspended particulate matter	SO	CO	NO
Industrial and mixed use	500	120	5000	120
Residential and rural areas	200	80	2000	50
Sensitive areas*	100	30	1000	30

(*Tourist resorts, places with monuments, game reserves and sanctuaries)

The Air Act, 1981 also prescribed emission standards for many industries, which signify the maximum amount of pollutants that can be discharged from a point source. Thus, air quality standard gives the maximum limit of a particular pollutant permitted in the air around us that is, in the ambient air. Table 5.4 lists the ambient air quality standards in India.

Units of Air Pollutants

The air quality parameters are expressed in Table 5.5 depending on the status of the pollutants, gaseous and particulate.

Table 5.5 Units of Air Pollutants

States of Pollutants	Units for measurement
Gases and vapours	ppm by volume or $\mu g/m^3$
Particulate matter (by weight)	$\mu g/m^3$
Particulate matter (by count)	Number per cubic metre
Visibility	Kilometres
Emission and sampling rates	Cubic meter per minute or m^3/min
Pressure	mm per mercury
Temperature	Degrees Celsius

Air Quality Control Technique

Irrespective of technical advancements, sustenance of all forms of life on earth through the control of gaseous and particulate pollutants in the atmosphere has become of prime importance in the current times.

(i) Mass awareness and education of the causes and effects of air pollution and importance of air pollution control.

(ii) Establishment of emission standards and ambient air quality standards.

(iii) Measurement and control of air pollution from different sources.

(iv) Monitoring ambient air quality and maintaining air pollutants below safe level.

(v) Developing and adopting new technology to create fewer pollutants and/or remove gaseous and particulate pollutants from the atmosphere.

The control systems for air pollution depend on the type of pollutants. The most common types of control systems for Suspended Particulate Matters (SPM) are settling chambers, inertial separators or cyclones, electrostatic precipitators, filters and scrubbers. Gaseous pollutants, on the other hand, can be removed through adsorption and combustion.

(i) **Settling chambers:** They are simple devices for collecting dust particles bigger than 100 mm.

(ii) **Cyclones/multiclones:** It works on the principle of dust separation by a centrifugal force. Particle-laden air enters the cyclone and swirls in it. Dust particles up to 10 microns get thrown on the periphery and clean air escapes from the centre of the cyclone.

(iii) **Electrostatic Precipitation (ESP):** ESP works on the principle of charging dust by a high voltage current to settle the particles down. The ESPs are simple in operation, cost-effective in longer sizes and very common in thermal power plants and industries.

(iv) **Filters:** As the name implies, these consist of filter bags through which dust-laden air is passed. The dust gets filtered and clean air escapes. These bags are removed periodically to clean the dust collected.

(v) **Scrubbers:** The process consists of absorbing the gaseous pollutants or particulate matter in appropriate liquid by scrubbing to bring the pollutants from the gaseous phase to liquid or solid phase. It is further treated before its disposal. Scrubbing consumes a large quantity of water and power. There are different types of scrubbers: venturi, spray and impingement scrubbers.

(vi) **Absorption:** In this process, a mass of gases is transferred to a suitable liquid solvent through the boundary layer of appropriate equipment for liquid-gas contact.

(vii) **Adsorption:** In this process, gases, vapours or liquids concentrated on a solid surface due to surface or chemical forces are termed physical adsorption or chemisorption, respectively. The internal surface area of solid adsorbents and the kinetic energy of the process determine the extent of adsorption. Bauxite, silica gel, activated aluminium or carbon, molecular sieves are mostly used as adsorbents in industries.

(viii) **Combustion:** Many organic pollutants can be decomposed into CO_2 and H_2O at high temperatures directly or indirectly or in the presence of catalysts such as Cu, Ni, V and Zn compounds for decomposition of pollutants at a lower temperature. Hence, there is combustion. It is a very satisfactory process of controlling gaseous air pollutants. Catalytic converters installed in vehicles of advanced design use the same principle to convert nitrogen oxides into N_2 before emission.

If the concentration of air pollutants is high, a combination of absorption methods and catalytic reaction is used in industries. The emissions from motor vehicles contain mainly carbon monoxide, unburned hydrocarbons and nitrogen oxides. Vehicular emissions are usually controlled through the use of better quality fuel, advanced design of the engine, good tuning, maintenance of the carburettor and installation of a catalytic converter. Figure 5.2 shows the air pollutants emissions at some cities of India.

High Volume Sampler

For the monitoring of ambient air quality, a wide range of particulate samplers are used for the collection of ambient particulate matter of 10 micrometres or smaller in size. These particulates enter the human respiratory system through inhalation.

The US Environmental Protection Agency (US EPA) formulated new laws for monitoring and controlling these particulates in the air, on July 1, 1987. We have ambient air standards for SO_2, NO_2, Suspended Particulate Matter (SPM), Respirable Suspended Particulate Matter (RSPM), lead, ammonia and carbon monoxide. The monitoring of PM_{10} and $PM_{2.5}$ is done with high volume samplers.

Historically, the most widely used instrument is the High Volume PM_{10} Sampler ($HVPM_{10}$). It is a standard high volume total suspended particulate sampler with a modified size-specific inlet and flow controller to allow any particles smaller than 10 micrometres to collect on a quartz microfibre filter.

A Dichotomous PM_{10} or Virtual Impactor sampler is used in non-compliance areas where particulate control strategies must be developed. These instruments provide information on coarse and fine particles and also facilitate chemical analysis of the collected particles on inert Teflon filters.

The PM_{10} Automatic Seven Day Sampler (PASS) is used for long sampling periods, where weekly transport of filters is followed. Depending upon the use, there are mainly two kinds of high volume samplers.

(i) High volume samplers associated with a cyclone sampler at the inlet.
(ii) Impactor separation-based sampler.

High Volumes Associated with Cyclone Separator

Ambient air, laden with suspended particulates enters the inlet to pass through the cyclone. Here, coarse non-respirable dust is separated from the air stream by centrifugal force acting on the solid particulates. The separated particulates fall through the cyclone's conical hopper to collect in the sampling bottles fitted at the bottom.

The fine dust (RSPM) of the suspended particulates passes through the cyclone and is carried by the air stream to the filter paper clamped between the top cover and filter adaptor assembly. The respirable dust (RSPM) is retained by the filter and the air which is free from particulates is blown out from the system through the blower.

Impactor Separation-Based Samplers

In this type of instrument, as the motor is started, the ambient air is sucked out at a certain flow rate. This sucked air is inserted into the sampler from the bottom of the lid and then passed through the impactor. The impactor arrests particulates larger than the desired size (PM_{10} and $PM_{2.5}$) as the pores of the filter paper used are necessarily smaller than the diameter of the particulates to be collected. The passed out air, which is free from particulates, is evacuated from the sampler through the exhaust pipe with the help of a brushless motor.

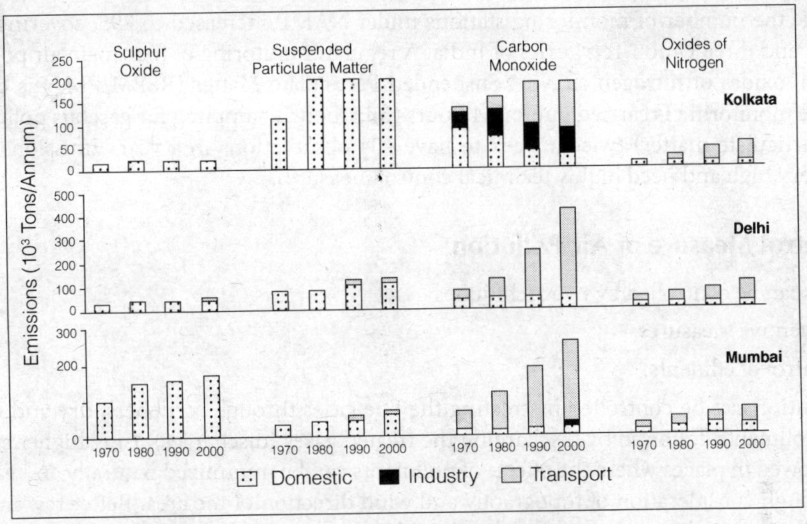

Source: WHO/UNEP, 1992 (Estimated/Projected Figures)

Figure 5.2 Air Pollutants Emissions at some Cities of India

National Ambient Air Quantity Monitoring (NAAQM)

The Central Pollution Control Board initiated the National Ambient Air Quantity Monitoring (NAAQM) programme in 1984 in seven stations at Agra, India and Ampara, Sri Lanka. Later, in 1998-1999, the programme was renamed National Air Monitoring Programme (NAMP).

The air quality management process adopted in India is shown in Fig. 5.3.

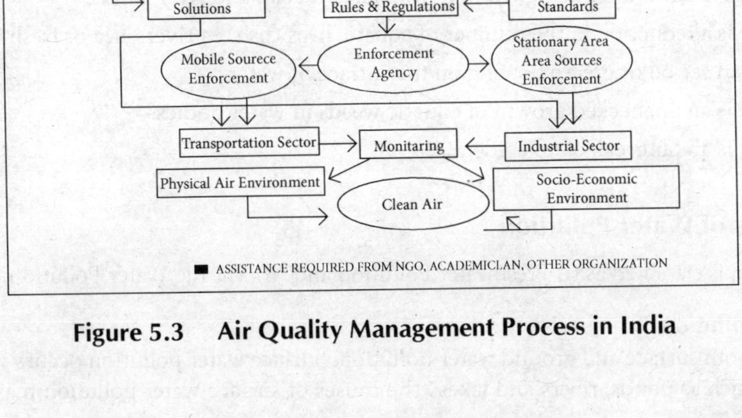

Figure 5.3 Air Quality Management Process in India

By 2000-2001, the number of monitoring stations under NAMP increased to 295, covering 98 cities/towns in 29 stations and three Union Territories of India. A regular monitoring of four main air pollutants, sulphur dioxide (SO_2), oxides of nitrogen as NO_2, Suspended Particulate Matter (RSPM/PM_{10}) is done in all these locations. The monitoring is carried out for 24 hours (four-hourly sampling for gaseous pollutants and eight-hourly for particulate matter) twice a week to have 104 observations in a year. Emissions from industrial sources are very high and need highly technical control measures.

5.2.6 Control Measure of Air Pollution

Air pollution can be controlled by two techniques

(1) Preventive measures.

(2) Control of effluents.

Air pollution can be controlled by treating the flue gases through scrubbers, dry and wet collectors to remove the pollutants, followed by discharging the treated gases (discharges) using higher stacks. Industries should be allowed in places where the effects of pollutants can be minimized naturally and easily. This can be achieved through consideration of topography and wind direction of the area. Better raw materials that will cause less pollution can be used. Automobiles should be provided with catalytic converters. The use of fossil fuels should be reduced with unconventional energy sources such as solar, tidal, wind and biogas. Last, but not the least, National Ambient Air Quality monitoring should be practised strictly by the State and Central Pollution Control Board.

5.3 Water Pollution

The naturally occurring dynamic equilibrium existing among environmental segments such as hydrosphere, atmosphere and lithosphere gives rise to pollution in water. The solvent property of water makes it very susceptible to pollution. Deviation of the properties of water in its pure condition as detected by the changes in its normal function and properties is known as water pollution. Water is used for washing, irrigation, flushing away waste, cooling and industrial use. All the uses of water led to its pollution. Human beings use rivers and lakes as dumping grounds by pouring highly toxic sewage and industrial waste into them.

Hence, water pollution can be termed as the addition of some excess material or heat to water making it harmful for all living beings. The following are indications that the water in question is polluted:

(a) Water has a bad taste or odour.

(b) Offensive odours are being emitted from lakes, ocean and river banks.

(c) There is a reduction in the number of aquatic lives (fish) in rivers, sea or fresh water.

(d) You can see oil or grease floating on the surface of water.

(e) There is an unchecked growth of aquatic weeds in water bodies.

(f) There is an outbreak of an epidemic.

5.3.1 Types of Water Pollution

Water pollution is classified as (i) Freshwater Pollution and (ii) Marine Water Pollution.

Freshwater Pollution

This includes both surface and ground water pollution. Surface water pollution occurs when pollutants enter water bodies such as ponds, rivers and lakes. The causes of surface water pollution may be either natural or man-made, or both. For example, silt in rivers and sedimentation on river-beds by the discharge of sewage and industrial effluents, blocks the free flow of water, causing floods. Discharge of industrial effluents into rivers also causes pollution due to discharge of harmful chemicals.

However, when pollutants find their way into ground water through an aquifer they pollute it too. As a result, the quality of ground water deteriorates. Excessive use of nitrogenous fertilizers causes seepage of water along with soluble nitrate salts into the aquifer and ultimately into underground stores of water. Using this water for drinking causes methaemoglobinaemia, particularly in infants. The ground water of industrial areas is contaminated with hazardous chemicals such as nickel, iron, copper, chromium, cyanide and arsenic in alarming concentrations.

Both surface and ground water pollution are categorized as freshwater pollution as the salt contents of both the types of water bodies is below five ppt (parts per thousand).

Marine Water Pollution

Marine water is easily polluted by contaminants carried by rivers and also by oil-spills and radioactive and industrial waste. Thermal pollution also occurs due to the discharge of hot water from different industries. Mining of polymetallic nodules also damages the ocean bed.

5.3.2 Sources of Water Pollution

When the source and place of pollution is readily identifiable it is known as point source of water pollution. For example, municipal and industrial discharge pipes. But when a source of pollution such as acid rain and mining runoff cannot be readily identified it is said to have a non-point source of pollution.

Surface Water Pollution

Different types of water pollutants along with their sources can be classified as:

(i) **Sewage and other Oxygen-Demanding Waste:** Discharging untreated or partly treated sewage of domestic and municipal origin into nearby rivers, lakes or seas has been a common practice. Domestic or public sewage mainly contains food waste, water from domestic use and water from lavatories. More accurately, this sewage contains pollutants that originate from different human metabolism and vital activities. Various bacteria, yeast, algae, mould and eggs of helmonth viruses are detected in this water. It also contains clay, mineral salts and sand as inorganic impurities. An average Indian family throws away about 50 kg of refuse per week.

These pollutants which are rich in carbonaceous organic material are oxidized by micro-organisms to carbon dioxide and water.

$$C + O_2 \longrightarrow CO_2$$

As a result, the Dissolved Oxygen (DO) level of water decreases, affecting aquatic life. When the amount of discharge in sewage is less, this biological degradation method helps to remove the waste from rivers. In case of strong sewage and when dissolved oxygen in water is less, the waste cannot be removed through biodegradation and ends up polluting the river. Water becomes septic when dissolved oxygen level is too low.

The level of these types of pollutants in water is measured in terms of Biological or Biochemical Oxygen Demand (BOD), Chemical Oxygen Demand (COD) and Dissolved Oxygen (DO). The amount of molecular oxygen required for aerobic biochemical decomposition of waste material is known as BOD. It is the amount of dissolved oxygen in water required for the aerobic decomposition of a sample of pollutant water in five days. In terms of milligrams, oxygen per litre is a measure of pollutants present in that sample.

Dissolved Oxygen (DO): Dissolved Oxygen is required for the sustenance of aquatic and semi-aquatic life. The optimum value of DO for good water quality that ensures a healthy aquatic life is 4–6 mg/l. Their survival depends on the ability of water to maintain this optimum DO value. Lower DO value leads to water pollution.

The amount of dissolved oxygen in water mainly depends upon:

(a) the rate of photosynthesis and respiration by aquatic life,

(b) the rate of re-aeration of water and

(c) oxidation of waste materials.

In the day time, as the process of photosynthesis takes place, the liberated oxygen increases the DO level of water. However during respiration, oxygen is continuously consumed. In aeration, the solubility of oxygen in freshwater decreases as the temperature increases. The process of photosynthesis, respiration and re-aeration results in the diurnal variation of DO as shown in Fig. 5.4.

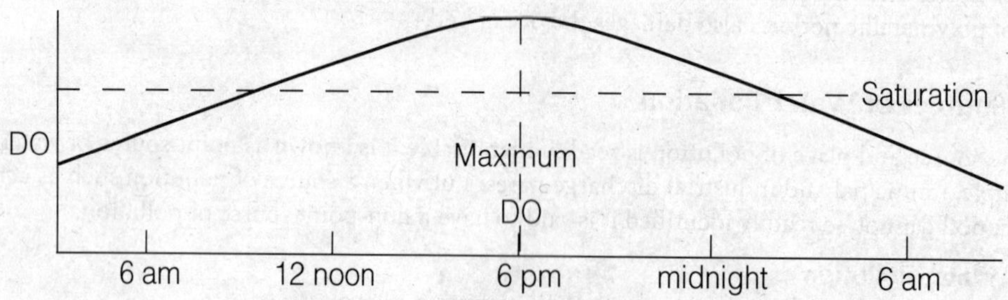

Figure 5.4 Diurnal Variation of DO

Water gets deoxygenated due to the presence of oxygen-demanding waste. Oxygen demand increases due to the death and decay of an explosive bloom of algae and related species and aerobic decomposition of organic matter by bacteria to CO_2 and water. Although none of the products are pollutants, the presence of oxygen-demanding waste causes depletion in Dissolved Oxygen. Sometimes heavy depletion of Dissolved Oxygen takes place, bringing it down to zero value. This is identified by a strong anaesthetic odour, production of scum and solids that make water unfit for recreational use. Consequently, deeper water in lakes becomes O_2 deficient, destroying fish and many other desirable aquatic species.

Biological (Biochemical) Oxygen Demand (BOD): Biological or Biochemical Oxygen Demand gives an empirical value to water quality and is a parameter for the organic matter in water. The respiratory demand for oxygen by aerobic organisms for the metabolism of organic waste and nutrients is known as Biological or Biochemical Oxygen Demand. Hence, it gives a measure of sewage strength.

BOD of a water body is measured by monitoring the decomposition of a known quantity of organic matter in a definite amount of water at 20°C for five days. This is represented as BOD. It is also observed that BOD increases with the amount of organic matter oxidized and vice versa.

$$\text{BOD} = \frac{\text{Volume of } O_2 \text{ required}}{\text{Volume of sample used}} = \text{parts of } O_2 \text{ in ppm}$$

BOD for very pure water is one ppm. However, water with a BOD level of up to three ppm can be considered pure. Table 5.6 denotes the BOD for different types of water samples.

Table 5.6 **BOD Level of Industrial Waste Water**

Type of water	BOD Level (ppm)
Municipal sewage	100–400
Dairy industry	2,000–15,000
Food processing industry	5,000–10,000
Pulp sewage	10,000–15,000

When DO level drops and BOD increases, bacterial decomposition shifts from aerobic to anaerobic type, producing more odorous and toxic products such as CH_4, NH_3, amines, H_2S , PH_3, and lower-valent phosphorous compounds. Thus, BOD value is very important for pollution control.

Chemical Oxygen Demand (COD): Chemical Oxygen Demand is also a water quality parameter depicting the amount of oxygen demanding substance present in water. Chemical Oxygen Demand (COD) is more scientific than BOD.

In this method, a water sample, after dilution to an appropriate level, is mixed with a measured amount of chemical oxidizing agent, such as potassium dichromate in concentrated sulphuric acid, and kept under reflux for two hours. After the completion of oxidation, the amount of unused potassium dichromate is estimated by titrating against ferrous ammonium sulphate. The difference in the values gives the amount of oxidants used, hence the amount of oxidizable organic material present in the sample. COD values are usually higher than BOD values because certain organic substances which are difficult to oxidize using common microbial oxidants easily undergo chemical oxidation with strong oxidizing agents.

Lower BOD impairs domestic and livestock water supplies as it affects the taste, odour and colour and may give rise to scum. Secondly, untreated or partially-treated sewage from domestic and municipal origin will stimulate algae growth spread dreaded water-borne diseases such as dysentery, typhoid and cholera in an epidemic form.

(ii) **Industrial Waste**: In India, almost all the industries discharge the waste by-products from the manufacturing processes into rivers or other water bodies. So, most rivers and freshwater streams are badly polluted with industrial effluents from industries such as paper and pulp, refineries, textiles, tanneries, distilleries, coal washers and steel industries. This industrial waste can be categorized as:

(a) Organic substances such as phenol and alcohol that increase the BOD by decreasing the oxygen content.

(b) Inorganic substances such as chlorides, nitrogen and carbonates that encourage the growth of micro-organisms and make the water unfit for use.

(c) Acids and alkalis which change the pH of water thereby posing a threat to aquatic life.

(d) Toxic substances such as cyanide, acetylene and heavy metals like mercury, lead and arsenic that cause extensive, irreversible damage to plant and animal life.

(e) Oil and other floating impurities that interfere with self-purification of water bodies.

(f) Colour-producing dyes that change the colour of the water with depleting oxygen content thereby affecting aquatic life.

(iii) **Agricultural Chemicals:** With population explosion and advancement of science and technology, the demand for more crops from the same land leads to excessive use of fertilizers and pesticides. This injudicious use of chemicals results in water pollution as these are washed off the lands through irrigation, rainfall and drainage and carried into rivers and streams.

Fertilizers: Excessive use of chemical fertilizers results in accumulation of nitrogen, phosphorous and water on land. These are washed off the land with water through rainfall and irrigation into water bodies such as streams and rivers. Excessive storage of potassium and phosphorous enhances a luxuriant growth of algae in water bodies which goes on to decrease the dissolved oxygen content thereby leading to eutrophication.

On the other hand, nitrogenous fertilizers get converted to nitrates and get deposited in water bodies. When these nitrates enter the human body they are transformed to nitrites by intestinal bacteria. These nitrites combine with haemoglobin (oxygen carrier in blood) to form methaemoglobin. Methaemoglobin prohibits the formation of oxyhaemoglobin and causes damage to the vascular and respiratory system and blue colouration of skin leading to skin cancer.

Pesticides: Pesticides are used to kill and control harmful pests especially those which severely affect crops. But these chemicals not only kill the targeted pests but they also affect the untargeted helpful organisms, besides their effects are long-lasting. For example, the effect of DDT is pronounced even after 15 years of its application. Indiscriminate use of pesticides results not only in accumulation of these harmful chemicals in land but also in the ultimate deposition of these chemicals in water bodies. Studies show that about 25 per cent of the total DDT used till date is likely to have been transferred to the oceans. BHC, PCBs and DDT have been detected in the Arctic region as well. The effect of the injudicious use of these pesticides on the environment can be summarized as follows:

(a) Pesticides attack both targeted and non-targeted species.
(b) The harmful effects are long-lasting.
(c) As they are easily transported to different places, they cause unintended effects like faunal displacement and increased disease susceptibility in host.
(d) Development of pesticide tolerance disturbing reproductive physiology.
(e) Food contamination.
(f) Bioaccumulation and biomagnification.

(iv) **Thermal and Radioactive Waste:** Different industries and nuclear power plants use water for cooling and discharge the heated water into nearby streams or water bodies.

Nuclear reactors, nuclear explosions, nuclear war, medicinal use and research laboratories are the main sources of radioactive waste. These are most toxic as their effect persists for a number of generations. The long-suffering people of Hiroshima and Nagasaki are a living example of radioactive pollution.

(v) **Biodegradable and Non-Biodegradable Pollutants:** Biodegradable pollutants consist mainly of organic matter from domestic sewage. They are decomposed by micro-organisms naturally, or can be decomposed artificially in chemical treatment plants. But when there is excessive biodegradable waste in the environment the dispersal or recycling capacity becomes a problem and poses a threat to the environment.

Non-biodegradable waste includes plastic bags, long chain detergents, aluminium cans, glass and phenolic chemicals which can neither be decomposed nor recycled. These are used only for filling lowlands.

(vi) **Eutrophication:** Newly formed water bodies such as lakes, ponds and reservoirs, whether natural or man-made do not support aquatic life as they are poor in nutrient supply or oligotrophic. Gradually, with the passage of time these water bodies become rich in nutrients through the deposit of domestic waste, agricultural residue (rich in nitrogen and phosphorous), land drainage and industrial waste. As a result, eutrophication that is enrichment of the water body through organic waste matter rich in nitrate and phosphate starts. Natural eutrophication is a very slow process, often taking more than 100 years. But artificial eutrophication is very fast as it depends on the input of organic waste matters. The aerobic decomposition of organic waste in the presence of oxygen by bacteria leads to eutrophication. The nutrient-rich water body supports the growth of algal blooms and other aquatic plants. As more plants grow due to the additional supply of nutrients more plants also die. Bacteria decompose these dead plants and organic waste using dissolved oxygen. As a result the BOD of water increases. Fish and other aquatic animals start dying due to the depletion of oxygen. Such a water body is said to be eutrophied. With an increase in BOD water starts emitting an offensive smell and the aesthetic and recreational importance of the water body decreases. In the advanced stage, eutrophication can be detected visually and also by smell. However, measurement of different parameters provides evidence of eutrophication.

The measurement of the total solid content or increase in calcium, sodium, potassium, sulphate, phosphate and nitrate content of the water body provides an indirect measurement of its eutrophication.

Direct eutrophication can be measured both by quantitative and qualitative methods. Table 5.7 illustrates the qualitative features of both oligotrophic and eutrophic water bodies.

Table 5.7 Characteristics of Oligotrophic and Eutrophic Water Bodies

Parameter	Oligotrophic	Eutrophic
Production (plant and animal)	Low	High
Variety	Many species	Few species
Distribution	To great depth	Trophogenic layer
Diurnal migration	Extensive	Limited
Water blooms	Very rare	Frequent
Plant nutrient flux	Low	High
Oxygen lower level	Present	Absent
Water quality for domestic use	Good	Poor
Presence of Salmon fish	Present	Absent

On the other hand, measurement of oxygen, biological productivity and the nutrient level in water provides a direct quantitative measurement for eutrophication.

Sometimes, oligotrophic lakes are purposefully turned into eutrophic lakes by controlled addition of fertilizers (nutrients) for fish cultivation.

Eutrophication can be controlled by:

(a) Limited input of nutrients through treatment of waste water before discharge into water bodies.

(b) Removal of dissolved nutrients from water bodies using physical, chemical or biological means.

(c) Harvesting algal blooms to check recycling of nutrient supply to water through their death and decay.

(d) Reduction in amount of nutrient solubized in water through stimulation of bacterial multiplication, to disrupt the algal food web.

Water pollution leading to eutrophication is a threat not just to fishermen but also to those who depend on recreational activities on water bodies.

(vii) Biological Magnification: The transfer of food energy from the primary producers (plants) to human beings through a series of organisms including repeated eating and being eaten is known as the food chain. Generally, this helps in the dispersion of food energy but sometimes the concentration of a particular chemical increases manifold inside the food chain. This is known as biological magnification.

For example, DDT is used in crop fields. It is washed away into water bodies by irrigation and is absorbed into the bodies of fish through their skin. When this fish is eaten, DDT deposition in the human body increases manifold as it tends to become stored in fat deposits. Excessive organochlorides in fat deposits in human bodies is very poisonous leading to sickness followed by death. A similar effect is observed with heavy metals and radionucleotides.

(viii) Heavy Metals: All metals with atomic numbers greater than 23 excluding rubidium, strontium, calcium, barium and francium are heavy metals. Some of the worst water pollutions are due to heavy metals. They have toxic and pathological effects on living beings. Domestic sewage and industrial effluents that cause the addition of metals to both surface and ground water are the sources of heavy metal pollutants. These metals are essential for plant and animal growth but when present in higher concentrations they are poisonous and severely affect plants and animals thereby disturbing the ecosystem. Heavy metals have great affinity for sulphur and attack the sulphur bonds in enzymes, protein, carboxylic acid and amino groups. They tend to precipitate phosphate bio-compounds and catalyse their decomposition. They also affect the transport process through the cell wall by combining with the cell membrane. Some of the trace elements and their adverse effects on the biosphere are discussed in Fig. 5.5.

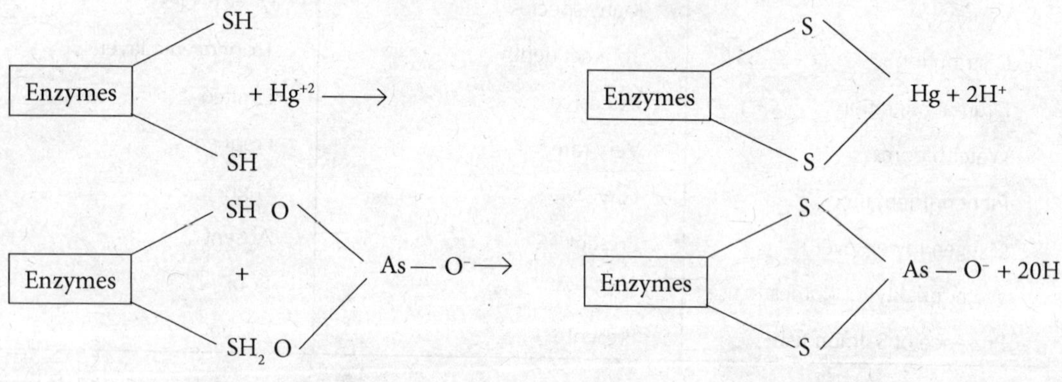

Figure 5.5 Trace Elements and their Adverse Effects on the Biosphere

(a) **Arsenic (As):** Arsenic occurs in nature as sulphide, regular (AS_2S_3) or orpiment (As_2S) and as an oxide (As_2O_3). Earlier, it was widely used as an insecticide, rodenticide, for wood preservation and medical preparations. Arsenic is also used for preparation of bad alloys for bullets and shots, pyrotechnic and boiler composition, paint pigment and opal glass enamels. Most of the anthropogenic inputs of water reach the water bodies. Different fungi and micro-organisms convert arsenic to dimethyl arsine in water. This methylated form of arsenic is detected in natural water, bird eggshells, seashells and human urine. The presence of arsenic in human blood results in vomiting, abdominal pains, dermatitis and bronchitis. This is carcinogenic to the tissues of the mouth, oesophagus, larynx and bladder leading to death. In India, the safety limit of arsenic in industrial water has been fixed at 0.2 ppm.

(b) **Cadmium (Cd):** Cadmium is used in industries to make fusible alloys and electroplating control rods in nuclear reactors. Cadmium enters water bodies from these industrial areas and also from zinc-mining areas. The concentration of cadmium in sea water is approximately 0.15 mg/litre of water whereas in river water it varies from 1 to 100 mg/litre of water. Cadmium is harmful to both fauna and flora as most of them absorb and store it in their tissues. The presence of cadmium adversely affects the heart, liver, lungs, reproductive organs and kidneys. The famous itai-itai disease of Japan was due to cadmium pollution. Patients affected with itai-itai showed signs of osteomalacia in bones and calcification and pyelonephritis in kidneys resulting in skeletal deformation and renal dysfunction.

(c) **Cyanide (CN):** Cyanide is a highly toxic chemical and its inhalation or ingestion leads to nausea and even to death. It is widely used in the extraction of gold and silver metals, metal plating, in pesticides and in some disinfectants.

(d) **Iron (Fe):** Iron is most widely found in the earth's crust and soil along with most of the minerals. The presence of iron in water severely affects industrial processes and it is disagreeable also in households. Diseases like thalassaemia get aggravated due to excessive iron in water. Water bodies become foul due to the an abundance of iron-oxidizing microbes.

(e) **Manganese (Mn):** Like iron, manganese is also abundantly available in nature and widely used in different industrial processes. It is very difficult to remove manganese completely. Most of the water around the globe, contains 0.005 to 1 mg of manganese per litre. Its tolerance limit is also fixed at 0.005 ppm because at a higher concentration, manganese causes cramps, tremors, hallucinations, manganic pneumonia and renal degeneration.

(f) **Lead (Pb):** Although lead is widely present in minerals, it is more abundant in the atmosphere because of its widespread use as an anti-knock agent (tetraethyl lead) in petrol. Through wet and dry deposition processes lead is removed from the atmosphere to enrich the roadside soil with concentrations of the order 1,000–4,000 mg/kg on busy streets which finally reaches the water bodies. Fig. 5.6 represents an average lead balance of a city resident.

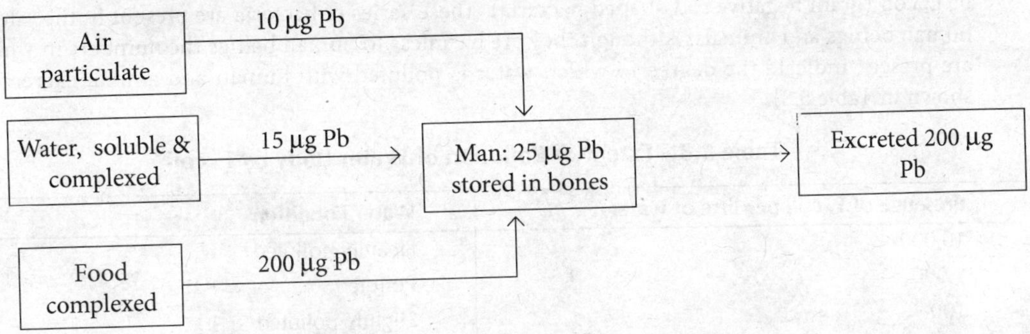

Figure 5.6 Daily Lead Balance of a City Resident

Lead inhibits several key enzymes and leads to anaemia, brain damage and loss of kidney function. A high level of lead of about 0.3 ppm in blood inhibits utilization of O2 and glucose for life-sustaining energy production.

(g) **Mercury (Hg):** The only liquid metal Mercury occurs as a trace element in many minerals and fossil fuels. It enters the environment mainly through human activities such as use in chlora alkali industries where mercury electrodes are used to manufacture electrical appliances and for use as fungicides. Mercury is discharged into water as an elemental divalent mercuric ion and also as phenyl mercuric acetate. All forms of mercury are not toxic to human body with the exception of methyl mercury. Almost all mercury in fish is methyl mercury which passes into human body and causes biomagnification. Mercury that enters into the cell membrane inhibits active transport of sugar across the membrane but allows potassium to pass into the membrane. This results in energy deficiency in brain cells and disorders in nerve impulse transmission. Mothers of new born babies with methyl mercury poisoning suffer irreversible damage of central nervous system, cerebral palsy, mental retardation and convulsion. Methyl mercury poisoning also causes segregation of chromosomes. The tolerance limit of methyl mercury in blood is 0.5 ppm. Minamata disease of Japan which killed more than 100 people and left thousands permanently paralysed resulted from methyl mercury poisoning caused by eating contaminated fish.

(ix) **Ground Water Pollution**: Water from seepage pits, refuse dumps, septic tanks and barnyards percolates into the layers of the earth, reaches the ground water table and pollutes it. It could be many years before the polluted water shows up in a well. Similarly, it would also take a long time to clean and restore contaminated aquifers.

In the industrial areas of Punjab and Haryana where bicycles and woollen garments are manufactured ground water samples show a high concentration of poisonous nickel, iron, copper, chromium and cyanide. Continuous intake of this contaminated water causes deposits of these minerals in human fat tissue. Being difficult to metabolize, they remain in fat tissue showing effects of biological magnification. On starvation when the body draws food from reserve sources these compounds are released into the blood stream producing toxic effects.

(x) **Biological Contamination of Water:** Nutrient-rich water is also rich in different types of bacteria, algae, protozoa, worms and other types of organisms. These micro-organisms use the carbon from various organic compounds to build up their bodies. Water bodies polluted with organic matter are a good medium for the growth of protozoa. Most worms and parasites enter through the food chain and infest the internal organs of man and animals thus making them sick.

The degree of pollution with pathogenic bacteria is measured by the presence of Escherichia coli or E. coli (gram negative rod-shaped bacteria). These series of bacteria are present in the intestine of human beings and animals. Although they are harmless to human beings the numbers in which they are present indicate the degree to which water is polluted with human and animal excrements (as shown in Table 5.8).

Table 5.8 Extent of Pollution of Water Body by E. coli

Presence of E. coli per litre of water	Water condition
10,000	Heavily polluted
1,000	Polluted
100	Slightly polluted
10	Satisfactory
3	Suitable for drinking

According to an estimate by the National Institute of Communicable Diseases, 30 per cent of the mortality and 50 per cent of the morbidity due to infectious diseases in India is from water-borne diseases. According to the World Health Organization, 90 per cent of deaths are avoidable through safe water drinking and proper sanitation.

Purification of Water

Due to the fact that water is a solvent, it is never found in a pure state. Therefore, rain water, surface water and ground water, has to be purified before being used for domestic purposes such as washing, cleaning and drinking. Depending upon the requirement, water can be purified on a small scale as well as on a large scale.

(a) **Small Scale Water Purification**: Water for domestic purposes becomes potable only after it becomes free from pathogenic agents and toxic chemicals. This can be done by boiling it or using chemical disinfectants or by employing filtration techniques.

Boiling: It is the simplest, cheapest and safest way to purify water. Boiling water for at least 15 minutes continuously kills all bacteria, spores, cysts, ova and harmful pathogens. Water should be boiled in the same container in which it will be stored to avoid contamination. The CO_2 dissolved in water escapes in the form of bubbles when water is boiled, calcium carbonate, on the other hand precipitates, thereby reducing temporary hardness.

$$2CaHCO_3 \xrightarrow{\text{On boiling}} CaCO_3\downarrow + CO_2\uparrow + H_2O$$

Use of Chemical Disinfectants: Water for drinking can be purified by using chemicals such as iodine, potassium permanganate or chlorine. Although iodine and potassium permanganate purify water they are incapable of killing all the pathogens. They are physiologically active (thyroid action) and costly too. In addition to this, they change the colour, odour and taste of water. So iodine and potassium permanganate are seldom used.

Chlorination is an effective way of water purification as it has few disadvantages. A dose of approximately 0.7 mg of chlorine is sufficient to disinfect one litre of water. Chlorine is added as bleaching powder $Ca(O\ Cl)_2$, or as chlorine solution, hydrochloride (per gallon) or chlorine tablets. The simplest way is to add a chlorine tablet to a glass full of water before drinking.

Filtration: Water filters are used in homes for safe drinking water. Water filters contain ceramic candles which remove suspended impurities and some of the bacteria present in water. In modern times, advanced technology ceramic candles have been replaced by ion-exchange resins and a combination of carbon filters with UV light. Carbon filters remove the odour and colour from water while UV light kills all types of pathogens. However, its cost is inhibiting and therefore many people cannot afford it.

Wells are still the main source of drinking water in villages. Adding bleaching powder to well water is the cheapest and most effective method to disinfect it. The volume of water in a well is calculated by measuring the diameter of the well and depth of water. A calculated amount of chlorine (bleaching powder) is taken in a bucket containing water and the supernatant solution is poured into the well thus purifying it.

(b) **Large scale Water Purification**: When a large amount of pure water is required for supplying in urban areas it is purified on a large scale through storage, filtration and chlorination.

Storage: Water is stored in large open tanks under direct sunlight. Within 24 hours, 90 per cent of the suspended material in the water settles down making it clear. Direct sunlight allows photosynthesis to release oxygen which is used for aerobic oxidation of organic matter. About 90 per cent of pathogenic organisms dry out within the first week of storage. Alum is added for effective sedimentation. However, storage for a longer period is avoided as it enhances algal growth producing odour.

Filtration: Filtration removes about 98 to 99 per cent bacteria and is carried out by biological filters, mechanical filters, slow sand filters or rapid sand filters. Out of these the last two methods are commonly used for their low cost and high-quality output. They are simple to construct and operate. The filter tank is made up of concrete, containing sand bed and gravel. It consists of four elements:

(i) Supernatant raw water
(ii) A bed of graded sand
(iii) An under-drainage system and
(iv) Filter-control valves.

The clean upper layer of water from a storage tank is led into the sand-graded bed. Water percolates through the sand layer at a rate between 0.1 and 0.4 cubic metre per square metre area per hour (m3/m2/hr). In the rapid sand filter alum is mixed and water is violently shaken before filtration. The rate of filtration at five to 15 m3/m2/hr, is fast. During filtration, a biological layer develops in the sand containing a lot of algae and bacteria which removes organic matter, oxidizes ammonical nitrogen to nitrates and yields bacteria-free water. The under drainage system made up of perforated pipes, functions as an outlet for the filtered water and supports the filter medium. Certain valves are present in the under drainage system for the smooth flow of filtered water.

Chlorination: Chlorination is a supplement for sand filtration. It is the addition of a calculated quantity of chlorine to water to kill pathogenic bacteria. Chlorine also oxidizes iron, manganese and hydrogen sulphide, destroys odour-producing constituents and controls algal growth. But chlorination is suitable for water in the pH range of 0.7 to 8.5. If not administered in proper doses it can be fatal to human beings. Recently, ozonation (adding ozone) and UV light has replaced chlorination.

Treatment of Waste Water

Waste water (sewage) from industrial, commercial and residential areas is collected through a network of underground pipes (sewers). This sewage is then transported through a water carriage system known as the sewerage system to treatment plants for purification. Sewage treatment or purification is carried out in three distinct steps as shown in Fig. 5.7.

(i) Primary treatment for removal of suspended particles.

(ii) Secondary treatment for aerobic decomposition of organic matter.

(iii) Tertiary treatment for production of water, free from pathogenic bacteria and harmful chemicals which can be discharged without any risk to health and environment.

(i) *Primary Treatment*: The sewage is first allowed to pass through large metal screens and then allowed to flow into the grid chamber. A metal screening removes all big, floating solid particles such as wood, rags, garbage and dead animals. In the grid chamber heavier solid particles such as sand and gravel settle down due to gravity. The sewage then passes into a big sedimentation tank. It is kept here for 6 to 8 hours and within this period, 50 to 70 per cent of the solids present in his sewage settle down. Decomposition of organic matter occurs to produce sludge which is mechanically removed. Primary treatment removes about 60 per cent of suspended solids, 30 per cent of oxygen-demanding wastes, 20 per cent of nitrogen compounds and 10 per cent of phosphorous compounds.

(ii) *Secondary Treatment*: It is a biological treatment process in which the total organic matter present in the sewage is removed. It is done either by the slow filter method or activated sludge process.

In the slow filter method, waste water is sprayed on the surface of a filter bed consisting of a top layer of coal followed by a layer of smaller silica granules and finally a layer of small grains and high-density sand. The sludge formed at the top is disposed off.

In the activated sludge process which is more recent in origin, effluents from a sedimentation tank are mixed with sludge collected after it finally settles in the tank.

Sludge being rich in bacteria is called activated sludge. The organic matter present in sewage is oxidized by aerobic bacteria and broken down to form carbon dioxide. Typhoid and cholera organisms are destroyed totally in this process.

The oxidized sewage from the secondary treatment chamber is passed to the secondary sedimentation tank. The activated sludge is collected here and the waste water now becomes free of bacteria by 95 per cent, free of phosphorous by 10 per cent and nitrogen by 50 per cent.

The volume and characteristics of activated sludge are modified through anaerobic digestion. This sludge serves as excellent manure.

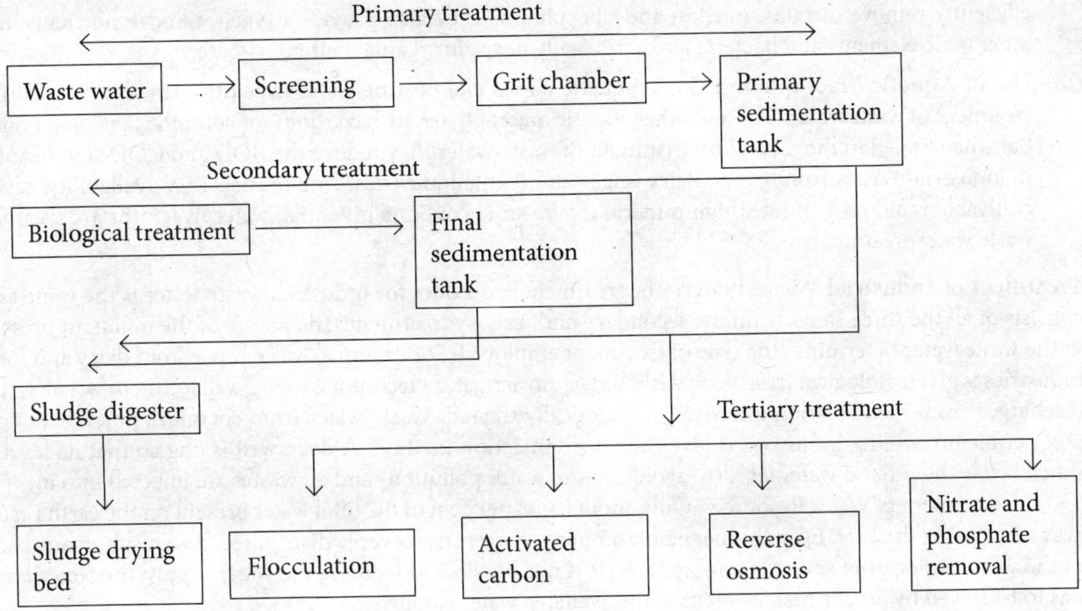

Figure 5.7 Flow Chart for Treatment of Waste Water

(iii) *Tertiary Treatment*: It is a combination of advanced physical and chemical methods. The water from the secondary sedimentation plant is subjected to tertiary treatment which consists of precipitation, reverse osmosis and/or chlorination methods so as to get pure water. Precipitation removes 90 per cent of suspended solids, phosphates and nitrates, while reverse osmosis removes organic and inorganic substances and chlorination is essential for killing disease-causing organisms.

Low Cost Waste Water Treatment Method: The conventional method of waste water treatment is costly and also requires regular expensive maintenance and adequate technical manpower. Underdeveloped and developing countries follow an alternative low-cost method which involves land treatment and use of algae and aquatic macrophytes.

(i) **Land Treatment:** In this method the partially treated waste water is applied in a regulated and controlled manner to the land surface for a designed degree of treatment. In this process, physical methods like filtration, absorption, photo-oxidation, evaporation, transpiration, osmosis and chemical methods like ion exchange, precipitation solubilization, oxidation, reduction, biological methods like carbon assimilation, proteolysis, enzymatic hydrolysis, respiration, catalytic decomposition, ammonification, nitrification, denitrification and phosphate mobilization work one at a time.

Application water for land treatment process is a versatile, efficient and an advanced process. It helps to improve the physical and biochemical qualities of soil in the wasteland.

The main advantages of this system are: (i) The water and nutrient components of waste water are recycled through crop irrigation. (ii) High crop-yield supports economic growth. (iii) Water pollution is controlled.

The success of the process depends upon the quantity and quality of water, nature of soil and nature of crop species.

(ii) **Use of Algae:** Use of algae-bacterial symbiosis instead of bacteria alone in the secondary treatment plant of waste water is extremely efficient. Algae in general and blue-green algae in particular, are used as they efficiently remove nitrates, nitrites, and phosphates. They also provide oxygen, remove nutrients and other toxic elements and bacteria and aerobically degrade organic matter.

(iii) **Use of Aquatic Macrophytes:** Green aquatic plants like floating water hyacinths can also be used for treatment of waste water. This and other aquatic macrophytes are hazardous for common lakes and ponds but when added in the secondary treatment of waste water they reduce the BOD and COD significantly in industrial wastes from paper, dairy, sugar and textile industries. Due to their easy availability, easily cultivable rapid growth rate, high mineral intake and economic potential, water hyacinths are used for waste water treatment.

Treatment of Industrial Waste Water: The treatment procedure for industrial waste water is the same and consists of all the three steps; primary, secondary and tertiary treatment. The nature of the pollutant present in the waste water determines the type of treatment employed. For example, waste water from dairy and food industries is given biological treatment. Although spray irrigation technique works well for them, waste water discharged from metal plating industries is chemically treated. Waste water from chemical, pharmaceutical and petroleum drilling industries is given the deep injection method. A deep well is dug so that its level is much below the ground water level (to avoid ground water pollution) and all wastes are injected into it.

Management of Water Resources: Only about 0.003 per cent of the total water present on the earth's crust is available as fresh water. Even this negligible amount of water is unevenly distributed. As a result, many parts of the world suffer from severe water crisis. As it is not possible to increase the water supply, the water crisis has to be solved by proper management of the available water resources.

Management of water resources means providing adequate supply of pure, fresh water for various uses without endangering the water resources. Water resources can be managed by: (i) Increasing the supply of usable fresh water, (ii) Reducing the wastage of water by proper methods and (iii) Reuse of water.

(i) **Increasing the Supply of Usable Fresh Water**: The supply of good quality water can be increased by:

(a) **Constructions of Dams and Reservoirs**: The rainwater and water from the melting of snow, flows through the rivers to the sea. Dams and reservoirs can control this wastage of water. They not only serve the purpose of water reservoirs, but also control floods, provide irrigation and generate electricity.

(b) **Water Diversion**: Water can be diverted with the help of canals, from areas of surplus water to those facing a water crisis. An existing example of this can be seen in America, where water is transported from water-rich north California to water-deficit south California.

(c) **Tapping Underground Water**: In areas with insufficient water supply, tapping underground water can help to meet scarcity. A geological formation with sufficient underground water which can be supplied to human beings is called an aquifer. Aquifers constantly gain water through percolation. By properly managing a constant supply of water (rainwater, domestic, municipal sewage) to aquifers through percolation, the ground water supply of fresh usable water can be maintained.

(d) **Desalination**: Saline marine waters can be converted to fresh drinking water through a process called desalination. It can be done through reverse osmosis or by distillation. It is an expensive process that requires a lot of energy. This method is used in the Arab countries where there is severe scarcity of fresh water but no paucity of finances. In India, desalination is used to obtain fresh water in Bhavnagar in Gujarat and Churu in Rajasthan. However, inexpensive solar energy is used for the purpose. A huge amount of salt is residual in the process of desalinization. Disposal of this salt is a problem.

(e) **Cloud Seeding**: In this method, a large suitable cloud is chosen and silver iodide powder is dispersed over it from an aircraft. The chemical particles serve as nuclei of condensation causing small water droplets of the cloud to coalesce and form large ice particles which finally precipitate on to the earth.

This method was successfully followed in the US bringing rain to 7 per cent of land areas. However, this method now evinces only theoretical interest because of the non-availability of rain clouds in dry areas and the harmful effects of the chemicals used on soil, wildlife and humans.

(ii) **Reducing the wastage of water by proper methods**: In India, a lot of water is wasted from leaking taps due to bad plumbing, lack of proper education and wastage during irrigation. Improving the flow of irrigation canals, use of sprinklers and deep irrigation techniques should be followed to check water loss.

As industries require huge amounts of water they can decrease the pressure on water resources by recycling their waste water. Therefore, preventing wastage of water from irrigation, industry and domestic use will help in the efficiently management of water as a resource.

(iii) **Reuse of Water**: The property of water as a universal solvent has made it reusable after proper treatment. This has reduced the wastage of water. Water can be reused (i) directly; (ii) indirectly or (iii) inadvertently.

(a) **Direct Reuse**: It involves the supply of well-treated and recycled water directly for use in industrial or agricultural activity.

(b) **Indirect Reuse**: When the sewage from domestic, commercial and industrial areas is transported through the sewerage system to the treatment plant, it is subjected to primary and secondary treatment followed by chlorination and natural filtration through soil. This water is then discharged into the river through a network of drains. It may also be used for irrigation of forests and wastelands.

(c) **Inadvertent Reuse**: In a city, town or industry, water is usually withdrawn from a river, used, treated and then discharged into the same river. Water from the same river is again withdrawn and the process is repeated. This practice however, suffers from the risk that discharged water may not be properly treated.

The symptoms of arsenic toxicity cannot be immediately identified and may take as long as eight to 14 years before they show up (depending on the quantity/volume of arsenic ingested, nutritional status of the affected person and his immunity level after he starts drinking contaminated water).

CASE STUDY

Arsenic Pollution: Arsenic is a naturally occurring element which forms compounds with not only non-metals such as oxygen, chlorine, sulphur, carbon and hydrogen but also with metals like lead, gold and iron. Organic compounds of arsenic are less toxic for human beings than its inorganic counterparts. Although in the elemental form, arsenic is insoluble in water but in an oxidized state it is water-soluble. Many a times, oral intake of arsenic has proven to be deadly. Earlier, it was used to treat skin diseases, worms and syphilis. A century ago, incidences of skin cancer were reported due to the intake of arsenic medicines.

Today, arsenic contamination has become a global problem. It occurs as a result of geological processes and human activities such as agriculture, manufacturing, mining and smelting. It also occurs due to contamination from food and beverages.

Arsenic toxicity in the human body is manifested in the following three manners:

(a) Arsenic in urine (for recent exposure)

(b) Arsenic in blood (for instant exposure)

(c) Arsenic in skin, hair and finger nails (for chronic exposure)

Large doses of arsenic can cause human death, whereas lower doses cause systemic effects.

The US Environmental Protection Agency (US EPA) has set the maximum concentration level of arsenic in drinking water as 20 mg/L. But according to a number of scientists, 20 mg/L is the threshold level for risk to human health. There is a consideration to lower this to 10 mg/L as per the WHO approved limit.

Dermal exposure or inhalation has lesser effects on human health. Oral intake may cause:

1. Skin abnormalities like diffused melanosis, spotted melanosis, peratosis of palms and soles which may progress to skin cancer.
2. Increased cancer risk mainly of the liver, bladder, kidney and lungs.
3. Irritation of the digestive tract leading to pain nausea, vomiting and diarrhoea.
4. Decreased production of red and white blood cells.
5. Damage to blood vessels, liver and kidney.
6. Abnormal heart function.
7. Impaired nerve functioning, feeling hot, and feeling pins and needles in hands and feet.
8. Foetus-damage during pregnancy.

Bangladesh Case Study: The first case of arsenic poisoning was identified on July 3, 1996 at a health camp at Paksi, in Pabna district of Bangladesh. It is also the largest mass poisoning case in the world. Large scale unplanned withdrawal of ground water for agriculture under the Green Revolution led to this incident. Of the 64 districts in Bangladesh, 53 are suspected to have high levels of arsenic in water and people of 34 districts have already been identified as suffering from arsenic poisoning. In India, the arsenic-affected areas of West Bengal are lying on sediments of younger deltic deposition extending eastwards towards Bangladesh covering at least 41 districts which lie mostly in the Atrai Meghna and Gangetic flood plains and tidal regions.

Scientists and support personnel of the Dhaka Community Hospital (DCH) and the School of Environmental Studies (SOES), Jadavpur University, Kolkata, India conducted laboratory tests on water samples collected from one lakh tube wells from 64 districts in Bangladesh, including those 53 districts which were suspected to be contaminated. From the tests, it was inferred that:

(i) The concentration of arsenic in ground water was above the maximum permissible limit (0.05 mg/l) in the 53 districts.

(ii) Arsenic contamination in Bangladesh is much more severe than in West Bengal. In Bangladesh it had affected 65 million people living across an area of 87,000 sq km, whereas in India, it has affected 38 million people covering an area of 38,000 sq km.

Smatler Case Study: According to a special joint study conducted by the DCH and SOES on the arsenic contamination of ground water in Smatler Village under Shirt Police Station in Jester, 91 per cent of the existing tube wells of the village were found unsuitable for human use.

If someone is suspected of having arsenicosis, a sample of the source from where he gets his drinking water has to be tested first. If it is found to contain arsenic concentration greater than the permissible limit (0.05 mg/L) then the use of water from that tube well for drinking purposes has to be immediately discontinued. An early stage of apparent oral contamination through drinking water can be cured by taking nutritious food rich in vitamins A, B, C and E. At an advanced stage of toxicity, although cure is difficult and medical consultancy is a must.

Methods to Combat Arsenic Contamination: These days' people have started drinking tube well water to avoid water-borne diseases. However, lately, even some tube wells have become shown signs of arsenic toxicity and have become a threat to human beings. To combat this arsenic contamination we have to:

(i) Go back to our age-old practice of using surface water for drinking after having discarded the sediments.

(ii) Test all tube wells for arsenic pollution and use water only from the arsenic-free tube wells for drinking and cooking purposes.

(iii) Use water from a well or pond in case all tube wells are found to be polluted.

(iv) Use pond or river water when there is no alternative.

(v) However, if there is no alternative to arsenic poisoned water, then it should be collected only in iron vessels and allowed to settle for 12 to 24 hours. Then three-fourths of the water has to be poured slowly into another clean container. For better results, water should be filtered through 4 to 5 layers of clean gauzy material as it reduces the arsenic content of the water by at least 70 per cent. This water must be further filtered before it becomes fit for drinking.

(vi) Alum nugget may be dipped for one second in the iron vessel containing contaminated water to ease its arsenic deposition process. Then the earlier method for water treatment can be followed.

The residual arsenic sediment rich water must be poured carefully into a dung-heap to neutralize its arsenic toxicity.

Conclusion: Arsenic pollution has taken a disastrous form due to the wanton extraction of ground water in Bangladesh and neighbouring West Bengal. The affected regions have huge wetlands, flooded river basins and high rainfall however, the major source of water there is ground water. In order to overcome this situation:

(a) Awareness campaigns have to be carried out at the grassroots level.

(b) Urgent watershed management must be undertaken.

(c) Safe tube wells should be marked properly and villagers of affected regions should be urged to drink water only from the safe tube wells and also to eat nutritious food.

(d) All tube wells and deep wells must be checked regularly.

(e) Arsenic contamination can be prevented from becoming a calamity, by using surface water and also by harvesting rainwater instead of using ground water.

Fluoride Pollution: Fluoride compounds are necessary in minute quantities and are also essential for prevention of tooth decay. But when these exceed the permissible limits, these compounds become toxic and cause harm to to the plant and animal kingdom.

The fluoride absorbed from air comes from chimney emissions, volcanic eruptions and insecticide spray. It settles on leaves, penetrates the stomata and migrates to the periphery of leaves. When it accumulates in larger amounts, it inhibits the metabolism leading to fluorosis. Turgid plants are more susceptible to fluoride accumulation than wilted ones. All leafy vegetables are more prone to fluoride accumulation, which in turn burns the tip and the sides of foliage. When cattle graze on these plants they also ingest contaminated fodder. In Orissa, the area surrounding the National Aluminium Company (NALCO), Angul, has already started showing signs of fluoride pollution. There is heavy damage to the crops in fields surrounding the plant.

In human beings and other animals, fluoride toxicity is mainly water-borne. Fluorine being the most electronegative element is universally present in nature in the combined form as it carries a negative charge (F^-) on it and is attacked by positive charged ions like calcium (Ca^{+2}). Bones and teeth have the highest amount of calcium in the body so they attract the maximum amount of fluorine, which gets deposited as calcium fluoroapatite. At the same time, the unbound calcium is also lost from the tissues. Incidents of fluorosis are higher in tropical and sub-tropical countries, probably due to more consumption of drinking water. The reported cases of endemic fluorosis in Haryana and Punjab were caused by the intake of well water which had higher concentrations of fluoride. Fluorosis affected people are found in Andhra Pradesh mainly in Prakasam, Nalgonda and Guntur districts. According to studies by the Defence Laboratory in Jodhpur, flourosis in Rajasthan has reached its threshold limit. The districts of Jodhpur, Bhilwara, Jaipur, Bikaner, Udaipur, Nagpur, Barmer and Ajmer have been crippled for life by fluorosis. The high concentration of fluoride in water sources in arid and semi-arid zones is the cause of fluorosis. Since the soil of these areas is also polluted with fluoride, the food grains cultivated on these fields also contain high levels of fluoride.

It is observed that in adults exposed to high-fluoride ingestion, the hydroxyl bonds of the hydroxyapatite material in the bone are partly replaced by fluorides. In order to immobilize fluoride from the circulating fluoride phase in the body (blood and cellular fluids), the body's defence mechanism fixes F– to the hydroxyapatite (HA) material of the bone by replacing OH– with F– (Teotia and Teotia 1992) irreversibly, till the exposure continues. In this process, the rate of bone material (HA) synthesis is considerably increased, leading to excessive bone formation or osteosclerosios, a basic symptom of subjects suffering from skeletal fluorosis. The effect of fluoride on the bone depends on the type of bone and its constituents. The following are the characteristics of the structural changes in fluorosis:

1. Increased bone mass and density.
2. Exostosis (bony out-growth) in bone surfaces.
3. Increased osteoid seam and resorption surface.
4. Increased trabecular bone volume.

Marine Water Pollution

When the salt content of a water body is equal to, or more than sea water, that is, 35 ppt, then it is known as a marine water body. Oceans, seas, lakes, estuaries, brackish water and salt marshes are all examples of marine water bodies. Estuaries are present at the tail-end of rivers and are full of lush green vegetation. They play the role of a buffer zone between coastal and river water and balance the ecosystem. As a result of the constantly increasing marine pollution this balance is getting disturbed.

Marine pollution can be defined as the direct or indirect introduction of matter or energy by mankind into marine life that is harmful to living resources, hazardous to human health, hinders marine activity and adversely affects sea water as well as reduces its amenities. As the definition suggests, marine pollution is purely a man-made activity. Any disturbance or damage caused by nature (like earthquake, volcanic eruptions, and tsunamis) is not considered marine pollution.

Everything that is carried away by rivers is ultimately dumped in the sea. This includes industrial waste, heavy metals, agricultural waste, mine trailings, urban and rural sewage. Sewers are also polluted by non-biodegradable waste, radioactive wastes and spillage of oil tankers or due to other causes. Statistics show about nine billion tonnes of solid waste; including three million tonnes of toxic matter produced by major industries is thrown into the ocean. The large amount of plastics thrown into the sea is also threatening marine life. About 15 per cent of the world's 280 species of sea birds are known to eat plastic. It causes a multitude of gastrointestinal problems such as reduction of appetite, sensation and development of stomach and intestinal ulcers.

Radioactive waste in ocean water totals about 0.1 per cent of the concentration. This waste gets disturbed by turbulence and ocean currents and is a great threat to civilization. Herman Miller, a Nobel Laureate has predicted reduction of the survival ability of the human species as several generations are exposed to radiation. Out of 2,500 million tonnes of oil produced per year, 1,700 million tonnes are transported in oil tankers across the ocean. An average of 80 to 200 litres of oil is spilled in the ocean per second per day. Besides this, oil pollution in marine water is also caused by maritime accidents, wars, drilling of rigs and so on. Oil pollution is an oxygen-demanding process. A drop of petroleum spread over a large area becomes more dark and viscous due to slow evaporation of lighter volatile fractions. Sunlight and oxygen polymerize oils to highly viscous tar balls. The oil is attacked by micro-organisms of the sea, which use hydrocarbons as a source of energy. Besides this, oil floats on water forming a film and inhibiting sunlight from entering the sea thereby prohibiting photosynthesis of aquatic plants and simultaneous oxygen production. Most surface-dwelling organisms are found near the shore from where they derive mineral and inorganic substances necessary for their vital process. Hence, shore-line pollution greatly affects aquatic life. Birds and furry marine mammals are frequent victims of oil pollution. The feathers of diving birds and the fur of marine mammals get soaked with oil displacing the air present there, interfering with their buoyancy and maintenance of body temperature (natural insulation). Once these animals are soaked, they either get drowned or die due to loss of body heat. Hydrocarbons from oil also enter the human body through the food chain and cause biological magnificence.

Control: In order to control such severe pollution, studies on marine environment have started. The Marine Pollution Convention (MARPC) has stipulated rules for reduction of pollution by ships and ports. Port authorities are also following anti-pollution measures.

5.4 Soil Pollution

Soil is the thin outermost layer of the earth's rocky surface. The major components of soil are air, water, minerals, organic and inorganic matter. The organic matter of the soil is decomposed by micro-organisms forming biomass, which is mixed in the soil and the inorganic matter is obtained from the weathering of parent rocks. Soil plays a vital role in supplying nutrients and water to the plant and animal kingdom respectively. The nutrient power of soil is a measure of its fertility. The productivity of soil is measured in terms of crop or animal biomass per unit area while the yield of biomass solely depends on soil and crop management strategies. Hence, there lies a dynamic equilibrium between the crop and the soil. The disturbance of this dynamic equilibrium leads to soil degeneration followed by degradation. This is called soil pollution.

Thus, soil pollution can be defined as the mixing of any form of matter or energy into the soil that will lead to modification of the features of soil or, more generally, of its chemical or biological balance that will have an adverse effect on plant growth and animal health.

The degradation of soil (soil pollution) is the result of both natural and human activities. Major human activities that have accelerated the process of soil pollution include, exposure of the soil by deforestation, overgrazing, intensive cultivation, mining, different developmental activities (like construction of dams and industries) and solid waste disposal.

Exposure of soil due to these causes, results in soil erosion leading to loss of soil fertility and deterioration in depth and quantity of underground water table. Soil erosion along the slopes of the Aravalli Hills and the Himalayas has made them prone to landslides. Soil erosion washes away the topmost layer of the soil and not only destroys the productivity of the eroded land but other productive land by getting deposited over it, thereby destroying it as well.

5.4.1 Mining Activities

With advancement in science and technology, our underground resources are now excavated through mining activities. For this, usually opencast mining and underground mining are followed. In opencast mining, underground resources are excavated directly by removing the top soil. As a result, the total area is destroyed and loses its productivity. However, in underground mining the minerals are mined through tunnels. When the mine is abandoned, the whole area collapses causing the formation of a big depression turning the area into a wasteland.

However, in both types of mining, the mined materials are dumped near the mine which causes unwanted occupation of land. Rain water washes away the mining waste (rich in sulphuric acid) to the nearby agricultural fields and pastures. The debris blocks the drainage system leading to water logging, as well.

Faulty agricultural practices such as unskilled irrigation, shifting cultivation, injudicious use of chemical fertilizers and pesticides are also responsible for soil degradation. If the drainage system of the agricultural field is not maintained scientifically, water logging may occur. Evaporation causes an increase in the salt concentration on the top soil, making the land more and more saline and finally, unsuitable for agriculture. Secondly, water logging closes the passage of air to the soil and stops the growth of soil organisms making the soil barren.

In shifting cultivation, practiced by almost all the tribal people all over the world, the forest is burnt to use the land for cultivation. The practice exposes the soil and makes it prone to soil erosion. In India, the *Podu* cultivation of Orissa or *Jhum* of Assam or *Dahya* of Madhya Pradesh are examples of shifting cultivation.

Inorganic fertilizers are used on a large scale, to restore soil fertility for increased crop yield. Inorganic fertilizers are mainly rich in nitrogen, phosphorous and potassium. Indiscriminate use of fertilizers increases the nutrient contamination, both in run-off water and ground water. Microbes of the soil reduce the nitrogen ions to nitrite ions which enter the animal body through food, crops or water and are directly absorbed in the blood stream and oxidize the oxyhaemoglobin (the O_2 carrier) to methemoglobin (which cannot carry O_2 any more). This leads to suffocation, breathlessness and ultimately the death of animals.

Although phosphorous is insoluble in soil, it goes into the soil and spreads as organic phosphate. Increase in the percentage of phosphorous in water causes eutrophication of nearby waterways.

Pesticides

Farmers use chemicals, in the form of herbicides, insecticides, fungicides and pesticides to protect their crop from herbs, insects, fungi and pests, respectively.

Earlier, plant-extracted organic compounds like pyrethrum, nicotine, rotenone were used to control pests. Subsequently, inorganic compounds like lead, mercury and arsenic were used as pesticides. These are known as first generation pesticides. Accumulation of these highly toxic chemicals in the soil not only inhibited the plant growth but also poisoned animals. Gradually, pests became resistant to them.

With the discovery of Dichloro Diphenyl Trichloroethane, (DDT) by Paul Muller, the second generation pesticides came into use. Some chemicals similar to DDT such as dieldrin, eldrin, heptachlor, chlordane, mirey, endusulfan were synthesized. Although, the potential insecticide properties of these organo-chlorides increased the crop yield with simultaneous reduction of pest damage it showed adverse effects with large-scale accumulation of residual organic chlorides in soil crops or animal tissues. Due to these environmental hazards use of DDT has been banned in most countries.

Different types of pesticides based on organo-phosphates and carbonates have been discovered. They affect the soil for a span ranging from one week to several months (in contrast to five to 15 years in case of organo-chlorides). However, soluble pesticides (like organo phosphates) can leach into ground water and affect the birds, animals and even human beings.

Injudicious use of chemical pesticides can create:

(i) Adverse environmental impact.
(ii) Development of pest resistance.
(iii) Resurgence and secondary pest outbreak.

(i) **Adverse Environmental Impact**: Large amounts of pesticides in soil interfere with the soil metabolic process by killing many non-targeted beneficial soil organisms such as earthworms. Hence the soil becomes infertile. Because organo-chlorides are non-biodegradable organic chemicals, they accumulate and magnify in the food chain (biomagnification) and interfere with the calcium metabolism of birds. As a result birds lay fragile, thin-shelled eggs. High concentration of pesticides present in fatty tissues of prey organisms can kill many forms of predators also. Thus, excessive use of pesticides can be described as a profit-oriented poisoning of the ecosystem.

(ii) **Development of Pest Resistance**: Pesticides kill almost all the pests. A few survive as their genes become resistant to a specific pesticides. The species that have survived can reproduce a large number of pesticide-resistant offspring within a short span of time despite frequent stronger doses and a switch-over to different chemicals. This will only help to generate increased number of pesticide-resistant species. Statistics show that about 462 different species of fungi, 50 species of weeds and 10 species of small rodents are already resistant to one or more chemicals.

(iii) **Resurgence and Secondary Pest Outbreak**: Resurgence is the return of some completely eliminated species of pests due to pesticide use. When pesticides are used along with the target pests, some beneficial, natural predators and parasites are also killed. As a result, some pests which were not a cause of concern earlier suddenly start multiplying and creating problems. This is known as secondary pest outbreak. Without sufficient natural enemies and with abundant food, these harmful species multiply alarmingly in a few days after being controlled. Now this problem is handled more scientifically in the following ways:

(i) Cultural practice.
(ii) Biological control.
(iii) Natural chemical control.
(iv) Genetic control.

(i) **Cultural Practice**: The usual practices of cultivation that is, crop rotation, adjustment of planting times, timely harvesting and multi-crop agriculture, inhibit the growth of pests to a large extent. Rotation of the crop every year will decrease the population of pests that feed on a particular crop. Timely plantation and harvesting will cut-off the food supply to the pests. As most pests are specific to certain crops, multi-crop farming will discourage their growth.

(ii) **Biological Control**: Unlike chemical methods, biological methods do not leave any toxic material in the soil. In the biological pest control method, some natural predators, parasites and disease-causing viruses and bacteria are introduced in the soil. These control the population of pests that are harmful to us. For example, the bacteria bacillus thurigiens is available as a powdery material. It effectively controls mosquitoes, gypsy moths and many leaf-eating caterpillars.

The biological method is economical and kills the target pests, while simultaneously decreasing their genetic resistance. Secondly, once the population of the predators is established, it becomes self-perpetuating, so no reintroduction of predators is necessary.

(iii) **Genetic Control**: This method was practised in the USA to eliminate the screw worm fly during the 1970s. In this method, male insects are raised and sterilized in the laboratory and released in the affected area. These sterile male insects mate unsuccessfully and this results in a decrease in population.

(iv) **Natural Chemical Control**: Hormones and pheromones are chemical compounds generated in the cells of organisms they control the development and sexual attraction/reproduction of the organisms. The use of natural or synthetic hormones and pheromones can control most pest problems. These agents are sometimes called third generation pesticides. However, they have not been developed and used to their full potential due to a lack of awareness, the attitude of farmers and because they are not favourable to the economic interests of the pesticide industry.

An overdose or underdose of natural hormones extracted from the species or chemically synthesized hormones, causes disorder in the lifecycles of the pests resulting in premature death and destruction of the species in the area.

Female insects that attract their male mates during the mating season secrete pheromones. Natural pheromones extracted from the insects and/or chemically synthesized pheromones can be sprayed in the pest-infected areas to attract male insects into one localized area away from the female insects. This will stop the growth of the pest population. The concentrated male population of the insects can additionally be destroyed by any other method for more effective pest control.

These natural methods of pest control can replace the present pesticide-based approach; this will ensure that the crops are not sacrificed and the harmful effects on ecology are also prevented. This carefully designed ecological approach to pest control, that integrates any number of natural controls and addresses sociological, economic and ecological factors is called Integrated Pest Management (IPM). Some of the sociological approaches of IPM and the ill-effects of industrial pesticides provide insurance coverage to the farmers for probable crop loss.

Another important cause of soil pollution is the solid waste disposal system. The garbage and trash generated through various activities in houses, commercial set-ups, hospitals, industries and agricultural fields results in millions of tonnes of solid waste material each year. Much of this waste is dumped on the road side which then floats into the open drains and streams. Although, most of the agricultural and organic waste is recycled to get fertilizers and biogas, the municipal waste from homes and commercial houses is used as landfill. If this land filling is not followed scientifically it will not only become a breeding ground for disease-carrying animals like rats but will also contaminate the aquifers. If solid waste is burnt in the open, it causes severe air pollution. Hence, solid waste has to be incinerated or placed in sanitary landfills. The discharge from the thermal power plant (fly ash) is used for brick manufacturing and also mixed with soil to get high crop yield. However, toxic solid wastes like petroleum oil, chemicals, biomedical waste, used batteries, e-waste and radioactive metals are hazardous to public health. Hence, they need special attention. The Love Canal case is a striking example of soil pollution in New York. Toxic chemical waste leaked from a former disposal site called Love Canal in Niagara Falls and resulted in cancer, respiratory problems and renal diseases. After the incident, about 58 per cent children located in the area were born with birth defects and finally the Health Commissioner declared Love Canal, a threat to human health.

Some major problems that result from improper solid waste disposal are the following:

(i) The dumping of solid waste on land leads to the emission of toxic gases that are detrimental to health.

(ii) The inconvenience caused to people due to the unpleasant smell and spread of insects that gather on this solid waste as well as smoke from burning of such waste.

(iii) Contamination of surface and ground water due to decomposition of solid waste.

(iv) Disfiguring solid waste.

5.4.2 Control of Soil Pollution

Soil pollution damages the thin layer of fertile soil covering the earth's surface. Nature takes hundreds of years to develop soil from its parent material, whereas man destroys it in a matter of few years. Thus, soil pollution has to be checked immediately. Soil erosion can be checked with planned afforestation and controlled construction in sensitive areas. The use of fertilizers and pesticides needs to be minimized. Instead, biofertilizers and natural pesticides should be used. A generation of solid waste can be controlled following the principle of the three Rs, namely, Reduce, Reuse and Recycle. Industrial discharge has to be treated physically, chemically and/or biologically until it becomes non-hazardous. For sanitary landfills, new storage areas have to be located. The deep well injection method should be followed for scientifically secure dumping.

5.5 Noise Pollution and Sound

In addition to other pollutants, noise also plays a major role in the degradation of the environment. Noise can be explained as any sound that is not desired by the recipient. So, it can also be defined as a wrong sound at the wrong place and at the wrong time.

Sound is measured in terms of sound pressure or sound intensity, that is, the quantum of sound energy which flows through unit area of the medium in unit time. It is measured in decibel (dB) units.

1 Decibel (dB) = 10 log 10 [Intensity measured/Reference intensity]

The reference intensity used is the threshold of hearing that is, the sound which can be first heard at a sound pressure of 2×10^{-5} Newton m^{-2} or sound intensity of 10^{-12} watts/m^2.

Noise as a potent menace is evaluated in decibels which is the noise level scale. However, the health hazard is a matter of Perceived Noise level Decibels (PNdB). According to the WHO, the optimum noise level for daytime is 45 dB, whereas for night it is 35 dB. Noise levels above 80 dB are hazardous to health. Noise sources with their respective intensities are listed in Table 5.9.

Table 5.9 Noise Source and their Intensity

Noise Source	Noise scale
Breathing	10 dB
Quiet conversation	20–30 dB
House in a quiet street	35 dB
Radio music	50–60 dB
Loud conversation	60 dB
Office noise	60 dB
Children playing	60–80 dB
Lawn mower	60–80 dB
Vacuum cleaner	80 dB
Traffic noise	60–90 dB
Sports car	80–95 dB
Heavy truck	90–100 dB
Thunderstorm	110 dB
Electrically amplified music	120 dB
Jet take-off (at 100 m distance)	120 dB
Aircraft noise	90–120 dB
Jet engine (at 25 m distance)	140 dB

5.5.1 Classification of Noise

Depending upon the source, noise can be broadly classified into:

(i) Transport Noise
(ii) Occupational Noise
(iii) Neighbourhood Noise

(i) **Transport Noise**: This can be subdivided into (a) road traffic noise; (b) aircraft noise; and (c) rail traffic noise.

(a) **Road Traffic Noise**: Vehicles on the roads produce non-stop noise pollution, particularly on urban roads where there are distinct traffic peaks in the morning and evening as people and children travel to and from work and school, respectively. The increase in the number of vehicles has increased the road traffic density, which in turn has increased the traffic noise steadily. Modern highways and traffic systems also encourage high speed so naturally the volume of noise increases with the increase in traffic speed.

Table 5.10 shows the permissible sound levels prescribed by the Central Pollution Board (India).

Table 5.10 Permissible Sound Levels

Area	Day	Night
Industrial	75 dB	65 dB
Commercial	65 dB	55 dB
Residential	50 dB	45 dB
Sensitive areas: up to 100 m radius around hospitals, educational institutions, courts etc.	50 dB	40 dB

(b) **Aircraft Noise**: Although, the volume of noise produced when an aircraft takes off or lands is very high, it differs from road traffic noise, in that it is not continuous. The peak frequency depends on the type of aircraft and also its operational height.
New models of aircrafts are being designed now to reduce the noise level.

(c) **Rail Traffic Noise**: This noise is generally of lower frequency than road traffic noise. As most of the railway tracks run through rural areas, the rail traffic noise does not pose greater health hazards than road traffic noise. Of course, the buildings situated close to the railway tracks are exposed to greater noise pollution. However, the introduction of diesel and electric locomotives has reduced the noise level considerably.

(ii) **Occupational Noise**: Heavy industrial machines are a source of occupational noise. Millions of people working in industries are exposed to this tremendous noise at the rate of eight hours per day, for six days a week for at least 30 years of their lives. This causes hearing loss and decrease in working efficiency and an increase in the probability of accidents. Hence, noise reduction is essential for industrial workers. Occupational noise also includes noise from domestic gadgets like vacuum cleaners, mixer-grinders and washing machines.

(iii) **Neighbourhood Noise**: This includes various noise sources that disturb and annoy the general public by interfering with their comfort and welfare. For example, noise from loud speakers in public functions, the loud volume of TV and radio sets and loud music late in the evening can disturb the neighbouring residents.

5.5.2 Effects of Noise Pollution

Noise is air-borne mechanical energy that strikes the eardrum. Noise affects hearing instantaneously, but in the long run too it affects human health in a number of ways. A noise of 2 dB can cause pain in the ear whereas the noise of 150 dB may even prove to be fatal. Prolonged exposure to lower noise levels can cause ear damage and hearing loss. The first symptom is in the form of temporary hearing loss and buzzing in the ears which eventually worsens to permanent hearing loss. Headache, fatigue, dizziness, inefficiency, high blood pressure, tension in muscles and abnormal heart rhythm are the side effects of prolonged exposure to noise. Effects that noise pollution can have on humans are shown in Fig. 5.8.

Children exposed to excessive noise show behavioural disorders and develop a destructive nature in later stages, followed by the development of neurotic traits as adults.

Noise effect on human beings

Noise hazards

Permanent hearing loss
Neural-humoral stress response
Destruction of artefacts

Noise nuisance

Comfort

Sleep interference
Invasion of privacy
Damage of artefacts interference
Habit of talking loudly

Efficiency

Mental stress
Frustration
Task interference
Irritability

Enjoyment

Concentration interference
Meditation
Recreational Interference
Temporary hearing loss

Figure 5.8 Effect of Noise on Human Beings

Noise pollution leads to cardiovascular problems and also causes malformation in the nervous system of a foetus. It also increases secretion of hormones such as adrenaline and alters blood circulation.

5.5.3 Control of Noise Pollution

According to the WHO, of all the environmental problems, noise is the easiest to control. This can be achieved by:

(i) Public awareness regarding the need to control noise pollution.

(ii) Reduction in exposure to noise by application of engineering control techniques such as alternation and modification of design to reduce noise, by construction of sound barriers or the use of sound absorbers.

(iii) Reduction in exposure to noise by making exposed personnel use protective ear plugs and decreasing the exposure time.

(iv) Construction of academic institutions and hospitals away from highways, railroads and airports.

(v) By creating vegetation buffer zones through large-scale tree plantation which absorb noise.

(vi) Apart from these, one can restrict the use of pressure horns and record players in the neighbourhood. The intensity of noise pollution can thus be decreased.

(vii) As per a decision by the Government of India in 1976, 'noise-induced hearing loss' became a noticeable disease under the Factories Act, 1948. Whenever detected, a medical practitioner has to report the case of noise-induced hearing loss in a worker along with all the information to the Chief Inspector of Factories. Failure to do so is a punishable offence according to Section 89 of the Factories Act. The permissible noise exposure level as per the Factories Act is a maximum of 115 dB, for a very short duration and a maximum of 90 dB for a duration of eight hours. Noise-induced hearing loss due to occupational exposure beyond the permissible levels is included in the Workmen's Compensation Act, 1923.

Noise pollution can be controlled if we spread consciousness through print and electronic media and social forums.

5.6 Thermal Pollution

Different industries and nuclear plants use water for cooling purposes and discharge the heated water into nearby streams or water bodies. As a result, the nearby water body is heated up and this in turn, affects aquatic life. This phenomenon is known as thermal pollution. It is also caused by:

(i) Random cutting down of shade-providing trees beside the water bodies.

(ii) Soil erosion due to poor agricultural practices, overgrazing, increase in solid suspended particles in a water body due to excessive recreational practices.

(iii) Natural phenomenon such as earthquakes can also cause thermal pollution.

Primarily, thermal pollution raises the temperature of the water body thereby decreasing dissolved oxygen levels. This results in suffocation of some of the aquatic life while accelerating the growth rate of others.

Secondly, with an increase in temperature most of the chemical and biochemical reactions proceed at an enhanced rate till the catalyst enzymes are denatured (enzymes can tolerate temperature change only up to a few degrees). In other words, from the ecological point of view, the increase in temperature will cause death of aquatic life by stopping essential biochemical reactions. For example, the mangrove forest density is decreasing day by day in the Sundarbans and in the Andaman bay areas in India. This loss of shade will accelerate the population of heat-tolerant species. This will upset the ecological balance of the water body. Hence, the effect of thermal pollution is summarized by scientists as:

(i) **Thermal Shock**: The sudden change in thermal status of the water body causes the fish and other aquatic life to migrate to a more suitable environment.

(ii) **Thermal Enrichment**: Warm water from different industries may be used for irrigation purposes during plant growth period. It accelerates the growth rate of fish and other aquatic life, making it beneficial for commercial purposes.

5.7 Nuclear Hazards (Radiation Pollution)

Radiation is a form of energy that unlike sound energy can also travel through vacuum. Radiation permeates the universe, the solar system and also the earth. The earth is constantly bombarded with solar radiation. Solar radiation consists of radiations of different wavelengths (energy). The earth's atmosphere can screen out much of the solar radiation. Some radiation is also trapped in the magnetosphere of the atmosphere, more particularly in the Van Allen belt. It is the region of two zones of high-intensity particulate radiation trapped in earth's magnetic field surrounding the planet, beginning at an altitude of about 800 kms (500 miles) and extending tens of thousands of kilometres (40,000 miles) in space. However, the total radiation can be broadly classified as: (i) Ionizing Radiation and (ii) Non-Ionizing Radiation.

(i) **Ionizing Radiation**: This type of radiation carries greater energy and can alter atoms to create ions. When ionizing radiation incidents on any living body, it passes through the protoplasm and produces positive or negative ion pairs. The damage depends upon the number of ion pairs produced in the absorbing medium. It can travel both as a wave, such as in the form of x-rays, gamma rays, or as particles like α and β rays.

(ii) **Non-Ionizing Radiation**: This type of radiation carries lower energy than ionizing radiation. Although this type of radiation can excite atoms, it cannot ionize them. Heat, light and radiowaves are examples of non-ionizing radiation.

5.7.1 Radiation Pollution

Radiation pollution is caused by the addition of more ionizing radiation to the environment and people are exposed to more radiation than they normally experience. Radiation interacts with living tissues and damages them.

Radioactive disintegration of atomic nuclei generates three types of radiation: alpha, beta and gamma which are very harmful to living systems. The sources of radiation may be natural, such as cosmic and terrestrial or artificial (man-made).

Natural Radioactive Sources

(i) **Cosmic Radiation**: These are high-energy charged particles (mostly protons) that enter the earth's atmosphere from outer space. While travelling they produce secondary radiation by colliding with other atoms such as oxygen and nitrogen, present in the earth's atmosphere. The extent of cosmic radiation increases with an increase in height from the earth's surface.

(ii) **Terrestrial Radioactivity**: This is another natural radioactive source. It originates from the deposit of radioactive minerals such as uranium, thorium, actinium and polonium. The amount of whole body radiation produced by them annually depends on the amount of mineral deposit and the distance of mineral deposit from the soil surface.

(iii) **Radio Isotopes**: These are the most damaging and dangerous pollutants. Some radioactive isotopes like K^{40}, Ca^{14}, Th^{232}, U^{238} are present in human body tissues, although their concentration varies with the geographical location and other factors. Radio isotopes are produced by the cosmic rays in the upper atmosphere. So far, 40 different types of radio isotopes have been discovered. Some radio isotopes are B^7, B^{10}, C^{14}, Si^{32}, H^3, Na^{22}, S^{35}, P^{32} and P^{33}.

$$_7N^{14} + {_0n^1} \longrightarrow {_6C^{14}} + {_1H^1}$$

$$_6C^{14} \longrightarrow {_7N^{14}} + {_{-1}e^0}$$

A specific example of natural radioactive contamination is the thorium-rich monazite sand deposits on the Kerala coast, the Chernobyl nuclear accident is an example of man-made contamination.

Artifical Radioactive Sources

(i) **Nuclear Power Plants**: The radiation from nuclear power plants beyond the boundary is negligible. They follow the radiation standards set by the US Atomic Energy Commission of a maximum radiation of 500 mrem/year at any point or beyond the boundary of a nuclear power plant. This limit was reduced to 5 mrem/year in 1971. According to this limit, a person present round the clock at the least favourable location (in the direction of wind blowing at the plant boundary) will receive less than 5 mrem/year.

(ii) **Peaceful or Constructive use of Nuclear Explosions**: Recently, nuclear explosions have begun to be used for constructive purposes such as, releasing natural gas from the underground region, harbour-building and for the construction of canals between oceans. These uses have led to radioactive concentration in air and water. For example, the burn up of an orbiting System for Nuclear Auxiliary Power (SNAP) generator in the atmosphere in April 1964 led to a detectable amount of Pu238 in rainwater. However, the benefits obtained from these peaceful and constructive uses outweigh the pollution risks through radiation.

(iii) **Radioactive Fallout from Nuclear Weapons**: This fallout is really dangerous to the living system. It has been found that testing of a nuclear weapon in the atmosphere results in the local fallout of radioactive fission products over the immediate area for about a day, followed by worldwide troposphere fallout for a month and a worldwide stratospheric fallout which continues for many years. This rapid transport of long-distance fallout has been detected within 22 days of a French nuclear test at 21^0 South latitude by the presence of radioactive I^{131} and Ba^{140} at 34^0 North latitude.

Radioactive isotopes with longer half-life periods are more harmful as they remain in the human body for long periods and cause damage. Many of the early field workers developed cancer due to constant exposure to radiation. Although, these tests are performed in unpopulated areas (oceans, deserts) to reduce their harmful effects, the long-term effects persist. Even after nearly 60 years, abnormally large number of cases of leukaemia are recorded among the survivors of the Hiroshima and Nagasaki nuclear blasts.

(iv) **Radiation from Miscellaneous Sources**: A person can also get exposed to radiation while watching television or smoking cigarettes.

A person watching television, which operates at less than 25 KV and is provided with a safety screen, at a distance of six to seven feet may absorb radiation of one milliard/year to the gonads whereas under the same situation if the set does not have a safety screen, the dose will increase to 1.1 milliard/year to ovaries and 7.5 milliard to the testes.

Smoking increases the risk of heart attack (by causing blockage of arteries) and lung cancer too. The top six inches of the soil contains approximately one gram of radium on an average, which decays to form another radioactive substance called radon. Being a chemically inert gas, radon diffuses in the atmosphere. When radon decays, it produces lead which is heavier, and comes out of air and settles on foliage such as tobacco leaves.

Some examples of other sources of man-made radiation include houses that are built with radioactive stones which give out 10 mrem/h or bathtubs that are glazed with uranium pigments give about 100 mrem/h radiation.

Units of Radiation

(i) **Disintegration per second**: Units of radiation can be defined as the number of radioactive disintegrations per second for a given amount of radioactive material. It is expressed as $N\lambda^0$; where N is the number of atoms present and λ is the radioactive decay constant.

(ii) **Curie**: It can be defined as the amount of radioactive material that gives 3.70×10^{10} disintegrations per second. One gram of natural radium together with its decay products amounts to one Curie.

1 ml Curie = 1 Curie $\times 10^{-3}$

1 micro Curie = 1 Curie $\times 10^{-6}$

(iii) **Roentgen**: It is a measurement of a quantity of ionizing radiation. One Roentgen of radiation produces one e.s.u of charge in one cm^3 of air at standard temperature and pressure. This amount of radiation produces 2.08×10^9 pairs of gaseous ions in air under standard conditions.

One Roentgen of γ radiation was found to result in the absorption of about 97 erg/gm in the soft tissue. Roentgen Equivalent Physical (REP), a new unit of radiation is defined to denote a dose of 97 erg/gm of body tissue.

(iv) **RAD**: Radiation Absorbed Dose is defined as the quantity of radiation that leads to the absorption of 100 erg/gm of the absorbing material.

For body tissue, one RAD is very closely equal to REP. The same units of Roentgen of radiation will correspond to different numbers of RADs in different materials.

(v) **RBE and REM**: Different types of radiation have different biological effects and Relative Biological Effectiveness (RBE) characterizes them. Roentgen Equivalent Man (REM) is the quality of radiation that produces the same biological damage in man as one REP of γ radiation.

While RBE gives a number of REMs per RAD, it depends on the effect, considering the location of the radiation source and other things. Table 5.11 shows the RBE of different types of radiations.

Table 5.11 Relationship between Different Types of Radiations and their RBE

Type of radiation	Energy	RBE
β radiation	> 50 KeV	1
β radiation	< 50 KeV	2 to 5
γ radiation	>50 KeV	1
γ radiation	<50 KeV	2 to 5
Neutrons and protons and natural α particles	0.5 to 10 MeV	10
Heavy nuclei and fission particles		20

Harmful Effects of Radiation

When ionizing radiation passes through the body tissue, it may split the molecules of the tissue into ions or free radicals and result in useless or reactive fragments and allow the formation of other reactive compounds. It is felt that radiation exposure to man from artificial sources is already capable of producing serious diseases such as leukaemia, bone tumours, genetic damages and infant mortality.

The harmful effects of radiation may be of the following two types:

(i) **Direct Effects**: Fragmentation of biologically important molecules such as DNA.

(ii) **Indirect Effects**: Fragmentation of biologically less-vital molecules with the formation of reactive ions or free radicals that can later affect more important molecules and impair their functions. Strontium 90, Caesium 137 and Iodine 131 have been the most serious pollutants in the general environment in view of one or more than one of the following criteria:

 (i) Whether nuclear fission or fusion, its formation is substantial.

 (ii) Its half-life has been in the range of 1 to 1,000 years (if it is shorter it decays rather quickly and if it is longer its activity is low).

 (iii) It is easily taken up by either man or some other component of the food chain of man.

 (iv) It is not easily eliminated from important body organs or in other words, it has a long biological half-life.

Atom bomb explosions at Hiroshima and Nagasaki in Japan in 1945 and the Chernobyl nuclear mishap on April 28, 1986 near Kiev in former USSR are examples of serious radiation pollution in the history of mankind.

The damage caused by radiation depends on the amount of energy that gets lost during radiation and which then becomes available for the ionization of matter. The amount of energy lost for unit travel is called Linear Energy Transfer (LET). Tables 5.12 and 5.13 show the effects of radiation on human beings.

Low LET radiation causes tissue-damage through indirect action. High LET radiation causes damage through direct action. A particle with high energy will have low LET initially, followed by an increase in LET while its energy decreases. Such a particle can be used in treating a deep-seated tumour if the particle range is just right for it to come to a stop near the tumour.

Iodine 131, in spite of its short half-life of eight days is considered a serious pollutant as it causes thyroid in man and animals. Fission from reactors and weapons produces a large amount of Iodine 131, which is consumed by cows from grass and hay and into milk and from there to the thyroid glands of man and animals.

Table 5.12 Effect of Different Types of Radiation on the Human Body

Types of radiation	Effects on body
α particles Particulate in nature	Can travel in air only a few centimetres and in living tissues only 30μm (that is, can cross about three cells), generally cannot penetrate the skin. Entry to body parts such as bones and lungs cause irreparable damage.
β particles Particulate in nature	Can travel in air about 8 cm. And in tissue about 1 cm. These can penetrate the skin but do not reach underlying tissues. They can cause damage to skin, causing cancer and eye cataract.
γ radiation	Can travel about hundred meters in air and can easily penetrate the body and pass through it.
X-rays	Can travel extremely far and pass through the body tissue, except bones. Causes damage to the molecules in cells
Ultraviolet rays	Have relatively lower energy than X-rays, can cause skin cancer.

Table 5.13 Effects of High Level Radiation

Dose	Effect
Dose of 650 RAD (high lethal dose)	Kills within a few hours or a few days.
300 RAD (lethal dose)	Kills one half of population within 60 days.
50–200 RAD (sub-lethal dose)	Not immediate deaths, suffer from radiation sickness.
Immediate effects after two to 14 days.	Fatigue, nausea, vomiting, diarrhoea and loss of hair, reduction in blood platelets, sore throat. Recovery is possible within a few days of exposure.
General delayed effects on mothers	Cancer, leukaemia, cataract, sterility, decreased life span. In mothers, increase in spontaneous abortion, still births and early infant deaths.

Safety Measures

The National Committee on Radioactive Pollution (NCRP) and Advisory Committee on Reactor Safeguards (ACRS) were formed in 1928. They have special recommendations on space flights and nuclear reactors, respectively to avoid radioactive pollution. The International Commission on Radiology Protection (ICRP) formed in 1947, has recommended that the Maximum Permissible Dose (MPD) for adults and members of the public, exposed to radiation should not exceed 0.4 RAD a year and cumulative dose over a period of 30 years should not exceed five RAD. Among man-made sources, X–rays are the most harmful. X–ray examinations should be avoided as far as possible. Protection of workers and improvement of radiation techniques must also be practiced. Better technology and suitable waste disposal method is also essential for all nuclear power plants. Nuclear power plants should be located far from populated areas and should be provided with a suitable radiation absorption zone around them to check the escape of radiation.

Radiation Episodes

In spite of the extreme danger of radiation hazards to mankind, test explosions of nuclear weapons by developed and developing countries, the explosion of two atom bombs in Hiroshima and Nagasaki by the USA during World War II, the explosion at the Chernobyl power plant in Ukraine in 1986 and such other incidents, have been taking place all over the world. These incidents have resulted in radiation leakage and damage to the fragile ecosystem. Chernobyl was the first officially reported nuclear accident in USSR and the world. The first nuclear reactor of the world was commissioned in USA in 1942 and the research reactor Apsara was commissioned in 1956, in India. It is reported that there have been more than 10,000 reactor accidents in USA alone and 200 serious accidents in India. Since then however, most of the incidents are kept a secret by the respective authorities. Some of the known radiation episodes of the world are described below briefly.

Atom Bomb Disaster in Hiroshima and Nagasaki

On Monday, August 6, 1945 at 8.15 a.m. USA detonated an atom bomb in Hiroshima using U-235 with half-life of 8.5 x 108 years at 580 m height, killing at least 1,00,000 people. Many others were injured and missing in the incident, in which a kind of blue storm covered all of Hiroshima followed by black rain with tremendous heat, dust, smoke and fire. The bomb virtually demolished all buildings and structures within an area of 15 sq km.

The second atom bomb, a weaker one with P4-239 having half-life of 24,000 years was detonated by USA at Nagasaki on August 9, 1945 at 11 a.m at 507 m in the air. In this 49,000 civilians were killed, injured or disappeared and a six to seven sq km area was devastated.

The first hydrogen bomb was exploded on Bikini Island in the Pacific Ocean. The effect of the radioactive emissions was such that it reached 150 km away, affecting the crew of Lucky Dragon, a Japanese fishing boat.

On March 31, 1993 at 3.31 a.m. a disaster was averted at Narora Atomic Power Plant in Uttar Pradesh in India, when a fire broke out due to a spark in the lens beam or arc in the cable, in Unit I of the Plant generating 190 mw. The fire damaged the turbine room badly, causing damage to equipment worth Rs 25 crores and a total loss of Rs 150 crores. The disaster was averted by activating the fast activating system to shut down the reactor automatically.

Radiation pollution is caused due to the addition of more ionizing radiation to the environment. When people are exposed to more radiation than they normally experience, it comes in contact with living tissues and damages them.

5.8 Solid Waste Management

Unwanted solid waste generated through various activities in homes, society, commercial places, industries and agricultural fields are known as refuse in the cities and litter in the countryside. Depending upon the locality and season, the composition of the solid waste differs. Depending upon its origin, solid waste is grouped as:

(i) Agricultural waste.
(ii) Animal waste.
(iii) Aquatic weed.
(iv) Community waste.

Whatever the nature of the refuse, it attracts rats, flies and other pathogens and becomes their breeding ground. Secondly, its accidental or spontaneous combustion causes severe air pollution. As it rains, the water passes through the garbage, fermenting the refuse and contaminating the surface and ground water aquifer. Finally, solid waste releases toxic chemicals which are absorbed by the soil and go on to become a threat to plants and human beings.

Therefore, management of solid waste is very essential for the sustainable growth of human civilization. This can be achieved by following the principle of the three Rs – Reduction, Reuse and Recycle.

The management of waste generation can be achieved through improvement and/or modification of the present process; reusing or recycling the waste to generate energy or other materials; disposing waste by burning in incinerators or dumping into inhabitable low lands. However, dumping or burning the waste, is no longer an accepted practice, both from the health and environment point of view. Today, disposal of solid waste is an integrated waste management programme.

Solid waste can be classified as non-hazardous or Municipal Solid Waste (MSW) and hazardous solid waste. The former has been discussed in detail in the following section:

Non-hazardous or Municipal Solid Waste (MSW)

Municipal solid waste commonly known as garbage, includes everything from leftover food, agricultural products, papers, tetrapacks, plastics, clothes, metals like iron, zinc and aluminum foil, glass bottles and rags. Previously, waste management was not a big concern because the amount of garbage generated was small and the variety was also not much. But with advancement of science and technology and introduction of new packaging systems, waste management has become a matter of great concern.

Control Measures for Urban and Industrial Waste

As already pointed out, the solid waste management programme has three components:

(i) Reduction

(ii) Reuse and recycling

(iii) Disposal.

Reduction at Source: This can be done by reducing the amount of consumption, for example buying materials of actual need. The quantity of waste can also be decreased by concentration through evaporation, precipitation or decantation method. Individually, one can reduce solid waste by reducing unnecessary items and avoiding plastic carry bags. Another way, in which solid waste can be reduced is through modification of industrial processes with a view to optimizing the use of raw materials. For example, if sulphate salt is substituted by a chloride compound in zinc electroplating with a slight process modification, it can eliminate the problem of pollution being caused by cyanide.

Recycling the Waste: This is an eco-friendly method of solid waste treatment. Many substances in the garbage are recyclable. They include cardboard and paper for making paper pulp, copper and other metal salt solutions used for metal recovery and glass and plastics.

Some waste material cannot be recycled due to technical or economical problems. For example, plastics are not very easy to recycle, as they are made of different polymer resins, which have their own physico-chemical properties. Hence, separation of different plastics before recycling is necessary. Secondly, recycled plastic or paper is not 100 per cent hygienic. The recycled plastic is unfit to be used in food containers.

But the biodegradable solid waste can be mixed with soil and decomposed by aerobic bacteria to be used as compost. When the process is done with the help of earthworms it is known as vermicomposting.

Composting: It is the conversion of solid waste to manure and biogas with the help of bacteria and fungi. Composting is widely practiced in rural areas to produce manure for use as fertilizer and biogas for the purpose of household cooking. Composting is a self-heating, managed microbial system. However, large-scale composting is expensive as it needs screening of different types of waste, such as metal, glass, plastic and organic matter. Many hazardous compounds are resistant to microbial degradation due to their complex structure, toxicity and compound concentration. Microbial growth also depends on moisture, pH of organic nutrients and particle size. In this method, micro-organisms such as bacteria and fungi convert the organic waste in the presence of air and heat.

Composting can be done by both the open system (land treatment) and closed system. Although, the open system is cheaper it suffers from the following disadvantages:

(i) Rate of biodegradation depends on temperature. (summer and winter rates vary)

(ii) It may suffer from oxygen-deficiency. To increase the efficiency of an open system, air passage is necessary. In the closed system, there is an air blower through which controlled air is supplied for the maintenance of microbial activities. This increases microbial growth followed by an increase in the rate of volatility of hazardous compounds. Consequently, the internal temperature increases, resulting in balanced evaporation of water and removal of toxic substances that are volatile through an exhaust outlet, thereby producing solid manure and generation of biogas.

Vermicomposting: In this method, composting is done with the help of earthworms. Earthworms play an important role in the turnover, aeration, decomposition, fertility and cycling of soil and increase of plant nutrients, especially nitrogen and maintaining soil-structure. Earthworms ingest organic matter with a wide C/N ratio and convert it to earthworm tissue with a lower C/N ratio, thereby increasing the nitrogen content of the soil. Significant research work is now being carried out in India and abroad on the food habits of earthworms, their biomass discharges, the relationship between inputs and outputs and tolerance to different synthetic agrochemicals. Scientists at IISC, Bangalore have recently developed methods for the decomposition of coconut coir through vermiculture. Composting of organic solid waste through vermiculture should be encouraged in an agriculture-based country like India.

Fly Ash

The main discharge from a thermal power plant and other industries is fly ash. It pollutes the air, water and land. In a highly polluted atmosphere, there is the presence of a bulk of mineral particulate matter which include oxides of metals and other compounds. Small ash particles of furnace fuel that emerge from the stock system in the absence of a collector device and enter the atmosphere are commonly known as fly ash. Various components of fly ash are listed in Table 5.14. This fly ash poses a health hazard to all living beings.

Table 5.14 Composition of Fly Ash from Coal Combustion

Component (As occurs)	Percentage
Silicon (SiO_2)	15.5–63.8
Aluminium (Al_2O_3)	9.8–58.6
Potassium (K_2O)	2.8–3.0
Iron (Fe_2O_3)	2.0–26.8
Carbon (C)	0.36–36.5
Sodium (Na_2O)	0.2–0.9
Sulphur (SO_2)	0.12–24.5
Calcium (CaO)	0.12–14.3
Phosphorous (P_2O_3)	0.07–47.5
Magnesium (MgO)	0.66–4.75
Carbonate (CO_3)	0.0–2.5
Titanium (TiO_2)	0.1–2.8

Now, this fly ash is being used for brick-manufacturing and making concrete. Research has revealed that fly ash mixed with soil gives high yield for crops like paddy, carrot, ladies finger and groundnut.

CASE STUDY

Recently, thousands of people living in areas around Kolaghat Thermal Power Station in West Bengal have been reported to be suffering from silicosis, a major respiratory disease. Silicosis is caused by very fine silica particles that float in the air and get deposited in the respiratory tract during breathing. The report also revealed the presence of lead, manganese, chromium and radioactive uranium besides silica in the fly ash generated from the thermal power plant.

The United States Department of Health, Education and Welfare, Division of Air Pollution Cincinnati, Ohio, in its report, published the following wide-range composition of fly ash emission from a variety of coal combustion units.

Mushroom Cultivation

Mushrooms have been known to human civilization since ancient times, for their culinary and nutritional benefits. People used to go to the forest to collect edible mushrooms, but, sometimes the mushrooms turned out to be poisonous and had harmful effects. In India, *Guchhi* mushroom from Kashmir was a delicacy known to people before the 1950s.

In the early 1950s, Shri S. S. Jain, the first Assistant Plant Pathologist of Himachal Pradesh, while travelling through apple orchards with a view to control diseases in apples, noticed a heavy growth of mushrooms in the rotting twigs, branches of apples and other fruit trees and amidst wheat straw in the barn along with cow dung. This led him to think of using waste material to grow purely edible mushrooms in this environment which seemed to be conducive. He started his laboratory at Solan in Shimla to grow edible mushrooms of Agaricus and other local edible species in laboratory conditions. He imported the spawn from France and Japan. Jain was successful in growing the edible mushrooms on the substrate prepared from rotting apple tree twigs, branches, cow dung and wheat straw. Later, the pioneer of mushroom cultivation in India transformed his laboratory into a training centre for mushroom cultivation for the poor farmers of Himachal Pradesh and other states of India.

Today, a huge factory in Maharashtra near Talegaon, Lonavala, solely meant for mushroom cultivation is an example of the development of mushroom cultivation in India.

Mushroom cultivation is a productive way of reusing biodegradable waste. The market for mushrooms is growing continuously due to its interesting culinary, nutritional and health benefits. Mushroom is a kind of fungi with a lifecycle which is very different from other green plants.

The choice of species to be raised depends greatly on the growth media (substrate) and also on the market demand. Oyster mushrooms, which grow on a variety of substrates and Shiitake mushrooms, are in great demand. Morels and Truffles are the only two Mycorrhizal mushrooms which are commercially cultivated. However, mushroom cultivation as a business is becoming popular day-by-day for its consistent nutritional quality and established buyers.

Disposal of Solid Waste

In the modern method, solid waste is first separated into different constituents depending on their size and then ferrous scrap is removed from it by employing the use of magnets. Material that can be recycled is then sent for recycling, whereas combustible refuse is treated in an incinerator at 150 to 250°C to dry and sterilize it. Dry garbage is used as fuel. A step-by-step management of solid waste is depicted in Fig. 5.9.

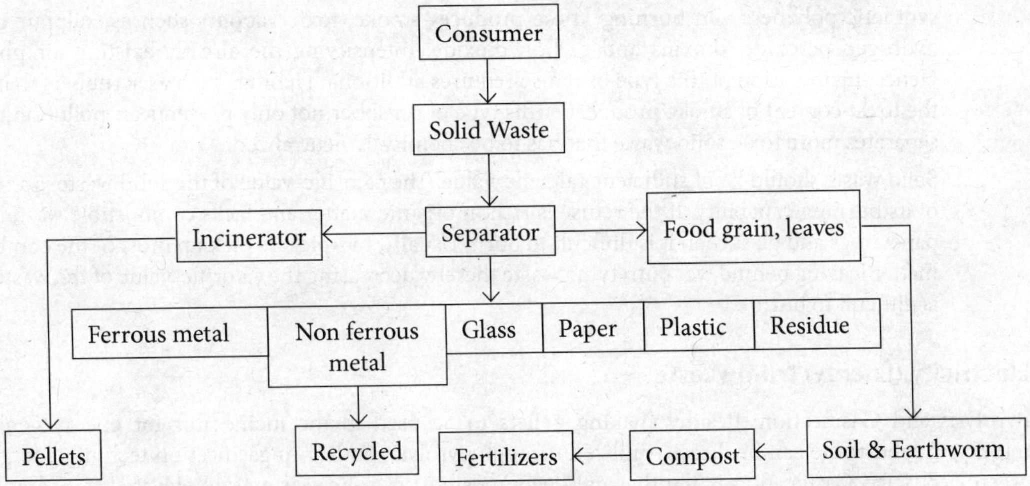

Figure 5.9 Management of Solid Waste

Dry garbage is screened to remove sand, plastic and metals. It is then passed through an air classifier. The lighter, combustible substances get separated and the heavier or the non-combustible portion is used as land fill. Rotten vegetables and their peel present in the refuse, act as a binder for the process of pelletization of the garbage. The pellets are dried to a point when they have eight to 10 per cent moisture content; they are then used in steam power plants.

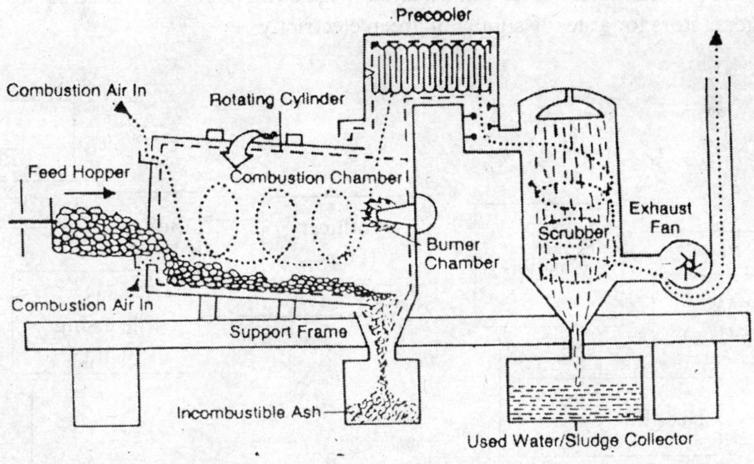

Figure 5.10 A Typical Waste Incinerator Unit.

Though the process of incineration aims to reduce the volume of solid urban waste and contamination of the environment with pollutants, it has certain limitations. A typical waste incinerator unit is shown in Fig. 5.10.

1. For incineration to be successful, the waste should not contain material that burns to generate air pollution problems. In developing and developed countries, solid waste is replete with plastic and

synthetic polymers. On burning, these produces smoke, toxic vapours such as, sulphur dioxide, hydrogen peroxide, dioxins and carbon dioxide, intensifying the already existing air pollution. Hence, incineration of this type of refuse requires additional facilities such as scrubbers that detain the toxic-content of smoke produced. This type of scrubber not only prevents air pollution but also separates more toxic solid waste that has to be chemically neutralized.

2. Solid waste should be of sufficient calorific-value. The calorific-value of the solid waste is a measure of its burning capability. If the refuse is rich in organic matter and lacks combustible waste such as paper, rags and glass then it is difficult to burn. Usually, rag-pickers pick up most of the combustible matter leaving behind wet putrefying waste thereby decreasing the calorific-value of the waste which is difficult to burn.

Electricity (Energy) from Waste

Pyrolysis and Gasification: Besides making pellets to be used in the incinerator for energy generation, municipal solid waste can be directly utilized through pyrolysis and/or in gasification technique to generate green electricity. Gasification is the thermal decomposition of solid waste at an elevated temperature in a controlled supply of oxygen, whereas in pyrolysis, the solid waste is thermally decomposed in the complete absence of oxygen. The process of gasification is shown in Fig. 5.11.

The solid waste is initially shredded so as to separate the combustible material from the incombustible material. The process is also known as fuel preparation. In either case, (pyrolysis/gasification), an initial heat supply is required to start the process. Once the temperature is reached, the processes is either self-sustaining or a minimal amount of recycled energy is required to maintain it. A large amount of a mixture of combustible gases such as methane, complex hydrocarbons, hydrogen and carbon monoxide are produced (producer gas or fuel gas) in the process. These gases can be directly used in boilers or cleaned-up and used in combustion turbines and generators for a steady supply of green electricity.

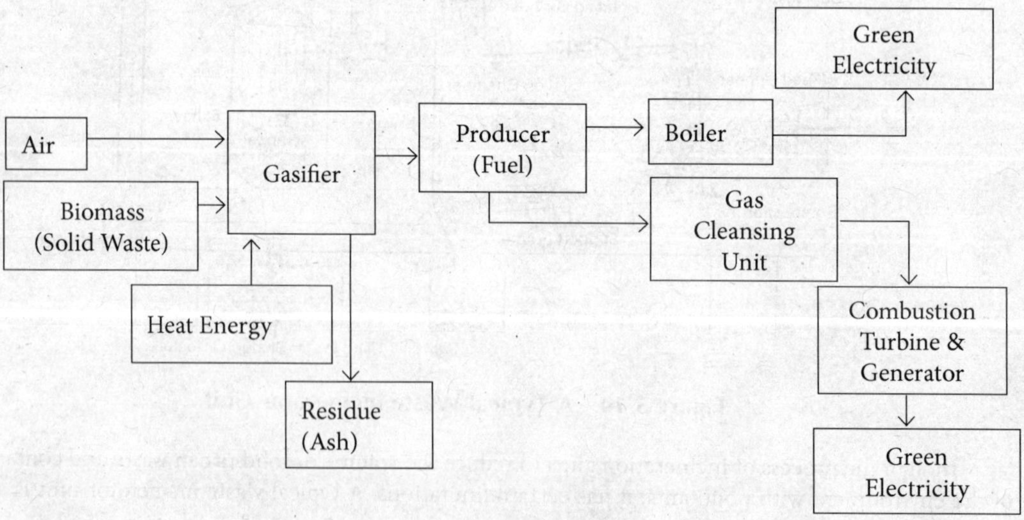

Figure 5.11 Flow Chart for Gasification Process

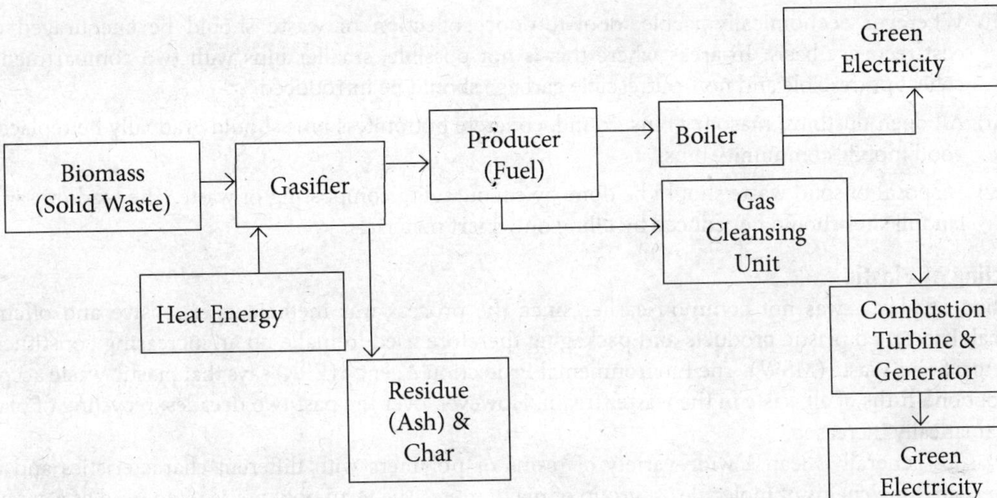

Figure 5.12 Flow Chart For Pyrolysis Process

In pyrolysis, thermal decomposition takes place in the complete absence of air. An energy-rich oil (tar) and combustible solid residue (char) are also produced along with fuel gas. In both the processes, emitted pollutants such as sulphur and heavy metals are retained in the ash and are not converted to the gas phase and therefore not released into the atmosphere. The flow chart for pyrolysis is shown in Fig. 5.12. The pyrolysis/gasification techniques are environment-friendly.

Compared to the conventional mass burn incinerators, the pyrolysis/gasification method is preferable because:

(i) Gasification/pyrolysis units can be built on a smaller scale, (with low capital) whereas mass burn incinerators always need large scale investment.

(ii) The start-up period of these methods being shorter than the incinerator method, the plant can be shut down at night or during weekends.

(iii) The pyrolysis/gasification units can be made to operate at 100 per cent or less fuel capacity, depending on the availability of solid waste. This is not usually possible in the incinerator method.

(iv) In pyrolysis/gasification method, harmful emissions are non-existent or much lower than the incinerator method.

Sanitary Landfill

Sanitary landfilling is a process of dumping solid waste in a scientifically designated land area, spreading waste in thin layers, compacting it to the smallest practicable volume and covering with soil on a daily basis. Methane-rich biogas is produced due to anaerobic decomposition of organic matter in solid waste. Garbage has a potential to generate about 150–250 cubic metres of biogas per tonne of waste depending upon its nature. An interactive workshop for the formulation of a national programme on sanitation and environmental hygiene was held at Vigyan Bhawan, New Delhi in April, 1995. Its recommendations are:

(i) Source segregation into two receptacles. Waste from households, shops, establishments, slums and squatter settlements to be collected in two different receptacles, one for putrescible and another for non-putrescible waste.

(ii) Wherever economically viable, door-to-door collection of waste should be encouraged on a cost-recovery basis. In areas where this is not possible, smaller bins with two compartments to collect putrescible and non-putrescible garbage should be introduced.

(iii) All open dustbins, masonry bins, round, concrete bottomless bins should gradually be replaced by good mobile community bins.

(iv) Disposal of solid waste should be done by encouraging composting of waste. The load on sanitary landfill sites should be reduced by filling only inert materials.

Recycling of Plastic

Recycling of plastic was not common earlier since the process was inefficient, expensive and often not practical. Discarded plastic products and packaging therefore used to make up an increasing constituent of Municipal Solid Waste (MSW). The Environmental Protection Agency (EPA) says that plastic waste accounts for about one-fifths of all waste in the waste stream. However, over the past two decades, recycling of plastics has dramatically increased.

Plastics generally mean a wide variety of resins or polymers with different characteristics and uses. Polymers are long chains of molecules, a group of many units. The term polymer is often used as a synonym for plastic. Many other types of molecules – biological and inorganic – are also polymeric. All polymers are not plastics, but all plastics are polymers.

Pure polymers are rarely useful and are most often modified or compounded with additives (including colours) to form useful materials. This compounded product is generally termed plastic. Polymers can be classified in many ways, based on how they are developed and how they perform. Thermoplastic polymers and thermoset polymers are the two basic types of polymers classified on the basis of their reaction to heat.

(i) **Thermoplastic polymers**: These can be heated and formed, repeatedly. The shapes of polymer molecules are generally linear or slightly branched. This means that the molecules can flow under pressure when heated above their melting point.

(ii) **Thermoset polymers**: These undergo a chemical change when they are heated, creating a three-dimensional network. After they are heated and formed, these molecules cannot be re-formed on heating again.

Hence, thermoplastics are much easier to recycle.

Polyvinyl Chloride or PVC is used for packaging and other short-life consumer products, furnishings and long-life goods, mostly construction material such as window frames and pipes. A major problem in recycling PVC is the high chlorine content of raw PVC – 56 per cent of the polymer's weight and the high level of hazardous additives added to the polymer to achieve the desired material quality. Additives may comprise up to 60 per cent of a PVC product's weight. Of all plastics, PVC uses the highest proportion of additives so these products cause serious waste problems, especially when incinerated. PVC bottles are almost similar to PET bottles. But one stray PVC bottle in a melt of 10,000 PET bottles can ruin the entire batch. It is therefore understandable why purchasers of recycled plastic want to make sure that the plastic is sorted out properly. In the case of plastic recycling there is often a need to identify which particular plastic material has been used for a given product. Most consumers recognize the types of plastics by the numerical coding system created by the Society of the Plastics Industry in the late 1980s. There are six different types of plastic resins that are commonly used to package household products. The identification codes are listed in Table 5.15.

Table 5.15 Classification of Plastics for Coding

Symbol	Type	Use	Recycling
1 PETE	PETE Polyethylene Terephthalate (PET)	Soda & water containers, some waterproof packaging	Recycling PET is similar to the polyethylenes (PE). Recycled products are used as polyester carpets, fiber filling for pillows, quilts and jackets, sheets or ribbon for VCR and audio cassettes and recycled bottles.
2 HDPE	HDPE High-Density Polyethylene	Milk, detergent & oil bottles, Toys and plastic bags	Recycled HDPE are plastic pipes, lumber, flower pots, trash cans, or formed back into non-food application bottles.
3 PVC	Vinyl/Polyvinyl Chloride (PVC)	Food wrap, vegetable oil bottles, blister packages	Generally not recycled.
4 LDPE	LDPE Low-Density Polyethylene	Plastic bags. Shrink wrap, garment bags	It is chemically similar to HDPE but less dense and more flexible. Recycling LDPE is very similar to HDPE, but less recycled. Recycled LDPE is used as plastic trash bags and grocery sacks, plastic tubing, agricultural film, and plastic lumber.
5 PP	PP Polypropylene	Refrigerated containers, some bags, most bottle tops, some carpets, some food wrap	Generally not recycled.
6 PS	PS Polystyrene	Throwaway utensils, meat packing, protective packing	Generally not recycled.
7 OTHER	OTHER Usually layered or mixed plastic		No recycling potential – must be land filled.

Uncoded Plastics

Plastic consumer goods not identified by code numbers are not usually collected. Plastic pipes, toys, computer keyboards and a multitude of other products simply do not fit into the numbering system that identifies plastics used in consumer containers. There are actually thousands of different varieties of plastic resins or mixtures of resins which are developed to suit the needs of particular products. There is limited recycling of some of these specific plastic products.

Glass, paper and metal cans can be used and recycled over and over again. With plastic recycling however, there is usually only a single-reuse. A recent development has been the bottle-to-bottle recycling of 'regenerated' pop bottles. Though, a 100 per cent recycled bottle is now technologically possible, there are serious economic questions and the environmental impact of the regeneration process in terms of energy used and hazardous by-products.

At present, only about 3.5 per cent of all plastic generated is recycled compared to 34 per cent of paper, 22 per cent of glass and 30 per cent of metals. Equipment to sort plastic is being developed but at present most recyclers are still sorting plastic by hand which is an expensive and time-consuming process. In plastic used as landfills, the decomposition process may take from 10 to 30 years. Recycling has therefore, become a reasonable solution to the landfill problem.

Plastic recycling is a five-step process: supply, collection, sorting and separating, reprocessing including the melting of used plastic and finally manufacturing of the melted plastic into new products.

There are four types of recycling processes that usually occur: primary, secondary, tertiary, and quaternary.

(i) The primary recycling process is the recycling of material and products that contain similar features as the original product. This process is only feasible with semi-clean industrial scrap plastic and is not widely used.

(ii) Secondary recycling allows for a higher mixture of combination levels in plastic. When the secondary process of recycling is used, it creates products like fenceposts and any products having low mechanical properties that can be used as a substitute for wood, concrete and metal.

(iii) Tertiary recycling process is widely used these days as it can adapt to the high level of waste contamination. The aim of the process includes producing basic chemicals and fuel from plastic.

(iv) Quaternary recycling process uses the energy from plastic by burning. This process is the most common and widely used in recycling because of the high heat content of most plastics. Most incinerators used in the process can reach temperatures as high as 900 to 1,000°Celsius. The use of incineration in the quaternary process is most beneficial because through the high temperature heating process the incoming waste is reduced by 80 per cent in weight and 90 per cent in volume. The material left over from this process is then placed in landfills.

The last two polymer energy systems offer a viable, economical and environmentally-responsible alternative to other methods of recycling plastic waste. These methods use pyrolysis and gasification methods to efficiently convert plastic waste into energy such as crude oil and green electricity.

The largest single use for plastics is packaging. The EPA says that in the early 1990s, about 80 per cent of all municipal solid waste was sent to landfills, 10 per cent was incinerated and 10 per cent was recycled. While more and more plastic is being recycled, the EPA estimates that plastics make up about 20 per cent of the solid waste that is landfilled.

The more serious problem with plastic waste is additives such as colourants, stabilizers and plasticizers, which include toxic components such as lead and cadmium up to the extent of two and 28 per cent, respectively.

A new technology for plastic recycling and powder production has been developed by the Department of Chemical Engineering, Northwestern University. It eliminates sorting by type or by colour. This technology, called Solid State Shear Pulverization (S3P), is a continuous one-step process for recycling unsorted pre- or post-consumer plastic waste. Unlike conventional recycling, S3P produces uniform powders that can be used to make a variety of high-quality products.

S3P subjects polymers to high shear and high pressure while rapidly removing frictional heat from the process to prevent melting. S3P can convert multi-coloured, unsorted (commingled) waste, industrial plastic scrap and virgin resins to a uniform, light-coloured, partially reactive powder of controlled particle size and particle size distribution. These powders are suitable for direct melt conversion by all existing plastic processing techniques. This energy-efficient process pulverizes plastic into powder of particle sizes ranging from coarse (10 mesh/2,000 microns) to very fine (635 mesh/20 microns). The resulting powder can be used in a variety of consumer goods and special products, such as automotive appliances to business equipment and furnishings, except for food applications.

Hazardous Waste Management

Any solid waste, other than radioactive waste, which by reasons of physical and/or chemical or reactive or toxic, explosive, corrosive and other characteristics, causes damage or is likely to cause danger to health or the environment, whether alone or by coming in contact with other waste and the environment is called hazardous waste.

In India, Hazardous Waste (Management and Handling) Rule, 1989 and its amendments regulate the various aspects of hazardous wastes as follows:

Schedule I – List of processes with waste stream

Schedule II – Concentration of hazardous component

The National Environmental Engineering Research Institute, NEERI, Nagpur has prescribed different aspects of hazardous waste management. The basic issues of hazardous waste management are as follows:

(i) Identification, categorization and qualification of hazardous waste.
(ii) Development of infrastructure and uniform procedures for hazardous waste analysis (industry-specific parameters).
(iii) Implementation of a combination of technological, legislative and economic instruments to promote minimization of hazardous waste.
(iv) Development and operation of common facilities for the handling and disposal of hazardous waste.
(v) Identification and investigation of abandoned contaminated sites.

Most of the industries generate hazardous waste or by-products. An indicative list of hazardous waste generated by industries is given in Table 5.16.

Table 5.16 Hazardous Wastes Generated By Industries

Industry	Hazardous Waste
Iron and steel	Acid tar, oily sludge, complex organic sludge, metal waste, flue dust, zinc dross etc.
Aluminum Smelter	Spent Pot line (Fluent and cyanide)
Fertilizer	Gypsum, spent catalyst
Sulphuric acid	Sulphur muck, spent catalyst
Sodium Dichromide	Leached solid (Cr^{+6})
Tannery	Chromium sludge
Solvent Trade	Spent solvent
Chloro alkali	Mercury contaminated sludge
Electroplating	Cyanide, heavy metal
Pharmaceuticals	Organic complex
Pesticides	Toxic Organic complex
	Date expired products

Industry	Hazardous Waste
Paint & Waste	Complex chemical sludge
Waste water treatment	Water treatment, sludge

Hazardous waste management depends on avoidance of hazardous waste generation and proper treatment of hazardous waste to minimize its effects.

The steps to manage hazardous waste are schematically represented in Fig. 5.13.

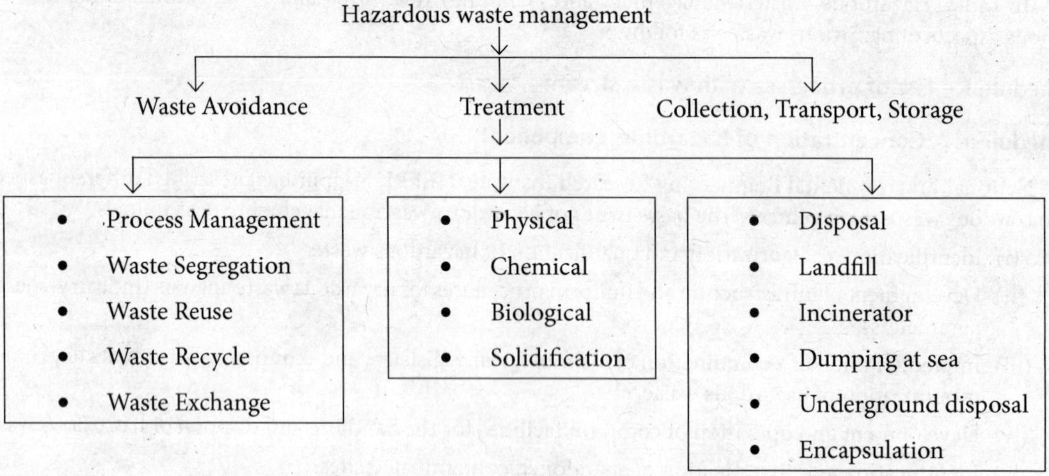

Figure 5.13 Various Methods of Hazardous Waste Management

Hazardous Waste Avoidance

Optimization of the consumption of raw materials or substitution of less hazardous raw materials avoids or minimizes the generation of hazardous waste, thereby compacting or reducing the volume of and helping in the segregation of hazardous waste from non-hazardous waste.

Hazardous waste should also be reused on-site or off-site to reduce hazardous waste generation.

With proper collection, transportation and storage, hazardous waste may be used as a raw material for other industries and such waste exchange may reduce generation of hazardous waste.

Treatment of Hazardous Waste

The type of treatment of hazardous waste depends on:

(i) characteristics of hazardous waste,

(ii) output desired and

(iii) economy and energy required for the treatment process.

Depending on the above factors, one of the following treatment options are selected:

(i) **Physical**: Phase separation, sludge-drying, compaction and lagooning.

(ii) **Chemical**: Chemical disintegration/modification of hazardous waste, oxidation/reduction, neutralization of acid/alkali, dilution of heavy metals precipitates/concentrates.

(iii) **Biological**: Microbial destruction of hazardous components.

(iv) **Solidification**: Conversion of hazardous waste into insoluble material (pre-treatment for landfill) mobilizes hazardous constituents (cement, lime, thermoplastic, organic polymer-based glass fixation). Various methods of hazardous waste management are shown in Fig. 5.13.

Collection, Transportation and Storage of Hazardous Waste

This is a very critical link in hazardous waste management. The collection network needs to be very effective and safe while transportation should be as per the rules of carriage of dangerous goods stipulated in the Motor Vehicles Act, 1988 with proper labelling and care. Finally, storage should be safe with a compatible container and provision of spill-containment.

Hazardous Waste Disposal

Disposal of hazardous waste, particularly engineering landfill and incineration needs special care as the effects of hazardous waste may harm us through various environmental pathways.

Engineering landfill needs to address three key aspects: waste characteristics, design and operation and site-selection. Waste should be compatible with the land selected for least propagation of its harmful effects and efficient design and operation. Leached collection and treatment or use of liners to avoid harmful effects through soil is also important in selecting landfill sites for hazardous waste. Hydrogeological and environmental laws, water use, population, transportation routes also play important roles in site-selection.

Dumping at sea or the underground disposal method can be used for hazardous waste disposal. Complex hazardous waste can be incinerated. In this, hazardous components are destroyed through high temperature oxidation without much air pollution and the incinerated remains of the hazardous waste (slag and ash) are disposed-off through landfill.

Treatment and disposal options for different hazardous waste are given in Table 5.17.

Table 5.17 Various Waste Disposal Methods and Their Merits/Demerits

Method of Disposal	Demerits	Merits
• Land Filling • Open land dumping • Burning/Incineration • Bioconversion into organic manure	• Restricted site availability cannot last longer • Contaminated water sources • Anaerobic gas production explosions • Environment pollution • Costly large area occupied • Increasing maintenance cost of open dumps • Ugly look to the cities & surroundings • Smoke and fire • Shifting of locations due to space becoming full • Smoke and gaseous contamination of environment • Temperature rise • Diesel costs higher • Capacity for incineration is a constraint	• Easy operation • Land gets leveled • Lower initial costs • Ease for ragpickers • Non-skilled job • Incineration is standard hygienic operation Burning is an easy operation • Highly useful product for land improvement for crop

Method of Disposal	Demerits	Merits
	• Technological constraints • Higher capital costs • Requires government support production • Value addition to waste resource • Sustainable approach	

CASE STUDY

The worlds most notorious and infamous soil pollution took place in Love Canal, New York in 1978. Hooker Chemical and Plastic Corporation had disposed of about 22,000 tonnes of mixed toxic chemical waste into the Love Canal as landfill in 1942. Due to heavy rainfall and snowfall in the area in 1975–1976, the toxic chemicals leached out of the storage drums into the soil contaminating the soil and surface water bodies. This resulted in residents suffering from various cancers, respiratory, nervous and kidney diseases. About 56 per cent of the babies born after the incident had birth defects. In April 1978, the Commissioner, Department of Health, New York declared Love Canal as a threat to human life and recommended temporary evacuation of pregnant women and young children from the first two rings of houses around the site. This was followed by President Jimmy Carter's order for total evacuation from the first two rings of houses around the site. The Love Canal environmental disaster raised public awareness regarding the importance of proper disposal and management of toxic chemicals and hazardous waste.

Biomedical Waste

According to Biomedical Waste (Management and Handling) Rules, 1998, amended in 2000, biomedical waste means any waste which is generated during the diagnosis, treatment or immunization of human beings or animals or in research work pertaining to the product or testing of biologicals, including categories mentioned in Table 5.18A.

The waste generated from healthcare units (hospitals, clinics and nursing homes) is more dangerous (hazardous) than others. The amount of waste produced in any healthcare unit depends more on in-house patients than outpatients. About one and a half kilograms waste is produced per day per bed, of which about 47 per cent is only biomedical waste. The human and animal anatomical waste along with items saturated or dripping with blood and body fluids, discarded medical equipment, soiled cotton, plasters, dressing, surgical and autopsy-waste is collectively known as biomedical waste. Being contaminated with disease-causing pathogens like Eschrichia, Salmonella, Vibria, Hepatitis Shigela, among others, biomedical waste provides a fertile breeding ground for bacteria, viruses and other micro-organisms to multiply. Hence, biomedical waste requires a proper waste treatment process for its safe disposal. The list of diseases caused due to improper disposal and treatment of hospital waste is endless, but a majority of them are deadly, such as AIDS, viral hepatitis, tuberculosis, bronchitis, gastroenteritis and other skin and eye-related diseases.

The Common Biomedical Waste Treatment Facility (CBWTF) which has been accepted in many developed countries is now legally introduced in India through Biomedical Waste (Management and Handling) Rules, 1998. The second amendment to the rules in June 2000 had further made CBWTF, a responsibility of the local authority, for providing a suitable site for its set up. However, for the establishment of a CBWTF unit, certain guidelines have to be followed for choosing the right technology, development of a CBWTF area and proper designing of a transportation system.

Land: A CBWTF unit has to be established at a place far away from residential and sensitive areas but as near as possible to its area of operation. In any area, a CBWTF unit is allowed for up to 10,000 beds for all healthcare units situated within a radius of 150 km. For an area where 10,000 beds are not available within an area of 150 km, another CBWTF unit may be allowed outside the said 150 km. For efficient performance of a CBWTF unit, a minimum of one acre of land is required. The location for establishment of a CBWTF unit is decided in consultation with the State Pollution Control Board (SPCB)/Pollution Control Committee (PCC). Enough space is required for the infrastructure of the CBWTF unit which includes:

(i) Treatment equipment room.

(ii) Main waste storage room.

(iii) Treated waste storage room.

(iv) Administrative room.

(v) Parking, loading and unloading area.

(vi) Washing room.

Collection and Transportation: All biomedical waste has to be segregated at source. It is the duty of the healthcare units to hand over segregated biomedical waste to the CBWTF operator as per Biomedical Waste (Management and Handling) Rules, 1988 in different colour-coded bags. The collected coloured bags are kept in similar-coloured containers with lids, in a fully covered vehicle and transported to the CBWTF unit for treatment followed by safe disposal. The total time limit from generation of biomedical waste to its treatment must not exceed 48 hours. Table 5.18A describes the classification of biomedical waste, its recommended treatment and disposal.

Table 5.18A Biomedical Waste and Recommended Treatment Options

Type	Waste description	Treatment & disposal
Human anatomical wastes and body fluids	Wastes consisting of anatomical human tissues, organs, waste body parts, body fluids, blood and blood products and items saturated or dripping with blood fluids removed during/after treatment, surgery or autopsy or other medical procedures.	Incineration
Animal waste	Wastes consisting of animal tissues, organs, body parts, carcasses, bleeding, fluid blood and blood products, items contaminated with blood and fluids, wastes from surgery treatment and autopsy and wastes of experimental animals used in research and waste generated by veterinary hospitals, colleges, animal houses and livestock farms.	Incineration
Microbiological wastes	Wastes from laboratory culture stocks or specimens of micro-organisms, live or attenuated vaccines, human and animal cell culture used in research and infectious agents from research and industrial laboratories, waste from production of biological dishes and devices used for transfer of cultures.	Incineration

Type	Waste description	Treatment & disposal
Waste sharps	Wastes consisting of sharps, such as needles, scalpels, blades etc. which include both used and unused sharps.	Chemical disinfection/ autoclaving followed by shredding.
Highly infectious wastes	Wastes containing highly infectious living and non living pathogens.	Incineration
Isolated wastes	Biological wastes from discarded material contaminated with blood exertion exudates or secretion from human and animals isolated due to communicable disease.	Incineration
Discarded medicines	Waste comprising expired, contaminated and discarded medicines.	To be sent back to the manufacturer for safe disposal.
Discarded glassware	Wastes generated from glass wares and glass equipment used.	Chemical disinfection/ autoclaving followed by shredding.
Soiled wastes	Wastes from soiled cotton, dressing, beddings etc.	Incineration
Disposable (PVC/plastics, cardboard and thermocol)	Wastes generated from disposable items other than waste sharps	Chemical disinfection/ autoclaving followed by shredding.
Biotechnology wastes	Wastes generated from activities involving genetically engineered organisms or products and their cultures not declared to be safe.	Incineration
Incineration wastes	Ash generation from Incineration of any biomedical waste.	Secured landfill as per hazardous waste (Management and Handling Rules, 1989)
Liquid wastes	Wastes generated from laboratory and washing, cleaning, kitchen house keeping and disinfection of the premises.	Effluent

All biomedical waste shall be segregated into colour-coded containers/bags (Table: 5.18B) at the point of generation, prior to its storage, transportation, treatment or disposal. The containers are also to be labelled accordingly. The untreated, segregated biomedical waste is transported to the CBWTF unit only in authorized motor vehicles and treatment is done as per Table 5.18A.

Table 5.18B Colour-Coding of Biomedical Waste and their Safe Disposal

Category no.	Colour and type of container	Type of biomedical waste	Nature of treatment
1	Yellow plastic bag	Human anatomical waste, microbial & biotechnology waste, soiled waste.	Incineration & deep burial

Category no.	Colour and type of container	Type of biomedical waste	Nature of treatment
2	Red plastic bag	Microbiology & biotechnology waste, soiled waste & solid waste	Autoclaving, microwaving, chemical treating
3	Blue/White translucent plastic bag/ puncture proof containers.	Sharp waste, solid waste	Autoclaving, microwaving, chemical treatment & destruction/ shredding
4	Black plastic bag	Discarded medicines, **cytotoxic** drugs, incineration ash, chemical waste	Disposal in secured land fill

Treatment and Equipment

As per Biomedical Waste (Management and Handling) Rules, 1988, about 90 per cent of the total waste in a healthcare unit is treated on non-burn technologies. It is mandatory to impart incineration/deep burial to anatomical and other types of waste coming under categories 1 and 2. If a secure landfill is not available, category 4 may also be incinerated. A common CBWTF shall have the following treatment facilities:

Incineration: It provides a complete, controlled combustion for waste. All harmful micro-organisms present in the waste are destroyed/denatured under high temperature. The design and construction of a biomedical waste incinerator should meet the following operating and emission standards:

(a) **Operating Standard**

Combustion Efficiency (CE) shall be at least 99.0 per cent, which is computed as:

$$\text{C.E.} = \frac{\%\ CO_2 \times 100}{\%\ CO_2 \times\ \%\ CO}$$

The temperature of the primary chamber shall be 800+50°C of the secondary chamber. Gas resistance time shall be at least one second at 1,050+50°C with a minimum of 3 per cent oxygen in the stack gas.

(b) **Emission Standards**

Parameter concentration mg/m3 at (12 per cent CO_2 correction.)

(i) Particulate matter 150.

(ii) Nitrogen oxides 450.

(iii) HCl 50

(iv) Minimum stack height shall be 30 m above the ground.

(v) Volatile organic compound in ash shall not be more than 0.01 per cent.

Autoclaving: It is a low-heat thermal process. Here, steam is used in direct contact with waste for sufficient duration to disinfect the waste.

Microwaving: Here, microbial inactivation of the waste occurs by the thermal effect of electromagnetic radiation spectrum of a frequency range of 300–300,000 MHz.

Hydroclaving: This method is similar to autoclaving but by indirect contact of steam with the waste. Indirect heating is conducted in hydroclaving by passing steam in the outer jacket. During the process the waste is also continuously tumbled.

Shredder: The biomedical waste is deshaped (cut) into smaller pieces for prevention of reuse and also as an identifier of disinfected waste.

Sharp Pit Encapsulation: It is used for dumping of treated sharps.

Effluent Treatment Plant: A suitable effluent treatment plant is installed to ensure that liquid effluent, generated during the process of washing containers and vehicles are disposed-off only after treatment. The treated effluent shall comply with the stipulated regulatory equipment.

Final Disposal: Properly treated biomedical waste is disposed-of in the following ways:

Incineration ash – secured landfill

Treated solid waste – municipal landfill

Plastic waste after the disinfection and shredding – recycling or municipal landfill

Sharps after disinfection – (if encapsulated) – municipal landfill

Treated waste water – sewer/drain or recycling

Oil and grease – incineration

E-waste

Used batteries, junk products with electrical cords, mobile phones, scanners, printers, calculators, different audio-video equipment, ovens, washing machines and electronic toys are known as electronic waste or e-waste. The disposal of this e-waste is becoming a global problem. Statistics show that the amount of annual e-waste in India is 1, 46,180 tonnes. For example, the average life span of a computer is about seven years however in India a computer is used for more than seven years. As a result, a large number of outdated computers and their parts are lying with individuals and corporations waiting to end up in landfills. However, the practice is harmful to the environment as this e-waste is composed of many other different substances along with metals such as lead, beryllium, mercury, cadmium, brominated flame-retardants and dioxin (a carcinogen and hormone disrupter) CFCs, which are hazardous to human life. According to the existing pollution control laws, it is the responsibility of the registered manufacturers to collect, recycle or destroy their hazardous waste. But there are no separate laws formulated regarding e-waste till date. As a result, most of the e-waste is making its way into the environment through improper disposal. However recently some of the State Pollution Control Boards have started working on this problem. The Karnataka State Pollution Control Board has instructed the IT companies of Bangalore to limit their e-waste within their premises. This is an important step towards e-waste management. The next step in this direction is to chalk out a suitable programme for proper disposal or recycling of e-waste.

5.9 Pollution Prevention

The first step towards pollution prevention is the reduction in waste generation. This can be achieved by enhancing the efficiency of a process which will increase production and generate the least possible waste. Efficiency of a process can be improved through process modification, better use of raw material, better management policy and overall cultural enhancements of society.

Pollution prevention as the name suggests, is associated only with reduction in waste generation; it does not deal with any method such as recycling of waste or diversification of the waste generated from processes such as volume reduction (incineration) for dumping or treatment of any kind (for reduction of toxicity of solid, liquid or gaseous waste) to comply with the Pollution Control Board's regulation.

5.9.1 Role of Individual in Pollution Prevention

Natural resources and energy help us to survive. Human need and greed are causing pollution and the fast depletion of natural resources. Therefore, each of us is individually responsible for the quality of the environment we live in. Our personal actions can either improve or worsen the quality of the environment. Instead of complaining about the deteriorating environmental situation and inadequate action on the part of the government we need to adopt habits in our daily life that are environment-friendly and save water, energy, soil, habitats and biodiversity. These include saving water in our bathroom, fuel in our kitchen, electricity in the house and fuel on the road and everything else that ensures prevention of pollution by changing the way we use the resources available to us.

Such actions not only improve the quality of the environment we live in they also provide immense economic benefits in terms of the environment.

Conservation of natural resources:

(i) Reduces soil erosion.
(ii) Generates less solid waste and landfills.
(iii) Ensures cleaner air and water.
(iv) Helps in the conservation of habitat and biodiversity.

The economic benefits include:

(i) Reduced water bill.
(ii) Reduced electricity bill.
(iii) Reduced conveyance expenses.
(iv) Reduced healthcare expenses.
(v) Increased property value.

What an individual can do and cannot do to prevent pollution is very important to note, preach and practice. Some do's and don'ts listed below are only illustrative and indicative not exhaustive and need very little elaboration.

(i) Plant trees and stop cutting them.
(ii) Conserve, reuse and recycle paper, packages and other wood products.
(iii) Don't disturb nature and be kind to it in all its forms, such as wildlife, animals and birds.
(iv) Advocate and participate in environment-friendly activities.
(v) Reduce the use of toxic pesticides and fertilizers.
(vi) Save water by consuming only as much water as is necessary.
(vii) Reuse water for toilet-cleaning and gardening.
(viii) Plug water leakages and avoid wastage.
(ix) Don't pollute sources of water or water bodies by throwing waste into them.
(x) Save electricity by switching off lights, fans, ACs and other electrical appliances when not required.
(xi) Use energy-efficient appliances only.
(xii) Use pressure cookers and follow other general tips for saving energy in the kitchen and in the refrigerator.
(xiii) Make do with natural air and light as far as possible.
(xiv) Try to use public transport, car pools and cycles instead of your own automobiles.
(xv) Own and maintain energy-efficient and pollution-free vehicles.

The natural processes with the help of solar energy renew the top soil, water, air, forests, grasslands and wildlife on which all forms of life depend. But we should not overuse these potentially renewable resources faster than they can be renewed. Natural processes also dilute, recycle and decompose some of our waste. They provide flood control and soil erosion control at no extra cost. We must therefore learn to value these resources and use them sustainably by preventing pollution individually and collectively.

5.10 Disaster Management

Natural phenomena such as earthquakes, cyclones, volcanic eruptions, tsunamis, wildfires, floods, landslides, droughts and accidents that occur due to human recklessness which cause grievous human casualties, economic and social loss and damage to the environment may be called disasters. One such recent disaster is the tsunami in South Asia on December 26, 2004 which killed more than 2, 00,000 people.
Some of the reasons for a disaster are:

(i) Traditional disaster threats such as natural phenomena and accidents caused due to human recklessness.

(ii) Wars and social violence.

(iii) Hazardous materials and substances.

(iv) Atomic and nuclear sources.

Hundreds of natural and man-made calamities occur around the world, year after year. While not all the calamities can be predicted or prevented, a state of preparedness and ability to respond quickly can mitigate the loss of life and property and reduce human suffering. Prevention, mitigation, preparedness and relief are the essence of disaster management. Earlier when a disaster struck, people relied on the government for relief and rescue. However now the approach has shifted to disaster management and risk-reduction based on the involvement and participation of the community. This ensures better safety to life and property.

5.10.1 Floods

Damage due to flooding tends to increase with increasing development in river basins whereas population pressure in flood-prone areas is often high. Full protection through large scale embankments of rivers is often not possible because of prohibitive costs or may not even be desirable because of environmental reasons. The majority of the Indian population lives in villages and depends on agriculture for their livelihood. Floods are one of the major natural disasters in India which result in loss of human life, cattle and agricultural crops every year. Measures for flood mitigation have been taken since 1950. As against the total area of 40 million hectares that is prone to floods, an area of about 15 million hectares has been protected by the construction of embankments. A number of dams and barrages have been constructed. The state governments have been assisted to take up mitigation programmes such as the construction of raised platforms. However, floods continue to be a menace mainly because of the huge quantum of silt being carried by the rivers emanating from the Himalayas. This silt-deposition in many rivers has raised the level of river beds making it higher than the adjoining countryside. Embankments have also given rise to problems of drainage with heavy rainfall leading to water-logging in areas outside the embankment.

Flood forecasting and early warnings are instrumental in saving lives and property. It also substantially contributes to improved flood management through more adequate reservoir operation. In order to respond effectively to floods a national disaster, risk management programmes have been initiated in all the flood-prone states.

Disaster management plans at the state, district, block/*taluka* and village levels are being planned and implemented. Awareness generation campaigns to sensitize all concerned on the need for flood-preparedness and mitigation and training of elected representatives and officials in flood disaster management are part of the National Disaster Risk Management Programme. Bihar, Orissa, West Bengal, Assam and Uttar Pradesh are among the 17 multi-hazard prone states; it is in these states that the programme is being implemented in collaboration with the United Nations Development Programme (UNDP), the United States Agency for International Development (USAID) and the European Commission (EC).

5.10.2 Droughts

Heat waves, forest fires and droughts also cause disasters and human suffering all over the world. In India, due to the erratic monsoon, both low and medium rainfall regions are vulnerable to periodic drought. About 68 per cent of the total area in India is drought-prone. Our experience has been that almost every third year there is a drought. However, in some states, there may be successive drought years which only go on to increase the vulnerability of the population in these areas. Heat waves, non-availability of drinking water, scarcity of basic human requirements, poor crop yield and damage of crops result in great human suffering. However, local communities have devised indigenous safety mechanisms and drought-oriented farming methods in many drought-prone parts of the country. From past experience of managing droughts, particularly the severe drought of 1987, the government has launched a number of programmes to mitigate the impact of droughts in the long run. These programmes include the Drought Prone Area Programme (DPAP), Desert Development Programme (DPP), National Watershed Development Project for Rainfed Areas (NWDPRA), Watershed Development Programme for Shifting Cultivation (WDPSC), Integrated Water Development Project (IWDP) and Integrated Afforestation and Eco Development Project Scheme (IAEPS).

5.10.3 Earthquakes

The devastating earthquakes on January 26, 2001 in the Bhuj area of Gujarat and the earlier one in the Latur district of Maharashtra on September 30, 1993 drew our attention to the large-scale destruction that an earthquake can cause. An earthquake, a sudden shaking or trembling of the earth due to release of energy from the core of the inside results in the physical damage of buildings and structures, heavy loss of life and property in densely-populated areas, heavy disruption of public health, transport water, electricity, communication, and medical services.

On December 26, 2003, an earthquake below the city of Bam in south-east Iran highlighted yet again the lethal connection between poor building quality and the number of victims. The earthquake killed as many as 40,000 people. The earthquake that hit Liaoning in the province of Manchuria in 1975 could be predicted by a series of tremors that preceded it and people were evacuated to safety. But the earthquake that hit Tang Shan in eastern China in 1976 came without warning and killed 2, 40,000 people.

In the seismically-active zones we are still unable to forecast or prevent earthquakes. We may however protect buildings against destruction though often at a high-cost. The main aim should therefore be to focus on the task of decreasing the level of risk and diminish the secondary consequences of the initial catastrophe. Examples of the most common chains of disastrous events triggered by earthquakes are as follows:

Earthquake – landslide – flood.

Earthquake – fire – conflagration.

Earthquake – contamination of potable water – disease epidemics.

Earthquake – tsunami – floods on distant coasts.

Physical planning must take into account all potential chains of events and must introduce appropriate measures to stop or limit their development.

Earthquake Risk Mitigation

In India, a comprehensive programme has been taken up for earthquake risk mitigation. Although, the Bureau of Indian Standards (BIS) has laid down standards for construction in the seismic zones these are not being followed. The building construction in urban and suburban areas is regulated by the Town and Country Planning Acts and Building Regulations. In many cases the building regulations do not incorporate BIS codes. Moreover the lack of knowledge among the architects and engineers regarding seismically-safe construction as well as lack of awareness among the population has resulted in non-compliance of BIS standards. In rural areas the bulk of the housing is non-engineered construction. The mode of construction in the rural areas has also changed from mud and thatch to brick and concrete thereby increasing its vulnerability. Population explosion has led to settlements in vulnerable areas close to river beds which are prone to liquefactions.

To address these issues the government has constituted the National Core Group for Earthquake Risk Mitigation consisting of experts in earthquake engineering and administrators. This core group has been assigned the responsibility of drawing up a strategy and plan of action for mitigating the impact of earthquakes. The salient points of the plan are as follows:

(i) Most casualties during earthquakes are caused by the collapse of structures. Thus, structural mitigation measures are the key to making a significant impact towards earthquake safety in our country for which the role of engineers and architects is most important. In view of this, the elements of earthquake engineering are being integrated into the undergraduate engineering and architecture courses.

(ii) Hospital-preparedness is crucial to any disaster response system. So each hospital should have an emergency preparedness plan to deal with mass casualty incidents. Hospital administrators and doctors are to be trained for the emergency.

(iii) The Planning Commission has already given a clearance in principle, to an earthquake mitigation project of an estimated cost of Rs 1,132 crore. The programme includes detailed evaluation for retrofitting of lifeline buildings (such as hospitals, schools, water and power supply units, telecommunication buildings, airport and its control towers, railway stations, bus stands, administrative buildings) in states of seismic zones IV and V and training of masons in earthquake-resistant constructions. Besides, assistance will be given to the state governments to put an appropriate technological regime in place. The Ministry of Rural Development under *Indira Awas Yojna (IAY)* and *Sampurna Grameen Rozgar Yojna (SGRY)* are providing small but compact houses to the vulnerable population.

5.10.4 Cyclones

Cyclones are violent storms caused by winds with a velocity of more than 25 kmph around a low pressure centre which sweep across the water and land surface with a velocity of 50 kmph. Cyclones take time to build hence satellite imagery can track their formation and movement. Very accurate predictions may not be possible but the Indian Meteorological Department issues warnings a few hours in advance. This is disseminated through SAT.com systems, radio, TV, telephone, fax, press and public announcement systems.

The east and south coasts of India which include West Bengal, Orissa, Andhra Pradesh, Tamil Nadu and Andaman and Nicobar Islands frequently experience cyclones. The supercyclone in Orissa on October 29, 1999 was the most devastating in the recent past. Cyclones cause physical damage to building structures, trees, transport, electricity supplies and communication networks. They also cause flooding and break down the public health system resulting in loss of human life and property.

Cyclone Mitigation

A project on cyclone mitigation has been drawn up in consultation with cyclone-prone states. This project envisages construction of cyclone shelters, coastal shelters, belt plantation in storm-prone surges, strengthening of warning systems, training and education. The Planning Commission has given its clearance for the project in principle and it is being taken up with the assistance of the World Bank.

5.10.5 Landslides

The sliding of mud, rock and earth debris down the mountain slopes or river banks due to gravitational action results in landslides, both gradual and sudden. They are often associated with earthquakes or volcanoes. Heavy rainfall and deforestation in steep-sloped areas also cause landslides. Though timely warnings of landslides may not be possible, the study of geology, hydrology, vegetation cover and past occurrences can help in identifying risk areas and hazards like the destruction and burial of settlements, destruction of transport, communication and electricity services, flooding of low-lying areas and loss of human lives and property. Casualties depend on the place and time of occurrence. Strong landslides may destroy many homes and kill thousands of people. The landslide in Peru in 1970 closely followed an earthquake and killed 18,000 people.

Landslide Hazard Mitigation

A National Core Group has been constituted under the chairmanship of the secretary, Border Management and comprising secretary, Department of Science and Technology, secretary, Road Transport & Highways and the heads of Geological Survey of India and National Remote Sensing Agency for drawing up a strategy and plan of action for mitigating the impact of landslides, providing advice and guidance to the state governments on various aspects of landslide mitigation, monitoring the activities relating to landslide mitigation including landslide hazard zoning and to evolve early warning systems and protocols for landslide risk reduction. The government has designated the Geological Survey of India (GSI) as the nodal agency responsible for coordinating/undertaking geological studies; landslides hazard zoning, monitoring landslides/avalanches, studying the factors responsible and suggesting precautionary and preventive measures. The states/UTs (Union Territories) have been requested to share a list of habitations close to landslide-prone areas in order to facilitate the GSI's ongoing assessment of such areas based on the Survey of India's toposheet and their existing database on landslides for the purpose of landslide hazard zoning being carried out by them. A national strategy for mitigating landslide hazard in the country is being drawn up in consultation with all the agencies concerned.

5.10.6 Tsunami

A tsunami is a wave train or a series of waves generated in a body of water by an impulsive disturbance that vertically displaces the water column. Earthquakes, landslides, volcanic eruptions, explosions and even the impact of cosmic bodies such as meteorites can generate tsunamis. Tsunamis can savagely attack coastlines causing devastation. Tsunamis have a low height while travelling over the deep ocean. Thus, vessels in the ocean that fall in their path may not even notice them. Tsunamis occur when shallow water reaches near coastlines.

The word tsunami is actually a combination of two Japanese words meaning harbour wave (Tsu-harbour, nami-wave). Tsunamis usually occur along the Pacific Ocean coasts of the Ring of Fire and a warning system has been in place since 1965. Tsunamis are rare in the Indian Ocean and the warning system there had not been a priority. Consequently, we were hardly aware of the devastating effect of tsunamis till December 26, 2004, when a mega undersea earthquake of magnitude 9.15 on the Richter scale struck at 07:58:53 local time in the Indian Ocean just off the western coast of Northern Sumatra, Indonesia resulting in a tsunami that left 2, 32,000 people dead and missing in 13 countries adjoining the Indian Ocean. It was one of the worst natural disasters in modern history. It was the largest earthquake on earth since the Good Friday earthquake of 1964 with a magnitude of 9.2 on the Richter scale and tied for the fourth largest since 1900. The tsunami devastated the shores of Indonesia, Sri Lanka, India, Thailand and other countries with waves as high as 15 m. The deadliest earthquakes and tsunamis in recent history are given in Table 5.19.

Table 5.19 Casualties in Historical Context

Types of Disaster	Place	Time	Casualties Reported
Earthquake	Tangshan, China	1976	2,55,000
Earthquake	Ashgabat, Turkmenistan	1948	1,10,000
Earthquake	Tsinghai, China	1927	2,00,000
	Great Kanto	1923	1,43,000
		1920	2,00,000
		1556	8,30,000

In the aftermath of the earthquake there has been a flurry of proposals to extend the tsunami-warning system to the Indian Ocean as well.

SUMMARY

Pollution can be defined as the accumulation of matter due to unknown activities in the environment that exceeds the capacity of the ecosystem to either neutralize or disperse them below harmful levels. Air, water and soil are essential for survival of life on earth but unfortunately they are being harmed by pollution. Some well known types of environmental pollutions are:

(i) Air pollution;
(ii) Water pollution;
(iii) Soil pollution;
(iv) Noise pollution.

(i) Air Pollution

Air pollution is one of the most dangerous chemical environmental pollutions of today. Due to natural and man-made activities, gases such as CO, SO_2, H_2S and oxides of nitrogen, mist particulates and aerosol etc. are continuously released into the atmosphere which gets distributed throughout the air. This disturbs the dynamic equilibrium of the atmosphere and causes severe health hazards. The pollutants are neither formed nor distributed locally or regionally, and the sources of pollutants are classified as point source and non-point source.

(ii) Water Pollution

Water is used for washing, irrigation, flushing away wastes, cooling, industrial use, etc. Water pollution can be detected from the following symptoms.

(i) Bad taste or odour of drinking water.
(ii) Oil or grease floating on water surface.
(iii) Unchecked growth of aquatic weeds in water bodies.

Depending upon the sink, water pollution is classified as (i) Fresh water pollution and (ii) Marine water pollution.

Both surface water pollution and ground water pollution are categorized as fresh water pollution since the salt content of both the types of water bodies is below 5 PPT (parts per thousand). In addition to contaminants carried by rivers, marine water is also polluted by oil spills, radioactive wastes, and industrial wastes, among others.

(iii) Soil Pollution

Soil plays a vital role of supplying nutrients and water to the plant and animal kingdom. Soil pollution is the result of both natural and human activities.

Pesticides

Farmers use chemicals to protect their crop from herbs, insects, fungi, pests, etc. that are known as herbicides, insecticides, fungicides and pesticides respectively.

Previously, plant-extracted organic compounds such as pyrethrum, nicotine and rotenone were used to control pests. Gradually, pests became resistant to them.

Injudicious use of chemical pesticides can create:

(i) Adverse environmental impact.
(ii) Development of pest resistance.
(iii) Resurgence and secondary pest outbreak.

Resurgence and Secondary Pest Outbreak

Resurgence is the rapid reappearance of a pest population in injurious numbers, a phenomenon that usually follows the application of a broad-spectrum pesticide which has killed the natural prey that normally keeps the pest in check.

Today resurgence is managed scientifically through:

(i) Cultural practices.
(ii) Biological control.
(iii) Natural chemical control.
(iv) Genetic control.

Cultural Practices

Traditional agricultural practices naturally inhibit growth of pests through crop-rotation, multi-crop farming and timely harvesting.

Biological Control

The biological method of introducing pest-eating bacteria is economical and kills the target pests only, while simultaneously decreasing the genetic resistance of the pests.

Natural Chemical Control

The application of natural or synthetic hormones and pheromones can control most pest problems through natural means. These agents are sometimes called third generation pesticides. Through the use of biofertilizers and natural pesticides, control of soil pollution is possible.

Genetic Control

The genetic control of pests involves releasing sterilized pests into the affected area. When these pests mate with the female of their species, there are no offspring, leading to a decline in the number of such pests.

Radiation Pollution

Radiation pollution is caused due to the addition of more ionizing radiation to the environment as a result of which people are exposed to a higher degree of radiation than they normally experience. Radiation interacts with living tissues and damages them.

Harmful Effects of Radiation

When ionizing radiation happens to pass through body tissue it may split the molecules of the tissue into ions or free radicals and result in useless or reactive fragments and allow the formation of other reactive compounds. It is felt that man's radiation exposure from artificial sources has already been sufficient to produce serious disease problems such as leukaemia, bone tumours, genetic damages, infant mortality, etc.

Plastic Recycling

Plastic recycling was not common earlier as processes were inefficient, expensive and often not practical. Plastic recycling is a five-step process namely supply, collection, sorting and separating process, reprocessing including the melting process of used plastic, and finally the manufacturing of the melted plastic into new products.

Hazardous Waste Management

The basic issues of hazardous waste management are: identification, categorization and qualification of hazardous wastes. Hazardous waste management depends on avoidance of hazardous waste generation, proper treatment of hazardous waste to minimize its effects and disposal of hazardous wastes. Engineering landfill and incineration need special care as the effects of hazardous wastes may harm us through various environmental pathways.

Biomedical Waste

The waste generated from health care units is known as biomedical waste. All biomedical waste has to be segregated at source. It is the duty of health care units to hand over segregated biomedical waste to a CBWTF operator as per Biomedical Waste (Management and Handling) Rules, 1988 in different coloured bags (as per the colour-coding) for recommended treatment and disposal.

E-waste

Used batteries, junk products with electrical cords, mobile phones, scanners, printers, calculators, different audio-video equipment, ovens, washing machines and electronic toys are known as electronic waste or e-waste. Statistics show that in India the amount of annual e-waste is 1, 46,180 tonnes. The Karnataka State Pollution Control Board has instructed the IT companies of Bangalore to limit their e-waste within their premises.

Disaster Management

Natural phenomena such as earthquakes, cyclones, volcanic eruptions, tsunamis, wild fires, floods, landslides, droughts and man-made major accidents which cause grievous human casualties, economic and social loss and damage to the environment may be called disasters. Prevention, mitigation, preparedness and relief are the essence of disaster management today.

(iv) Noise Pollution

Noise can be explained as any sound that is not desired by the recipient. So, it can also be defined as a wrong sound at the wrong place and at the wrong time.

Noise as a potent menace is evaluated in decibels which is the noise level scale. However, the health hazard is a matter of Perceived Noise level Decibels (PNdB).

Classification of Noise

Depending upon the source, noise can be broadly classified into:

(i) Transport Noise
(ii) Occupational Noise
(iii) Neighbourhood Noise

ESSAY TYPE QUESTIONS

1. What is the meaning of the term environment and what is the genesis of the environmental problem?
2. Discuss air pollution and its characteristics.
3. Write a note on common air pollutants and their sources and effects on human beings.
4. Write a note on particulate air pollutants mentioning their chemical composition.
5. Write a short note on sources of water pollution and effects of water pollution on human health.
6. What is ground water pollution? What are the detrimental effects of toxic metals present in water?
7. Write a note on the small-scale water purification process.
8. Give the outlines of large scale water purification process.
9. Write a short note on treatment of waste water.
10. What are the toxic substances present in industrial effluents and how are these industrial effluents treated?
11. What is low-cost waste water treatment method?
12. Write a short note on management of water resources.
13. What is soil pollution and what are its different sources?
14. What is soil pollution, give emphasis on its detrimental effects?
15. How can soil pollution be controlled?
16. How does degradation of pesticides take place in soil?
17. What are the effects of using synthetic pesticides?
18. What is marine water pollution? Explain the ways to mitigate it.
19. What are the sources of oil pollution in sea water?
20. Distinguish between sound, noise and noise pollution.
21. What are the sources of noise pollution and how can noise pollution be controlled?
22. Write a note on the detrimental effects of noise pollution on human beings.
23. Write a note on e-waste and its mitigation.
24. What is thermal pollution? What are its sources?
25. How does thermal pollution affect aquatic life? Explain the means to control thermal pollution.
26. Write a note on control of nuclear hazards.
27. Write a short description on the Chernobyl Disaster.
28. Write the main principles (components) of solid waste management.
29. Write a note on the main composition of Municipal Solid Waste (MSW) in urban areas and industrial solid waste.
30. Explain the process of incineration. What are its drawbacks?
31. Give the composition of fly ash obtained from coal-fired power plants and explain how it is utilized these days.
32. How is vermicomposting better than conventional composting used for biodegradable solid waste?
33. Write a short note on biomedical waste.
34. Write a short note on hazardous waste disposal.

35. Describe the role of an individual in the prevention of pollution.
36. What is a disaster? Briefly explain how human activities act as a catalyst in a natural disaster?
37. Discuss the effects of disasters on the environment.
38. Write short notes on the following, describing their causes, effects and ways of mitigation:
    (i) Flood.
    (ii) Drought.
    (iii) Cyclone.
    (iv) Earthquake.
    (v) Landslide.
    (vi) Tsunami.
39. Briefly explain the steps taken in India to fight these disasters.

SHORT-ANSWER TYPE QUESTIONS

1. Define pollution. What are pollutants and how do they affect the environment?
2. What are the various sources of air pollution?
3. What is fossil fuel? Write a note on the detrimental effects of fossil fuel combustion.
4. What are the different types of air pollutants? Explain in brief.
5. Explain the effect of particulates on the environment.
6. Explain how the extent of air pollution depends on meteorology?
7. Explain in short about different air quality control techniques.
8. Write a short note on high volume air sampler.
9. What do you mean by National Ambient Air Quantity Monitoring (NAAQM)?
10. Write short notes on:
    (i) Population explosion.
    (ii) Effect of particulate matter on the climate.
    (iii) Primary pollutants.
    (iv) Effect of secondary pollutants on the biosphere.
11. What is air pollution? Define clear air.
12. What is the meaning of vehicular pollution? Describe anthropogenic sources of this pollution.
13. How do you define water pollution? What are the different types of water pollution?
14. What are the different sources of water pollution?
15. What is BOD? How does it help to measure the pollution of a water body?
16. How does industrial waste cause water pollution? Write a short note on biodegradable and non-biodegradable pollutants.
17. How do agricultural chemicals cause water pollution?
18. Write a note on Eutrophication.

19. What is biological contamination of water?
20. Write short notes on:
    (i) Desalinization.
    (ii) Cloud seeding.
    (iii) Tapping of ground water.
    (iv) Composting.
21. Explain the difference between BOD, COD and DO of a water body.
22. What is soil? How is it formed?
23. Write a note explaining how mining activities accelerate soil degradation.
24. Explain the role of pesticides in soil degradation.
25. Write short notes on:
    (i) Biological control of pests.
    (ii) Natural chemical control of pests.
    (iii) Cultural practice of pest control.
26. What is marine water pollution? How does it affect marine life?
27. What is noise pollution? How can it be controlled?
28. What are the different sources of noise pollution?
29. What is thermal pollution? What are its sources?
30. Write a note on cooling towers.
31. Write short notes on:
    (i) Recycling of waste.
    (ii) Vermicomposting.
    (iii) Hazardous waste management.
32. Write a note on the Love Canal case study.
33. Write short notes on:
    (i) Landslide hazard mitigation.
    (ii) Cyclone mitigation.
    (iii) Earthquake risk mitigation.
34. What is the difference between natural and man-made disasters? Name two natural disasters that occurred recently in India.

MULTIPLE CHOICE QUESTIONS

1. Depending upon the generation of different air pollutants, they can be classified as
    (a) point source or non-point pollutants.
    (b) primary or secondary pollutants.

(c) natural or anthropogenic pollutants.
(d) None of the above.

2. PAN is a secondary pollutant that
   (a) forms when hydrocarbon radical reacts with nitrogen oxide.
   (b) causes photochemical smog.
   (c) may cause respiratory diseases in humans.
   (d) All of the above.

3. Stock emission and ambient air quality are
   (a) modes of causing air pollution.
   (b) modes to determine the ambient air quality.
   (c) used to determine various air pollutants (in ppm) present in the atmosphere.
   (d) None of the above.

4. ________ gives an empirical value to water quality and is a parameter for the organic matter present in water.
   (a) BOD
   (b) COD
   (c) DO
   (d) All of the above

5. More scientific method than BOD to determine water quality parameter–
   (a) COD.
   (b) DO.
   (c) Both of the above.
   (d) None of the above.

6. The itai-itai disease of Japan was caused due to
   (a) arsenic pollution.
   (b) cyanide pollution.
   (c) cadmium pollution.
   (d) lead pollution.

7. Fluorosis results in
   (a) exostosis.
   (b) increased bone mass.
   (c) increased trabecular bone volume.
   (d) All of the above.

8. Hormones and pheromones are
   (a) first generation pesticides.
   (b) second generation pesticides.

(c) third generation pesticides.

(d) fourth generation pesticides.

9. Behavioural disorders, development of destructive nature in later stages and neurotic traits as adults are the outcome of

(a) noise pollution.

(b) thermal pollution.

(c) radiation pollution.

(d) soil pollution.

10. One Curie is equal to

(a) 3.7×10^{10} disintegrations/second.

(b) 2.08×10^{9} pairs of gaseous ions.

(c) $2 \times 10^{-5}Nm^{-2}$.

(d) $10^{-12}w/m^2$.

11. The common conversion of solid waste into manure & biogas, using bacteria & fungi is called

(a) recycling of waste.

(b) disposal of waste.

(c) composting.

(d) incineration.

12. The process of conversion of electricity (energy) from waste is called

(a) pyrolysis.

(b) vermicomposting.

(c) sanitary landfill.

(d) plastic recycling.

13. The only two mycorrhizal mushrooms which are commercially cultivated are

(a) morels and truffles.

(b) oyster & morels.

(c) shitake & oyster.

(d) truffles & shitake.

14. CBWTF stands for

(a) the Common Biochemical Waste Treatment Forum.

(b) the Council of Biomedical Waste Treatment Forum.

(c) the Common Biomedical Waste Treatment Facility.

(d) None of the above.

15. The main discharge from a thermal power plant is

(a) smog.

(b) dust particulates.

(c) fly ash.
(d) smoke.

16. Vermicomposting is a method of composting that involves
(a) tapeworms.
(b) silkworms.
(c) earthworms.
(d) leeches.

17. Process of dumping solid waste in a scientifically designated land area is called
(a) solid waste disposal.
(b) waste dumping.
(c) sanitary landfill.
(d) None of the above.

18. Prevention, mitigation, preparedness & relief are
(a) principles of pollution prevention.
(b) objectives of hazardous waste management.
(c) the essence of disaster management.
(d) modes to combat exploitation of natural resources.

19. Used batteries, mobile phones, calculators and other such stuff are better known as
(a) hazardous waste.
(b) solid waste.
(c) industrial waste.
(d) electronic or e-waste.

20. The supply of usable, good quality water can be increased by
(a) cloud seeding.
(b) use of aquatic macrophytes.
(c) chlorination.
(d) activated sludge process.

ANSWERS

1 (b) 2 (d) 3 (b) 4 (a) 5 (a) 6 (c) 7 (d) 8 (c) 9 (a) 10 (a) 11 (c)
12 (a) 13 (a) 14 (c) 15 (c) 16 (c) 17 (c) 18 (c) 19 (d) 20 (a)

UNIT VI

Social Issues and the Environment

LEARNING OBJECTIVES

After reading the chapter, students will be able to:

- Define social issues and their impact on the environment.
- Explain the processes needed to transit from unsustainable to sustainable development.
- Describe urban problems related to energy.
- Define water conservation and describe methods such as rain water harvesting and watershed management.
- Explain resettlement and rehabilitation of people and outline the accompanying problems and concerns, with case studies.
- Define environmental ethics, and explain the issues and possible solutions.
- Define climatic changes, global warming, acid rain, ozone layer depletion, nuclear accidents and holocausts.
- Extrapolate on these terms with appropriate case studies.
- Explain wasteland reclamation.
- Define consumerism and waste products.
- Describe the issues involved in the enforcement of environmental legislation.
- Explain the need and importance of public awareness.
- Outline and explain the provisions of the Environment (Protection) Act, (Prevention and Control of Pollution) Act, Water (Prevention and Control of Pollution) Act, Wildlife Protection Act, and Forest Conservation Act.

6.1 Elementary Idea on Sustainable Development

It has already been discussed in the previous chapters that the advancement of human civilization from the hunter gatherer to modern industrial society through agriculture has resulted in the overuse and abuse of natural resources. Besides air and water pollution, the many consequences of this abuse include the depletion of the earth's protective ozone layer, global warming due to increase in carbon dioxide content by burning of fossil fuels, the Greenhouse effect, acid rain, decrease in forest cover, soil erosion, the increasing rate of extinction of species and the deposit of toxic and nuclear waste in vulnerable areas of the earth. This overuse and abuse of natural resources threatens to lead the world into an era of famine. It has already started in Ethiopia.

Our present behaviour with the environment is no longer life-supporting. To make our relationship with the environment sustainable, we have to take care of it by meeting our needs and aspirations of the present without compromising those of the future.

6.1.1 Concept of Sustainable Development

The concept of sustainable development has been defined by the World Commission on Environment and Development (WCED) 'as a process of change in which the exploitation of resources, the direction of investments, the orientation of technological development and institutional change are all in harmony and enhance both current and future potential to meet human needs and aspirations.' Thus development should not endanger the natural systems that support life on earth. Rather it should provide a framework for the integration of environmental policies and developmental strategies which can be implemented from the grassroots at home to the international level.

There are two schools of thought regarding our attitude towards Mother Nature: Frontier Mentality or the Throwaway Society and the Sustainable Society.

Frontier Mentality

The Frontier Mentality is a human-centred view mostly believed by economists and technocrats. According to this group, the continuation of the present trend of economic growth and technological advancement will make the earth a less crowded and less polluted place. Most of the people will be rich in both physical and material health with greater longevity. This concept is based on the following ideas:

(i) The world has an unlimited supply of resources for human use.

(ii) Humans are apart from nature.

(iii) Nature is something to overcome.

(viii) Technology will advance with these ideas but only at the cost of our natural resources which are very limited.

Sustainable Development

The other group, which consists mainly of environmentalists and conservationists, believes in Mahatma Gandhi's philosophy: 'Earth provides enough to satisfy every man's need but not any man's greed.' According to environmentalists and conservationists, the continuation of the present trend will make the world both more populated and polluted. The rich will be richer and the poor will be poorer. This will lead to greater political and economical instability increasing the threat of nuclear war and also an ecological crisis. In order to prevent an ecological crisis and for a new sustainable society, the strategies to be followed are:

(i) Conservation and reduction of excessive use of resources.

(ii) Recycling and reuse of materials.

(iii) More use of renewable resources (solar energy) than non-renewable resources (oil, coal) for energy.

Sustainable development is defined as the development that meets the needs of the present without compromising the ability of future generations to meet their needs. It believes in equality between countries and continents, gender, age, race and class. Sustainable development maintains a balance between social and economic development along with protection of the environment. It strengthens the interdependence of human needs and environmental requirements.

In the current strategy of economic development, the natural resources of the world are being used up at a much faster rate without any consideration for our future generation and the environmental degradation being caused.

This type of developmental strategy is simply unsustainable for the long term development of the world. At the Rio de Janeiro conference in 1992, several documents were created for the United Nations Conference on Environment and Development (UNCED). These documents conclusively pointed out to the need to care for the earth as the environment was closely connected to development.

In the sustainable development strategy each activity that is expected to bring about economic growth must consider its environmental impact through a scientific Environmental Impact Assessment (EIA). Thus it is consistent with long-term development and growth.

For example, large dams, highways and the mining industry are all essential for our social and economic development. At the same time, this developmental work means the use of natural resources, loss of forest cover, depletion of biodiversity and increase in pollution. Thus, for every project, according to the sustainable development strategy, a scientific EIA is done. The economic benefit of a project is weighed against the possible environmental costs. The government has also made it compulsory to publish the summary of the EIA report and conduct a public hearing. All responsible citizens should read, evaluate and respond to such public hearings and make comments on the matter so that our environment is not degraded further. For the sustainable development of the world we have to think globally but act locally.

6.2 Urban Problems Related to Energy

Energy is one of the major pillars of economic development of the society. Economic growth along with a growing population will obviously consume a lot of energy.

Houses in urban areas are now made of more heat sensitive materials such as metals (iron, steel and aluminium) glass and concrete instead of heat insensitive substances such as wood and brick. To make these houses comfortable, we use air conditioners or room heaters run by electricity. High rise buildings need energy to operate lifts and electrical energy for lighting. Most urban people use their individual transport rather than a public one. Similarly, each and every step in an urban centre needs energy in some form or the other. To meet the enormous energy needs and for long term sustainability we should be more specific about the most efficient and cost-effective manner of energy use.

This can be achieved through the use of more renewable energy resources than non-renewable energy resources in addition to steps such as:

(i) Urban planning for more efficient energy utilization.

(ii) Change of lifestyle to increase community involvement, which means using car pools.

(iii) In India, the importance of the development of renewable energy sources as an alternative to fossil fuels for a sustainable energy base has been recognized since 1970. Since then, considerable effort has gone into the development, trial and introduction of a variety of non-conventional energy technologies in the industry and for domestic use.

The Ministry of Non-Conventional Energy Sources, Government of India is involved in the implementation of these programmes for the development, demonstration and utilization of renewable energy-based technologies such as solar and thermal, and green technologies such as:

(i) Solar photovoltaic.

(ii) Wind power generation and water pumping.

(iii) Solar power.

(iv) Geothermal energy.

(v) Energy recovery from municipal and industrial waste.

(vi) Chemical source of energy.

(vii) Fuel cell.

(viii) Alternative fuel for transportation.

(ix) Biomass combustion.

(x) Hydroelectricity.

Today, India is in a position to solve its own energy crisis and offer these green technologies to other developing countries too.

Efforts are being made to meet the water crisis by building dams, reservoirs and digging wells to conserve water.

6.3 Water Conservation

In ancient days women used water economically as they had to cart it for their household from long distances and with much hardship. The used water was fed to the kitchen garden and not disposed off. Water was stored in small tanks called talaabs or jheels for domestic and agricultural use.

Over the years, with increase in population, the demand for more food crop and water has also increased. At the same time, growing industrialization and demand for more agricultural land has led to deforestation. But with deforestation, the surface run-off increases and the ground water table thus drops as water cannot percolate into the ground with no vegetation. The perennial rivers are also becoming seasonal due to the lack of forest cover. With the advancement of science and technology, ground water is also constantly withdrawn to meet the water demand. Hence, the water table is continuously receding. Now it is time we recharge the ground water table for our own sustenance. Efforts in this direction are made by rain water harvesting and watershed management techniques. Emphasis should also be laid on the conservation of water by employing modified techniques such as using the drip-irrigation method to water the plants near their roots in agricultural fields rather than the traditional method. Proper pipes should be used and leakages should be checked periodically. Water should be used economically and judiciously. This can be easily achieved, by using water from a bucket rather than using it directly from the tap and also utilizing used water (water used after washing vegetables, cereals and clothes) for the kitchen garden, cleaning floors and also by collecting rain water in buckets during the rainy season.

The surface water bodies are also responsible for recharging the ground water table of the area. Thus, the need of the hour is to recharge the aquifers, thereby recharging surface water bodies and hence, increasing the water table. We can also enrich the environment by maintaining some ground covered with vegetation, around our house. This will allow for easy percolation of rain water.

The sustainable use of rain water can only lower the demand for ground water. It can also augment local water supply through the recharge route. Ground water is a dependable source of fresh water. But, its continuous consumption has resulted in the drastic lowering of the water table. The result is the drying up of surface water bodies such as tanks, wells and ponds.

6.3.1 Rain Water Harvesting

Rain water harvesting is an age old method of collecting rain water on the roof tops and then using it directly or storing it for use when the rainy season is over. This method is widely practiced particularly in dry areas all over the world.

Extensive rain water harvesting apparatus existed in Palestine and Greece about 4,000 years ago. In ancient Rome, every house was built with its own cistern and paved courtyards for harvesting rain water. The farmers of Baluchistan and Kutch were also storing rain water in specially built irrigation dams to be used in farming.

In the present age, concrete houses, well-built roads, footpaths and well-concreted courtyards have left few open grounds. With the decrease in natural forest cover, increase in concrete jungles and the decrease in exposed earth; very little open ground is left for water to soak in and thereby increase the ground water table. So, artificial recharging of the ground water is extremely essential. It is done through rain water harvesting. For the purpose, rain water is collected at the roof top or in an open well and then carried down for immediate use or it is directed into the aquifier.

Several techniques are available for efficient rain water harvesting. Traditionally, rain water is stored in open storage bodies such as lakes, ponds and tanks. It is still practiced in villages. In rural areas the harvested rain water is stored in underground tanks. Underground storage tanks have an additional advantage over surface water storage tanks as there is no loss of water through evaporation.

Rain Water Harvesting Techniques

There are two main techniques for rain water harvesting:

(i) Storage of rain water on the surface for future use.

(ii) Recharge of ground water.

Storage of rain water on the surface for future use is a very old practice. Recharge of ground water is a recent concept and the structures used for the purpose are:

(i) **Pits**: For recharging a shallow aquifer, recharge pits are constructed. These are one to two metre wide and three metre deep and backfilled with boulders, gravels and coarse sand to aid filtration before percolation to the ground.

(ii) **Trenches**: About 0.5 to one metre wide, one to 1.5 metre deep and up to 20 metre long trenches are constructed where a permeable stream is available at a shallow depth. The trench is also backfilled with filter material just as it is done in the case of pits.

(iii) **Dug Wells**: Existing dug wells may be utilized as recharge structures. The excess water should be allowed to pass through a filter bed before putting it into the well.

(iv) **Hand Pumps**: Existing hand pumps may also be used as recharge structures for aquifers. The water should pass through the fitter media before percolation.

(v) **Recharge Shaft**: For recharging shallow aquifers located near a clayey surface, recharge shafts of 0.5 to three metres diameter and 10 to 15 metres deep are constructed and backfilled with boulders, gravels and coarse sand.

(vi) **Lateral Shafts with Bore Wells**: A similar technique is used to recharge upper as well as deeper aquifer levels.

(vii) **Spreading Technique**: If the permeable strata starts from the top, then water is allowed to spread in streams/nullahs, making check dams, nullah bunds, cement plugs or in a percolation pond.

In addition to these, in urban areas roof top and road top collection of rain water is also used to recharge aquifers.

In India, all the traditional houses, forts and palaces have a roof top rain water harvesting system. The collected rain water from the roof top is stored in underground tanks. In the hilly areas of North Bengal (Darjeeling), all the houses have a roof top rain water harvesting system. They use the collected rain water immediately or store it for a very short period. They do not have an underground storage system. These days, the Central Ground Water Board along with the civic authorities is encouraging artificial recharging of ground water through rain water harvesting. Laws are being framed to make the rain water harvesting system compulsory in all the newly-constructed buildings.

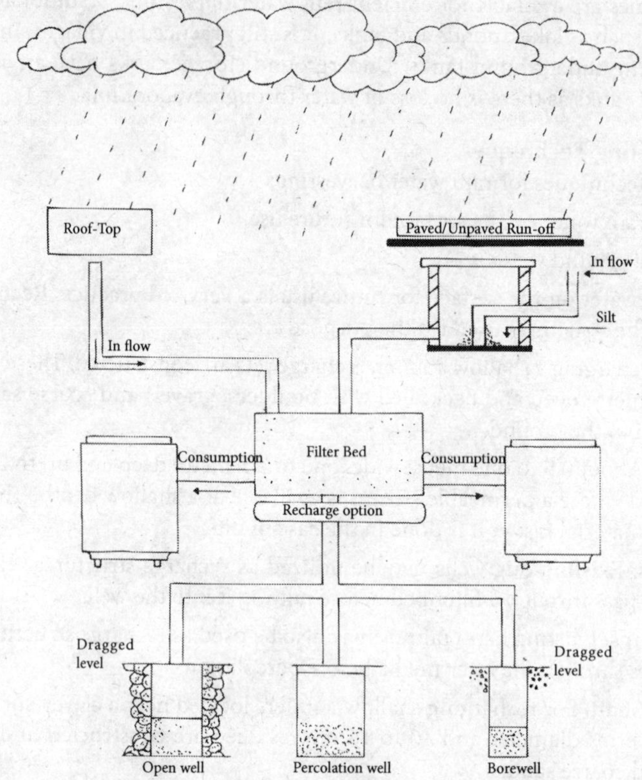

Figure 6.1 Rain Water Harvesting Techniques

6.3.2 Watershed Management

A watershed is a geographic unit (a piece of land) that collects, stores, and releases water. Collected water comes from rain, snow and fog. This water is stored in lakes, ponds, sub-surface soil and geological formations. The stored water is then released through rivers, streams and ground water flow. Thus, a watershed can be described as a land from where water drains into a particular stream, river, lake, wetland, estuary and even the ocean. Watershed is a natural system and all lands (and all humans, wildlife and activity on that land) are part of one or other watershed. Being a natural system, watersheds provide substantial benefits to people and the neighbourhood when they are kept in good condition. But individual or collective action directly affects it.

Depletion in the ground water table, drying of small streams after the monsoons, floods, and the spread of desertification are some of the consequences of an affected watershed. The watershed management programme was started in India in 1999 to control these adverse effects. It provides scientifically-based education to maintain the viable natural resource management of agriculture production while conserving, protecting and restoring watersheds to secure water both in quality and quantity for drinking, agriculture and sanitation.

The method includes mapping of the watershed area, construction of a series of long trenches and mounds along hill contours to hold rain water and allowing it to percolate into the ground.

Plantation of trees and grass is done to enhance seeping of water to the ground and to prevent water contamination, torrents and landslides. In this way, the ground water storage gets fully recharged. The next step is to make *nullah* plugs and check dams and streams to hold back large amounts of water and prevent it from rushing down the hillside. In this way, not only is an underground aquifer filled up but also water bodies such as streams and rivers flow for the whole year. The main benefits of watershed management can be listed as follows:

(i) It recharges the ground water, potentially reducing water storage during dry spells.

(ii) It reduces severity of floods downstream by retaining water and releasing it during the dry periods.

(iii) It protects stream banks and shore lines from erosion.

(iv) It improves water quality by breaking down, removing, using and retaining nutrients, organic waste and sediments carried to the wetlands with run-off from the watershed.

(v) It provides food and other products such as commercial fish and shell fish for human use.

(vi) It provides fish and wildlife, including numerous rare and endangered species, food, habitat, breeding grounds and resting areas.

(vii) It increases opportunities for recreation, bird watching, photography and outdoor education.

6.4 Wasteland Reclamation

Wasteland reclamation is the process of turning barren, sterile wasteland into something that is fertile and suitable for habitation and cultivation. India has shown an alarming rate of decline in the man-land ratio from 1.25 hectare per capita in 1921 to 0.48 hectare per capita in 1986. The formation of wasteland is nothing but the result of increasing overuse and abuse of our land resource.

Wasteland land can be classified as:

(i) Barren and uncultivable land.

(ii) Fallow land.

(iii) Fallow and cultivable wasteland.

Apart from these, we also have saline and alkaline land and water eroded wasteland. All types of wastelands should be reclaimed for productive use for:

(a) Meeting the constant demand for agricultural land for our ever growing population.

(b) Afforestation to maintain ecological balance.

(c) Protection of natural resources.

However, such reuse depends on the type of wastelands. By reducing the salt content of soil through leaching and flushing and using gypsum, urea, potash and compost before planting, crops can convert fallow and cultivable wasteland into a usable one.

Agro forestry is used in fallow lands for multiple use of trees, agricultural crops or livestock management. This is done by planting trees and cultivating crops in between or under trees to form an integrated system of biological production within a specified area.

In India, the National Wasteland Development Board was established in 1985 to formulate action plans to arrest further land degradation. In 1992, the Board was transferred under the Ministry of Rural Development with the formation of a new department of Wasteland Development.

Five districts in the country (under the state ministry) — Sundargarh (Orissa), Purulia (West Bengal), Almora (Uttar Pradesh) Bellary (Karnataka) and Durgapur (Rajasthan) — have been chosen for the implementation of the plan for reforestation and reclamation of degraded lands by the Ministry of Environment and Forests. Moreover, detailed maps of 146 districts, including every state of India had been prepared for identification of wastelands followed by action plans for their reclamation.

CASE STUDIES

1. *Jatropha curcas* Cultivation:

Biodiesel from Wastelands: Plantation of *Jatropha curcas* is a unique case of wasteland reclamation for study and adoption. *Jatropha curcas* which produces oil-rich seeds is known to survive in eroded land and requires limited amount of water, nutrients and capital inputs. The *Jatropha curcas* trees not only grow in uncultivable wastelands but also yield vegetable oil suitable for conversion into biodiesel. The concept of substituting biodiesel produced from plantations from eroded soils for conventional diesel fuel has gained widespread attention in India. The Government of India as well as many state governments has agreed to use marginal lands unsuitable for food production to be used for plantation of Jatropha trees.

National Bank for Agriculture and Rural Development (NABARD), an apex body of planning, policy and operation on agriculture and rural finance is extensively involved in extending credit support for renewable energy development in the rural areas. *Jatropha curcas* trees can be planted through joint forest management. It is hardy in nature, has a short gestation period, high rate of recovery and high quality of oil. The average production cost is Rs. 25,826 per hectare. Besides the use of *Jatropha curcas* trees as substitute of diesel, the following are some of the other uses of the Jatropha tree:

- To make soap, lubricants and candles.
- It is a raw material for plastic and synthetic fibres.
- It is a raw material for medicines.

The benefits to cost ratio is 1:19.

2. The City of Dawn: Auroville, the city of dawn, is situated about five kilometres away from the sea, 10 km north of Pondicherry and 150 km south of Chennai. It is designed as a universal town where people from all over the world live in peace and progressive harmony. The main purpose of Auroville is to realize humanity. About 35 years ago, Auroville was a desert land caused by 200 years of deforestation and overuse. This barren land was exposed to wind and water erosion with the fall of subterranean water tables, decreasing crop yield, denudation of vegetation cover, and a deeply furrowed erosion channel, sometimes in the form of several meters deep canyons. In spite of an average annual rainfall of 1,200 mm and 25 million CBM of rain water per annum, the area was gradually converting into a desert with the red laterite strata exposed.

Under these circumstances, it was not possible to build Auroville, a city for 50,000 people without any vegetables, water and food source. Herald Kraft, an expert in water conservation and watershed management of Berlin, Germany, studied the situation and said that if water could be conserved then construction of the city is possible with the available 25 million CBM rain water for a population of 50,000 per annum.

After studying the monsoon water flow of the place, Kraft built two watersheds. Starting from the top of the watershed, bunds were made around individual fields so that no water could overflow to another field. All the fields were protected from cattle grazing by planting thorny plants along the bunds. Small dams were also made to stop the flow of water into canyons. Plantation was done first on land, which was not entirely degraded. Indigenous tree species were used in seriously degraded lands. As the vegetation started growing, the microclimate also changed.

In this way, with effective watershed management, Auroville has been transformed from a desert into a lush and verdant jungle supporting a thriving diverse community of people, plants and animals.

The drinking water problem of the area was solved through rain water harvesting. The cheapest and best storage of water was found to be underground favoured by the geology of the existing clay layer just beneath. Water was kept in big lakes and ponds. It evaporates to the atmosphere and percolates to the soil. The best place to infiltrate water is just opposite the highest point, the Matri Mandir Gardens.

Dirty water running off the streets is processed here and converted into drinking water. After the final treatment, the dirty water is pumped and passed through the lake. In the process it gets purified and collected before it leaves the city and flows into the sea. All the roofs of the houses are connected to a common system for collection of rain water. All the roads also have a surface from where water that infiltrates the ground is treated. An artificial basin is created and all the water flows by gravity into the lake. For the first three years, the lake was completely sealed and the overflow was controlled with wells. The lake water remains crystal clear and is fit for human consumption.

It is clear that Auroville is a success story of sustainability and environmental restoration. Innovative developments in appropriate technologies of sustainable agriculture and rural regeneration brought it international acclaim and unanimous endorsement from the United Nations Educational, Scientific and Cultural Organization (UNESCO), which called it the "city of earth needs". Auroville is a classic case of wasteland reclamation through watershed management and rain water harvesting that should be studied and followed for implementation across the world.

6.5 Resettlement and Rehabilitation

Across the globe nearly 10 million people per annum are affected by forced displacement due to infrastructural projects such as dams, mines, roads, industries and power plants. They lose their shelter and livelihood and some of them die due to hunger and poverty. In India, planned development in growth sectors such as power, mining, heavy industry and irrigation, immediately after Independence had already displaced about 30 to 50 million persons. Only about 25 per cent of this number was resettled.

Uprooting people is a serious issue. The rights of the tribal people are also threatened with displacement. The developmental projects come into existence after a fairly long period of planning and awareness of displacement caused by such projects. Despite awareness about issues relating to the resettlement and rehabilitation of the displaced persons, very little attention is paid to them. Rather these projects focus on the economic efficiency and not on the person to be displaced from their land, livelihood and their socio-cultural life. Under the new economic policy, expecting large scale displacement, the Committee of Secretaries, Ministry of Rural Areas and Employment drafted the Land Acquisition Bill in 1998 for the rehabilitation of displaced persons. According to this Bill, people eligible for rehabilitation should apply to claim it. However, following a number of revisions, finally a National Policy of Resettlement and Rehabilitation on Project-Affected Families 2003 was gazetted on February 17, 2004 by the Ministry of Rural Development. In spite of this, resettlement and rehabilitation for development work in India is not adequate, uniform and consistent. Resettlement and rehabilitation, as per the Indian Constitution, is the responsibility of the individual states concerned. But only three states — Karnataka, Maharashtra and Madhya Pradesh have separate laws and only two public sector undertakings, the National Thermal Power Corporation (NTPC) and Coal India Ltd have separate policies for rehabilitation. No development work can be justified if a section of the society is pampered. Often, rehabilitation benefits are also discriminating. For example, a person displaced by canal or irrigation work gets lesser benefits than those affected by the construction of a dam.

Environmental Ethics

Ethics deal with moral duty and obligations and gives rise to a set of values, which in turn are used to judge the appropriateness of a particular conduct or behaviour. These are the basic principles by which a society should be guided in its decision making and activities. The entire world is laying greater emphasis on responsibilities to the environment. The basis of ethics is truth, honesty, justice, trustworthiness, competence and accountability.

Consciousness of ecological consequences has led the society to adopt environmental ethics. The new environmental ethics give as much importance to revitalising growth as they do to sustainability, that is, to developmental processes that last. Sustainability requires at least a constant stock of natural capital constructed as the set of all environmental assets. It consists of:

(i) Justice in respect of the socially disadvantaged.
(ii) Justice to future generations.
(iii) Justice to nature.
(iv) Aversion to risk arising from our ignorance about the nature of the interaction between the environment, economy and society and the social and economic damage arising from low margins of resilience to external 'stock'.

Some of the ethical guidelines on environmental protection are enlisted below:

(i) The earth is the habitat of all living species and not of human beings alone.
(ii) Natural resources and energies are depleting fast. We must protect them.
(iii) Keep yourself informed about ecological changes and developments.
(iv) Involve yourself in the care of the earth and experience nature.
(v) Respect nature, you are a part of it.
(vi) Observe austerity, reserve scarce resources for the future and the future generations.
(vii) Become involved in the environmental movement and bring about a change in the attitudes and acts towards nature of people known to you.
(viii) Think of the global cause and act for local protection.
(ix) We must be cooperative, honest, affectionate and polite to society and nature.

6.6 Air Pollution and the Biosphere

Biosphere denotes the domain of living organisms and their interaction with the atmosphere, lithosphere and hydrosphere collectively known as the environment.

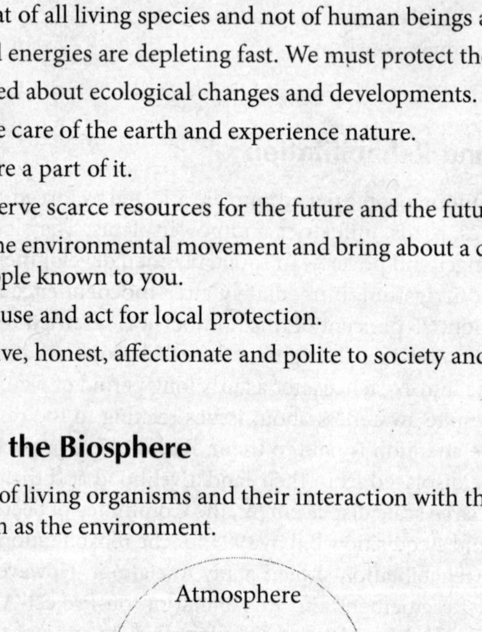

Figure 6.2 Biosphere

To a great extent, atmospheric pollutants are mainly present in the troposphere and lower stratosphere. Depending on the extent of air pollution, the atmosphere can be divided into three layers. The first layer, up to 100 metres from the ground is highly polluted, in general and in urban and industrial areas, in particular. A part of the pollutants present in this layer are absorbed by the vegetation, water surface and buildings. The next layer 100–2,000 metres high, contains lesser amount of pollutants as turbulent air currents, drizzle, rain and fog divert and dilute the concentration of pollutants there. The third layer containing large amounts of water vapour and clouds extend up to the troposphere. Here, cloud water droplets dissolve some pollutants and bring them back to the earth as rain. The pollutants present in the stratosphere due to volcanic eruptions, nuclear explosions and products of photochemical reactions, stay there for a long time due to the lesser movement of air currents in the stratosphere.

Large quantities of particulates and smoke from fossil fuels, particularly over urban and industrial areas, increase atmospheric turbidity and reduce visibility. These particulates absorb and reflect incident solar radiation, which causes a 20 per cent decrease in solar flux in the urban areas compared to rural areas. Particulates also participate in cloud formation resulting in smog over urban areas, which leads to erosion and corrosion of metals and materials, as well as damages plants.

6.6.1 Climate Changes

Global warming is now one of the most important environmental issues. Our every day activities are leading to changes in the earth's atmosphere that significantly alter the planet's heat and radiation balance and warmer climate. International efforts to address this problem have been on since the 1980s. The Earth Summit in 1992 was an important launching point.

The knowledge that CO_2 in the atmosphere was a Greenhouse gas dates back to the 19th century. Not only CO_2, but other Greenhouse gases such as methane, nitrous oxide and water vapour too help to retain the incoming heat energy from the sun, thereby increasing the earth's surface temperature. However, CO_2 is the most important Greenhouse gas that is being affected by human activities ranging from burning any kind of fuel containing carbon to human respiration. The concentration of CO_2 in the earth's atmosphere was 280 ppmv and 358 ppmv (parts per million by volume) in 1750 and 1994, respectively and is steadily increasing at a rate of 1.5 ppmv per year. The concentration of other Greenhouse gases such as methane and nitrous oxide has also been rising at a fairly rapid rate. Earth has warmed an average 0.3 to 0.6°C since the late 19th century and temperatures would rise by 1 to 3.5°C by the year 2100 and global mean sea levels by about 15 to 95 cm. It is likely that changes of this magnitude and rapidity could pose severe problems for many natural and man-made ecosystems as well as important economic sectors such as agriculture and water resources. These changes also threaten extinction of habitation in many low-lying areas.

Most of the ill-effects of climate change are linked to extreme weather events such as hot and cold spells of temperature or wet or dry spells of rainfall or cyclones or floods. Prediction of nature and distribution of such events in a changed climate may not be accurate also resulting in the intensification of the ill-effects. Global warming has often been described as one of the most serious environmental problems ever to confront humanity. Its effect covers the entire globe over a time scale of decades or centuries and is inextricably linked with economic development of the nations. Since Greenhouse gases are generated by burning fossil fuel in power plants, factories and automobiles, it is not easy to reduce emissions. Virtually every facet of our life is intimately tied to the consumption of energy and any serious attempt to cut emissions will have clear and immediate consequences.

To respond to climatic changes, negotiations began in 1991 under the United Nations, so as to formulate an international treaty on global climate protection. This was finalized at the Earth Summit at Rio de Janeiro in June 1992. The convention has a few binding requirements and calls for nations to limit carbon dioxide and other Greenhouse emissions by addressing anthropogenic emissions by some and removal through sinks of Greenhouse gases. It does not set out specific targets or time frames for reducing emissions.

The effect was found inadequate and the Kyoto Protocol to the United Nations Framework Convention on climate change decided in December, 1997 in Kyoto, Japan to commit specific, legally-binding emission-reduction targets for six Greenhouse gases — carbon dioxide, methane, nitrous oxide, hydro-fluorocarbons, perfluorinated compounds and sulphur hexafluoride by industrialized nations. The protocol was opened for signature on March 16, 1998. Developing countries have since then been arguing that developed countries which are mainly responsible for the present and past atmospheric changes should bear the financial responsibilities of change.

The mechanisms suggested to improve the Kyoto Protocol include IT, CDM and ETS. They are explained below:

IT: It is a voluntary project between a developed and developing country to reduce Greenhouse gas emission or increase carbon sequestration through afforestation.

CDM: It is IT in a new form — Greenhouse gas reduction projects between Annexure-B (developed) and non Annexure-B (developing) countries that result in Certified Emission Reductions (CERS).

ETS: Emission Trading Scheme is a market-based policy instrument that allows the Greenhouse producing corporate to reduce its emissions at a minimum cost.

For India, climate change issues have several ramifications.

First, India does not have current obligations to reduce emissions. But in the future there will be international pressure on India for necessitating correct emission inventories in reduction measures.

Second, we need to develop our emission target basis of norm and measurement and reliable measurement devices.

Third, we need to develop our disaster mitigation plan since global intervention; global warming, climate change and extreme meteorological events are not ruled out.

Greenhouse Effect of Global Warming

The Greenhouse effect is the phenomenon of the progressive warming up of the earth's surface due to the excessive presence of some gases such as carbon dioxide and methane in the atmosphere.

The temperature of the earth's surface is due to exchange of energy in the form of radiation between the surface and atmosphere and of physical characteristics of the earth's surface. Among all the constituents of the atmosphere, carbon dioxide and water vapour strongly absorb infrared radiation (14,000 to 25,000 mm) and effectively block a large fraction of the earth's emitted radiation. The radiation thus absorbed by carbon dioxide and water vapour is partly re-emitted to the earth's surface. As a result, the temperature of the earth's surface continues to increase.

The increase in carbon dioxide content in the atmosphere raises the heat (temperature) in the earth's surface or the Greenhouse effect. On the other hand, the carbon dioxide content of the atmosphere directly depends on deforestation and burning of fossil fuels.

A large amount of photosynthesis occurs in forests and they also maintain a vast amount of readily oxidisable stock of carbon as wood and humus. In other words, forests help to maintain the atmospheric balance of carbon dioxide.

Continuous increase in burning of fossil fuels increases the carbon dioxide content of the atmosphere. It has been estimated that about 50 per cent content of atmospheric carbon dioxide is from combustion of fossil fuel. The ocean also contains a large amount of dissolved carbon dioxide as bicarbonates.

Excluding carbon dioxide, other gases such as methane, chlorofluoro carbon and ozone released from industries, also play an important role in the temperature regulation process. Hence, these gases are known as Greenhouse gases. Just like the glass panes of a greenhouse they allow most of the solar radiation to enter the earth's surface but prevent some of the re-radiated infrared or heat to escape into space. As a result, temperature on the earth goes on increasing.

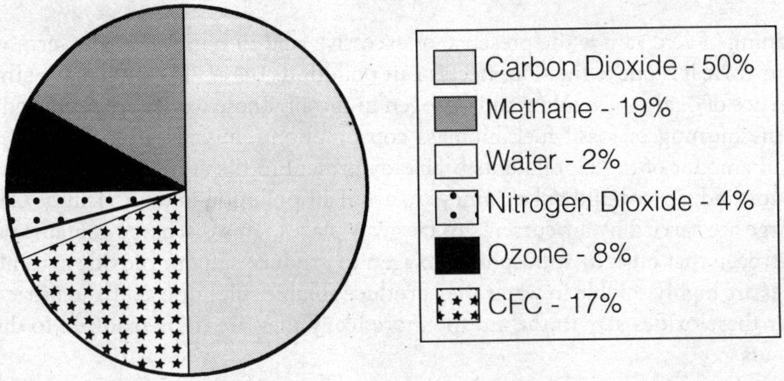

Figure 6.3 Relative Contribution of Radioactively Active Greenhouse Gases to Temperature Rise

Consequences of the Greenhouse Effect: Increase in surface temperature due to increase in the concentration of Greenhouse gases will increase the evaporation of surface water thereby raising the temperature further.

The Greenhouse effect will increase the surface temperature at the rate of 2–3°C for doubling the carbon dioxide content. Increase in surface temperature even by 1°C will adversely effect the world food production by shifting the wheat-growing zone from the northern latitude (Canada and Russia) to the poles. It will also reduce the biological productivity of the ocean which will in turn reduce the nutrient transportation from the bottom to the surface by vertical circulation.

The Greenhouse effect will raise the temperature at the poles resulting in partial melting of glaciers and ice caps. This will raise the water level by as much as two metres in the ocean. This rise in sea level will decrease the ice-gap of Greenland and will threaten the coastal countries. It will submerge densely populated places like Bangladesh, Shanghai and Maldives and coastal cities in India like Chennai and Goa may face the same fate.

An increase in Greenhouse gases, mainly carbon dioxide, will result in increasing the number of cyclones and hurricanes and the melting of snow in the high mountains to cause floods.

An increase in the percentage of atmospheric carbon dioxide content will adversely affect the earth by changing its climate. At present, carbon dioxide controls the global temperature to life, sustained at 15°C by trapping the heat radiation. It is true that without carbon dioxide our earth will be as cold as the moon but if its quantity goes on increasing then the day is not too far when the surface temperature of the earth will be as high as 450°C.

Control Measures: Increasing global warming due to an increase in the Greenhouse effect can be controlled by adopting the following measures:

(i) Use of solar energy and biogas instead of fossil fuels on a large scale.

(ii) By putting a check on population explosion mainly in densely populated countries, which will help in decreasing the quantity of CO_2 and chloro fluoro carbons released.

(iii) Afforestation programme should be followed on a massive scale and deforestation should be banned so as to maintain forests in at least in one-third of the geographical area, as forests regulate the balance in atmospheric CO_2.

By attaching anti-pollution devices such as catalytic converters and using unleaded petrol, emission of Greenhouse gases from automobiles can be reduced.

Acid Rain

The literal meaning of acid rain is the presence of excessive acid in rain water. The term was first coined by Robert Angus in 1872. It is one of the side effects of air pollution. The acidification of the atmosphere is mainly due to the presence of oxides of sulphur and nitrogen in the air. These oxides are produced mainly by human activities like the burning of fossil fuel, biomass consumption, emission from industries and automobile exhaust. A small amount of these pollutants is already present in the atmosphere due to volcanic emissions and swamps. Most industries build tall chimneys to avoid air pollution in their vicinity. Chimney smoke and gaseous discharge are carried by air currents to far away places. This discharge contains particulates such as sulphur and nitrogen that burn in atmospheric oxygen to produce sulphur dioxide and nitrogen dioxide. As these two oxides are highly soluble in water they produce soluble sulphuric acid and nitric acid.

The longer these oxides stay in the air, the more likely they are to be oxidized to different oxides and ultimately to acids.

$$S + O_2 \longrightarrow SO_2$$
$$2_2 + O_2 \longrightarrow 2NO$$
$$NO + ½O_2 \longrightarrow NO_2$$
$$S + O_2 \longrightarrow SO_2$$
$$2SO_2 + O_2 \longrightarrow 2SO_3$$
$$2SO_3 + 2H_2O \longrightarrow 2H_2SO_4 \text{ metal Sulphuric acid or soot particle}$$

In case of nitrogen

$$N_2 + O_2 \longrightarrow 2NO$$
$$NO + ½O_2 \longrightarrow NO_2$$
$$NO + O_2 \longrightarrow NO_3$$
$$NO_2 + NO_3 \longrightarrow N_2O_5$$
$$N_2O_5 + H_2O \longrightarrow 2HNO_3 \text{ Nitric acid}$$

All rainfall is slightly acidic (having a pH of about 5.6) because rain water reacts with atmospheric carbon dioxide to produce carbonic acid. Hence, rain with a pH below 5.6 is called acid rain.

$$CO_2 + H_2O \longrightarrow H_2CO_3 \text{ Carbonic acid}$$

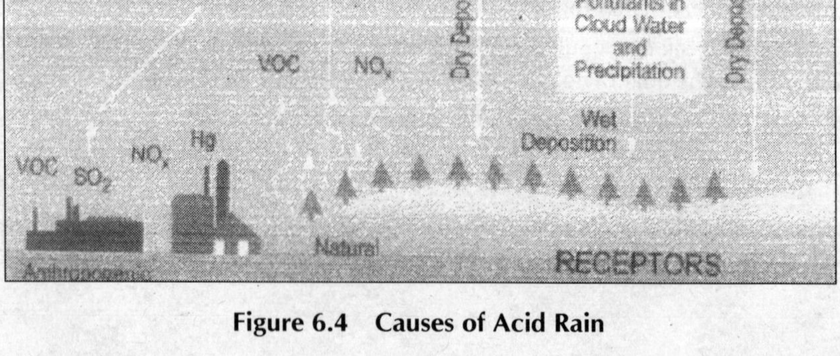

Figure 6.4 Causes of Acid Rain

These acids can get deposited in wet or dry form. Acid rain, snow, fog, frost, dew are different forms of deposition. In dry form, these acids may get deposited on the earth as dust. However, deposition of these acids as rain is more common and the pH of this rain water can be as low as 2.0.

The Effects of Acid Rain

(i) Acid rain increases the acidity of the soil and thereby leaches out all its nutrients and reduces agricultural productivity.

The industrial emissions of northern USA have already destroyed the crops and forests of North America, Europe and Canada by acid rain precipitation. In West Germany, about 10 per cent of the forests died and about 18 million acres of forests are severely damaged.

(ii) Acid rain also affects the aquatic ecosystem. It increases the acidity of lakes, ponds and other water bodies. As a result, fish and other aquatic population have decreased markedly. A decrease in the variety of species in the food chain has also been noticed. About 237 lakes in Adirondack have a pH below 5.0. They have become graveyards for fish and other aquatic life and disrupt the ecological balance.

(iii) Acid rain causes extensive damage to buildings, marble sculptures, limestone, slate and mortar in the following way:

$$CaCO_3 + H_2SO_4 \longrightarrow CaSO_4 + CO_2 + H_2O$$

(Comes out as soluble sulphate)

As a result, these materials become pitted and mechanically weak. This is termed as stone cancer. The Taj Mahal at Agra is also attacked by stone cancer by the pollutants (SOx) released from the Mathura Refinery. Many statues and sculptures in Greece and Italy have been affected by acid rain.

Acid rain has been a great threat to the environment in Britain, as it mostly experiences acidic snowfall, which, on melting, will destroy the whole of Britain. Hence, it can be suitably termed as a pollution time bomb.

(iv) Excessive acid rain will affect the nervous, respiratory and digestive systems of human-beings making a person a neurological patient.

Acid rain moves downwards from the source and spreads widely. Even the gaseous pollutants produced in one country are capable of producing severe acid rain problems in other countries. In short, acid rain is one of the most serious threats to the environment and it has both macro as well as micro biological effects on animals and plants.

Steps to Prevent Acid Precipitation: Earlier, farmers used to add lime to the fields to maintain the pH of the soil. New York has been adding lime to lakes and ponds since 1959. Recently, the United States limed the Adirondack lakes to prevent eradication of trout fish. But it is not a cost-effective method, as it requires nearly US$ 40 per acre with volunteers and vehicular assistance. If aeroplanes and helicopters are used, the cost rises to US$ 200 per acre. So, on a large scale, this type of air pollution is reduced by reducing the acid-forming emissions. These days, scrubbers are installed in coal-fired power plants and other industries to scrub acidic air pollutants and reduce their emission in the atmosphere.

Ozone Layer Depletion (Ozone Hole)

Ozone is a pale blue gas. It is continuously being produced and destroyed. The stratospheric pool of ozone is commonly known as ozonosphere or the ozone umbrella.

Although the presence of ozone in smog impairs vision and breathing but its presence in the ozonosphere filters out the sun's harmful radiation like ultraviolet rays and helps to sustain life on earth. In the absence of this protective umbrella of the ozone layer, all the ultraviolet rays of the sun will enter the earth's atmosphere and the biological furnace of the biosphere will change into the blast furnace.

Naturally, ozone is formed in the atmosphere through a photochemical reaction. Oxygen present in the lower mesosphere absorbs the sun's radiation to produce two oxygen atoms, one in the ground state and the other at the excited state, respectively.

$$O_2 \xrightarrow[175\text{mm}]{hv} O + O^*$$

Excited oxygen atom (O^*) being unstable, cannot stay in the thin upper atmosphere for a long period. It is deactivated by collision with another oxygen molecule, which produces a single molecular oxygen.

$$O^* + O_2 \longrightarrow O + O_2$$

Finally, the ground state oxygen atom combines with the oxygen molecule to form ozone in the excited state.

$$O + O_2 \longrightarrow O_3^*$$

The ozone formed in the excited state is stabilized in a third body by various ways.

$$O_3^* + M \longrightarrow O_3 + M$$

M can be O_2 or nitrogen or any other inert molecule which removes excess energy of recombination that can cause decomposition of ozone.

However, solar radiation (210–290 mm) can cause decomposition of ozone to molecular oxygen but it reforms again resulting in no net loss of ozone.

$$O + O_2 \longrightarrow O_3$$

$$O_3 \xrightarrow{kv} O_2 + O + O + \text{heat}$$

Ozone can also get dissociated by collision with a free oxygen atom.

$$O_3 + O \longrightarrow 2O_2$$

This dynamic equilibrium of ozone is naturally maintained in the ozonosphere. All these reactions together are known as the Chapman Reaction. As the height increases, the rate of dissociation of the O_3 molecule increases and the rate of formation of the same decreases. The concentration of ozone is a balance between these competing reactions.

During the past 15 years, this protective layer of ozone has reduced over the South Pole due to some natural and man-made activities. It has been estimated that one per cent depletion in ozone concentration will lead to a six per cent increase in skin cancer ailments. A study by scientists of the British Antarctic Survey revealed that a decrease in ozone concentration by more than 40 per cent was found over Antarctica. However, calculations at Antarctica stations indicated that depletion in the ozone layer over Antarctica is between 50 per cent and 90 per cent at some places.

The increase in air traffic of supersonic aircrafts that fly at the ozonosphere altitude for their required low air resistance, releases a large amount of nitrogen oxides. Atmospheric nuclear explosions also emit a significant amount of oxides of nitrogen which directly enter the atmosphere. The cumulative effect of increase in nitrogen oxides increases the rate of oxygen decomposition which in turn reduces the ozone layer as follows:

$$NO + O_3 \longrightarrow NO_2 + O_2$$

$$O_2 + hv \longrightarrow O + O$$

$$NO_2 + O \longrightarrow NO + O_2$$
$$O_3 + O \longrightarrow 2O_2$$

Chloro fluoro carbons (CFCs), fluorochloro methane (Freons), difluoro dichloro methane (CF_2Cl_2) and fluoro chloroform ($CFCl_3$) are used as coolants in refrigerators, air conditioners and propellants in pressurized aerosol cans since the 1930s with technological advancements. They found increasing applications until the 1980s when chloro fluoro carbons were found responsible for ozone layer depletion. They release chlorine by ultraviolet radiation in the atmosphere. Estimate shows that one molecule of chlorine from chloro fluoro carbons can destroy one lakh ozone molecules when they diffuse to the stratospheric level and 6.5 per cent of the total ozone layer depletion is reported to be due to chlorine alone.

$$Cl + O_3 \longrightarrow ClO + O_2$$
$$ClO + O \longrightarrow Cl + O_2$$
$$O_3 + O \longrightarrow 2O_2$$

Depletion of atmospheric ozone level reduces the screening effect of the sun's lethal radiation entering the earth's biosphere. An increase in ultraviolet rays will not only adversely affect aquatic organisms but will also reduce global food production. Excess ultraviolet rays of the sun will cause skin cancer, retardation in tissue growth, albumin coagulation and ecological disturbances.

Control Measure: Ozone layer depletion or ozone hole formation is serious for the existence of mankind. The control measures will be through mass awareness. Statutory regulations on the restricted use of CFCs and advanced research to plug the ozone holes that have already been formed.

Smog or Smoke-Fog

Smog, an odd combination of smoke and fog is prevalent in London. It is commonly known as sulphurous smog or London smog formed by the presence of sulphur dioxide. This smog affected London city badly for several centuries after the introduction of coal as fuel and hence the name. The mixture of components is chemically reducing in nature so they the smog is also called reducing smog. It occurs mainly in the early morning hours of the winter season. During this period, the relative humidity is high and also air near the ground is cool by conduction from cooled surfaces (particularly on a clear night), while higher up in the troposphere the air is not that cool. In the absence of wind, the warm air lies over cooler air and temperature inversion takes place. Under this particular atmospheric status, the following chemical reactions take place leading to the formation of sulphurous (London) smog.

$$S + O_2 \longrightarrow SO_2$$

In the presence of sunlight and particulate matter SO_2 so formed gets oxidized to SO_3 as follows:

$$SO_2 + \tfrac{1}{2}O_2 \longrightarrow SO_3$$
$$SO_3 + H_2O \longrightarrow H_2SO_4$$
$$H_2SO_4 + 2NH_3 \longrightarrow (NH_4)_2SO_4$$

The chemical reaction in this type of smog is not photochemical in nature. London smog causes throat irritation and difficulty in breathing.

The other type of smog is due to the presence of oxides of nitrogen (NO_2) in the atmosphere which mainly forms as a result of photochemical reactions of automobile exhaust. It is called photochemical smog or Los Angeles smog. The most severe smog of this type was caused in Los Angeles in 1944 and hence the name.

photochemical smog occurs in warm sunny days (sunlight is essential to carry out photochemical reaction) both in summer and winter when the sky is clear and humidity is low. A subsidence inversion is a perfect condition for formation of photochemical smog. Chemically, photochemical smog is formed as follows:

(i) Reactive hydrocarbons (RCH_2 having C=C) present in automobile exhaust react with O_3 of troposphere to form RCH_2^*.

$$\text{Automobile exhaust } (RCH_2) \xrightarrow{O_3} RCH_2^*$$

(ii) RCH_2^* immediately reacts with O_2 and yields the free radical $RCH_2O_2^*$.

$$RCH_2^* \xrightarrow{O_2} RCH_2O_2^*$$

(iii) $RCH_2O_2^*$ reacts with NO to produce NO_2 and another free radical RCH_2O^*.

$$RCH_2O_2^* + NO \longrightarrow RCH_2O^* + NO_2$$

(iv) RCH_2O^* then reacts with O_2 to yield a stable aldehyde (RCHO) and hydroperoxyl radical (HO_2^*) which reacts with another molecule of NO to yield NO_2 and HO^*.

$$RCH_2O^* + O_2 \longrightarrow RCHO + HO_2^*$$
$$HO_2^* \xrightarrow{NO} NO_2 + HO^*$$

(v) HO^* being extremely reactive, rapidly reacts with RCH_3 (a stable hydrocarbon) and regenerates the free radical RCH_2^*.

$$HO^* + RCH_3 \longrightarrow RCH_2^* + H_2O$$

(vi) The aldehyde (RCHO) so formed reacts with HO^* to form Peroxy Acetyl Nitrate (PAN) as a by-product of photochemical smog.

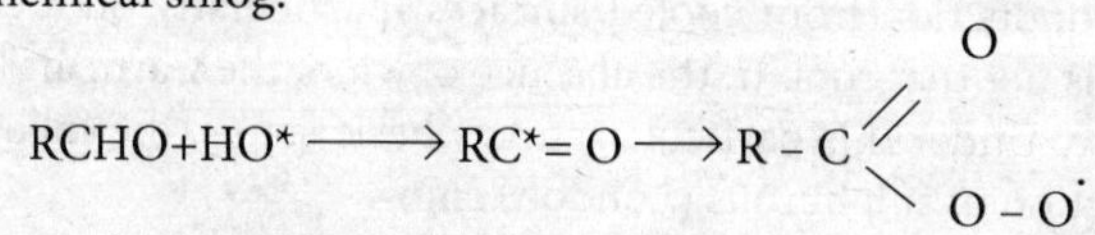

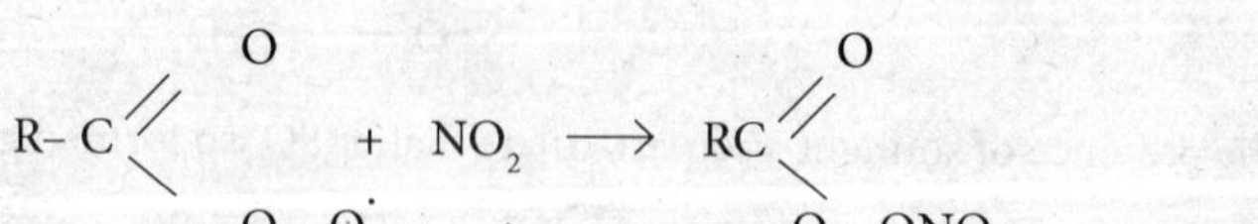

Peroxyacyl radical   nitrogen dioxide   Peroxyacylnitrate

The majority of the harmful effects of hydrocarbon pollution result not from hydrocarbons themselves but from the photochemical reactions of the hydrocarbons. Photochemical oxidants are formed in the atmosphere where large quantities of automobile exhaust is trapped by an inversion layer of stagnant air masses and exposed to intense sunlight. Photochemical smog consists of brown hazy fumes. It irritates the eyes and lungs leading to cracking of rubber and causes extensive damage to plant life.

6.7 Nuclear Accidents and the Holocaust

If used with proper care, nuclear energy can be a boon to the human civilization or else it can be a bane for all of us. Nowadays, radio isotopes are used for detection of ulcers and treatment of cancer. X-rays are a very useful tool in medical science for the diagnosis of fractures in our body. Laser rays are used for bloodless operations in serious patients. Nuclear energy is used for the generation of electricity. The role of nuclear energy for the betterment of human civilization is endless as long as it is used within limits.

But the same nuclear energy is responsible for the atomic and hydrogen bombs used for the destruction of Hiroshima and Nagasaki in Japan in 1945. The effects of radiation from these nuclear bombs can still be seen in the form of cancer and genetic mutation in the survivors of the incidents. Excessive exposure to the same nuclear energy that is used for treatment of patients can also produce serious diseases like bone tumors, genetic damage, infant mortality or even leukemia. The Chernobyl nuclear mishap on April 28, 1986 near Kiev in USSR is also an example of a nuclear holocaust in the history of mankind.

Radiation Episodes

In spite of extreme danger of radiation pollution hazards to mankind, test explosions of nuclear weapons by developed and developing countries, explosion of two atom bombs in Hiroshima and Nagasaki by the USA during World War II, explosions at the Chernobyl power plant in the Soviet Ukraine in 1986 and other such incidents have been taking place all over the world. These accidents/incidents had resulted in radiation leakage and damage to the fragile ecosystem. Chernobyl was the first officially reported nuclear accident in USSR and the world.

The first nuclear reactor of the world was commissioned in USA in 1942 and India's research reactor, Apsara in 1956. It is reported that there had been more than 10,000 reactor accidents in USA alone and 200 serious accidents in India. Since then, however, most of the incidents are kept secret by the respective authorities. Some of the known radiation episodes of the world are described below in brief.

6.8 Consumerism and Waste Products

Modern society has made mankind consumerist for comfort. Today, people consume more products and services and thus more resources and energies of the earth, which are fast depleting. The life cycle of the products and services has been shortened and even they have become 'disposable' for single use such as injection syringes, paper cups, cans, bottles, plastic cartons and condoms. Further, with a rapid increase in population, the amount of resources and energy consumed is increasing which results in a corresponding increase in waste disposal, both in quality and quantity.

Large quantities of solid, liquid and gaseous wastes generated by modern societies, mineral refuse, e-waste, biomedical waste, agriculture waste fertilizer and pesticide overuse, dead animals, human and animal excreta, increasing landfills, burning of waste and waste dumping in the rivers and oceans all pose a serious threat to environment management.

Careful and judicious utilization of resources and recovery of used material by recycling of waste are essential today. We must minimize the generation of waste at the source itself by improving the process efficiency with minimal use of resources. Reduce, reuse, recycle— the three Rs principle must be adopted by everyone to make a low waste or no waste society.

Plastic has become an indispensable part of modern life — cars, computers, baby bottles, telephone, clothing and packaging–it is used everywhere. Although plastic is non-biodegradable, its versatility in physical properties and unique formability has made plastic recyclable. Plastics can be recycled mechanically, chemically or thermally to produce new articles. In India, the Department of Science and Technology and Indian Oil Corporation are jointly working on the method of converting used plastics into petrol, diesel and LPG and expect to be India's first plant for recovering petrol, diesel and LPG from waste plastics.

Environmental Movements

Environmental deterioration is a by-product of rapid industrialization and developmental projects. It has become so great that immediate steps for conservation and regeneration of environmental resources are urgently necessary.

Initially, the environment referred only to the physical, chemical and biological aspects. But it has changed with time to include man-made social, cultural, economic and technological aspects too. The first social awareness about environmental problems was started by an 'elite group' of people in the 1950s. It gathered momentum in the 1980s and 1990s through scientific and technological research and development of organizations, writing textbooks on environmental science for teaching, educating the general masses through various media like posters, slides, audio-visual media and special exhibitions. Now, the world has entered the 21st century when all its natural renewable resources have been pushed closer to their limits. Drastic planning is the need of the hour to bring the environment back from non-sustainable to sustainable development.

NGOs

Various Non-Government Organizations (NGOs) have already been involved in educating people about environmental awareness. There are about 187 NGOs, of which 129 have been involved in environmental education and awareness, 56 in conservation of nature, 47 in pollution control, 46 in afforestation and social foresting, 28 in floral and faunal studies, 11 in rural development and 10 each in wildlife conservation and waste utilization and nine in eco-development. These NGOs work mainly at the grassroots level and also act as the ears and eyes of the Government.

(i) They give necessary aid and advice to the government

(ii) They educate people and create general awareness for conservation.

In addition to the efforts of the Government in preventing and controlling environmental pollution, these NGOs play a significant role in environmental protection.

One of the oldest cases of environmental pollution is between the *Society for the Protection of the Silent Valley* vs *Union of India* in the Kerala High Court for seeking a writ for banning the proposed hydroelectric project at the Silent Valley. The construction of the project could have endangered the famous tropical evergreen forest and its diverse flora and fauna and would have thereby adversely affected the climate condition as well.

Although the court's verdict was against the society, ultimately the project was abandoned by the timely intervention of the then Prime Minister Mrs Indira Gandhi. However, this not only saved the environment but also aroused active awareness about ecological imbalance amongst people.

Subsequently, the Court responded positively to a number of PILs on environmental issues. The PIL by the Rural Litigation and Entitlement Kendras, Dehradun against the Uttar Pradesh government for environmental degradation due to limestone mining in the sub-Himalayan terrain led to the court ordering the closure of half the mines and an award of Rs 10,000 towards the cost to be paid to voluntary organizations by the government. Other examples of PIL cases are: the Sriram Fertilizer Factory case, Oleum Leak case, Union Carbide case, and the Rajgangpur cement factory case all of which received a positive response.

The NGOs have also worked against so called developmental projects that appeared to have threatened the local inhabitants with ecological imbalance.

Appiko Movement

Dating back to 1983, this movement was begun in protest against the forest department's felling of trees in Salkane Forest in Sirsi district in Karnataka. *Appiko* means, to hug. Volunteers do not put a total ban on tree felling; rather they put some rules and regulations on the felling of trees. According to them, local people should be consulted before marking the trees for felling and there should be a total ban on tree felling within 100 metres of a water source and on a slope of 30° or above.

***Kalpavriksha* (KV)**

The organization, Kalpavriksha, started in 1979 to oppose the destruction of Delhi's green areas in developing workbooks for environmental studies at school levels and also conducting research on the same issue.

Worldwide Fund for Nature (WWF India)

This organization acts through its 200 volunteers and 10,000-subscriber supporters for the conservation of the country's natural heritage through support to research, field projects, education and training.

The contribution of Maneka Gandhi for the protection of the environment or the dedication of Medha Patkar in the *Narmada Bachao Andolan* and of Suman Sahani and Vandana Shiva for advocating development without destruction are unforgettable. Our then Prime Minister, Mrs Indira Gandhi was awarded with the International Union of Conservation of Nature for her sincere efforts for a cleaner environment.

Movement in Orissa

The NGOs and environment groups have also been very active in Orissa. The following environmental movements need special mention:

(i) Orissa *Krushak Mahasangha* and others agitated in 1991–1992 against the introduction of intensive shrimp farming in the waters of Chilika and sought discontinuation of the proposal.

(ii) *Gandhamardhan Yuva Parishad* led a movement in Western Orissa to make the state government agree to close down the Bauxite project in the famous Gandhamardhan Hill.

(iii) The Orissa Environmental Society launched the movement to declare Similipal forest as a biosphere reserve. It was declared the 15th biosphere reserve of India.

(iv) Southern Orissa spearheaded a movement to conserve Mahendra *Giri* in Eastern ghat, a rare forest of rich biodiversity in Gajapati district.

(v) Orissa Environmental Consciousness Society is actively engaged in spearheading and popularising environmental education in Orissa, which started functioning from Jyoti Vihar-its present headquarters.

Constitutional Provisions

The advancement of human civilization from a hunter gatherer society to the modern technologically advanced industrial society has induced man to tamper with the natural environment for creating an economic, social and cultural environment of his choice. This has resulted in the depletion of natural resources and led to water, air and soil pollution. Technological development has generated noise pollution. The negligence in the handling, storage and transportation of different hazardous and radioactive chemicals has resulted in mass causalities. To protect ourselves, plants, animals and the environment, several laws have been enacted at the national and international levels.

The United Nations Conference on Human Environment, 1972 at Stockholm was attended by the representatives of 113 world governments to discuss this issue. At the conference it was proclaimed that, 'The protection and improvement of human environment is a major issue which affects the well-being of people and the economic development throughout the world and it is the duty of all the governments and its people to exert common effort for the preservation and improvement of human environment, for the benefit of all the people and their posterity.'

Thereafter, many countries introduced control mechanisms by amending and enacting special criminal laws to prosecute most flagrant offenders. India is the first country to impose a constitutional obligation on the state and citizens to protect and improve the environment as a primary duty. Article 48A of the Indian constitution says:

The state shall endeavour to protect and improve the environment and to safeguard forests and wildlife of the country.

Article 51A also provides:

It shall be the duty of every citizen of India to protect and improve the natural environment including forests, lakes, rivers and wildlife and to have compassion for living creatures.

6.9 Laws Relating to Environment, Pollution, Forest and Wildlife

Numerous laws have been enacted at the international, national, state and municipal levels. The following national acts provide environmental legislation:

(i) The Factories Act, 1948.

(ii) The Insecticides Act, 1968.

(iii) The Water (Prevention and Control of Pollution) Act, 1974.

(iv) The Air (Prosecution and Control of the Pollution Act, 1981.

(v) The Forest (Conservation) Act, 1980.

(vi) The Wildlife (Protection) Act, 1972.

(vii) The Environment (Protection) Act, 1986.

The main provisions of some of these national acts are:

The Factory (Amendment) Act of 1987

This Act is aimed at declaring the information of hazardous processes taking place inside the factory to its workers, local residents and government officials. This rule also allows the employees to inform the Factory Inspectorate directly regarding violation of safety rules during plant operation. According to the second amendment of the Environment (Protection) Act, 1992 all these industries are covered by Water and Air Pollution Act and also by Hazardous Waste Handling Act. According to this amendment, every industry is to submit an environmental audit report to the State Pollution Control Board on or before May 15 for the financial year ending on March 31. In order to reduce the pollution burden and optimum utilization of natural resources, a comparative statement from the previous year is to be submitted.

The Insecticides Act, 1968

This is one of India's first national environmental laws enacted in 1968 and enforced from 1971 to regulate import, manufacture, sale, transport, distribution and use of insecticides to prevent risk to human beings and animals. Different agencies like the Central Insecticide Board, Pesticide Registration Committee, Pesticide Environment Pollution Advisory Committee, Central Insecticide Laboratory Committee were created for effective enforcement of the Act and to ban or restrict the use of pesticides.

The Water (Prevention and Control of Pollution) Act (1974)

This was the first national law of India on pollution control. This act defines water pollution and determines penalties. The Water Pollution Board at the Central and State levels has been set up for the prevention and control of water pollution in all sources of water under its jurisdiction. They also determine the permissible and impermissible levels of water pollution. This act was amended in 1977 and 1978 to improve the financial condition of the state boards through implementation of a water tax for use of water by industrial and local authorities.

The Air (Prevention and Control of Pollution) Act 1981

This Act regulates and controls emission from automobiles and industrial plants. The Central Board for Prevention and Control of Water Pollution is authorized to implement and enforce this act also. This body determines the standards of air quality. The Central Board is empowered to coordinate the activities of the

State Boards. After consultation with the State Board, the state government may declare an area within the state as:

(i) Air pollution control area.

(ii) Prohibit the use of any fuel other than the approved one in the area of air pollution.

(iii) Nobody can operate any industrial plant in air pollution control area without the consent of the State Board.

The Forest (Conservation) Act, 1980

The India Forest Act of 1927 was amended in 1980 to prohibit the state governments from declaring any reserve forest as non-reserve without prior approval of the central authority. This amendment has restricted the efforts of the state governments to use forest land for non-forest purposes.

The Wildlife (Protection) Act, 1972

This Act was enacted in 1972 to provide necessary protection against serious threats to wildlife by the expansion and advancement of agriculture, industry and urbanization. The Indian Board of Wildlife was set up in 1952 to protect the wildlife resources and prevent gene erosion. Various wildlife parks and sanctuaries were created; projects were launched to save endangered species such as lions (1972), tigers (1973), crocodiles (1974) and brown antlered deer (1981). India became a party to the Convention of International Trade in Endangered Species of Fauna and Flora (CITES) in 1976 and started a national component of the UNESCO's Man and the Biosphere Programme (1971).

The Wildlife Protection Act, (1972), was enacted to provide protection to wild animals and birds by:

(i) Constitution of wildlife advisory board in each state.

(ii) Regulating the hunting of wild animals and birds.

(iii) Specifying the procedures for declaring areas of sanctuaries and national park.

(iv) Regulating the possession, acquisition and trade of wild animals as well as animal products.

The Environment (Protection) Act, 1986

After the Bhopal gas tragedy in 1984, a series of new environmental laws were enacted. Through the Environmental (Protection) Act 1986, the Central Government has acquired the following powers:

(i) Laying down standards for emission or discharge of pollutants from various sources and for environmental quality.

(ii) Restricting area in which industries operation may or may not be carried out subject to certain safeguards.

(iii) Laying down safeguards for the prevention of accidents and remedial measures in case of accidents.

(iv) Procedures and protection for hazardous substance handling.

(v) Issuing directions to any person, officer or authorities for closure, prohibition or regulation in electricity or water supply or any other services.

This act also empowers a person to complain to the court regarding violation of the provisions of the Act after having given a notice of 60 days to the prescribed authorities.

A violator of the Environmental Protection Act can face imprisonment up to five years and a fine of up to Rs one lakh.

Role of the Judiciary: Issues Involved in Enforcement of Environmental Legislation

Regulatory measures in the form of legislation check the degradation of the environment and lead to the enacting of laws at the national or international levels to prevent pollution. The role of the judiciary in protecting the environment lies in formulation and enforcement of effective laws to protect the environment.

The Supreme Court of India's judgment sanctioning US$ 470 million towards compensation by the Union Carbide to the victims of the Bhopal gas tragedy is an example of what the judiciary can do to protect the environment.

The Constitution of India has provisions to make environmental legislation. Many legislations have already been enacted to protect the environment. Judiciaries with their limited resources, try to enforce such laws. But the judiciary alone cannot improve the environment unless the states and citizens do their duties and obligations to protect and improve the environment.

For successful implementation of an environmental legislation, collection of relevant data, its processing and final submission to the enforcement agency has to be done honestly and effectively. Violation of any law or rule by an individual or institution has to be punished legally. Information must reach the law enforcement officials from the concerned person or people. If no cognizance is taken, the affected or interested person must file a Public Interest Litigation (PIL) for the protection of the environment. Thus, the general public must be careful of any irregular practice that is likely to have an adverse effect on our national environment.

6.10 Environmental Impact Assessment

Environmental Impact Assessment (EIA) is a widely recognized study to assess environmental impacts of developmental projects. Decisions on proposed developmental projects are made based on the findings of an EIA study. An extension of an EIA could be a cost benefit analysis where the value of economic gains versus environmental losses or vice versa can be suitably compared to establish a sound basis for the choice of a particular project.

EIA is the analysis of any possible alteration of environmental conditions, adverse or beneficial, caused by the project under consideration. It is the tool of environmental protection to test the compatibility of the environment before taking any decision to construct public, private or governmental projects and industries which could deteriorate the natural resource. It is anticipatory in nature and attempts to answer the following questions:

(i) Are these physical/health hazards associated with the locations of the proposed projects?

(ii) Will these be significant disturbances on the existing community?

(iii) Are there potential conflicts with economic interests?

(iv) Will the project require major additions to the existing physical infrastructure including transport, water and housing?

It is a process of identifying likely consequences for the natural environment and man's health and welfare, of implementing particular activities and designs to identify, predict, interpret and communicate information at a stage when their decision may materially affect those responsible for sanctioning the proposal.

Undesirable environmental effects caused by man's deliberate and inadvertent activities can be prevented through EIAs. When EIAs are adopted, planners are bound to consider side effects of any proposal over a time frame of short, medium and long term. Possible alternatives and the proposal may be dropped or the most acceptable alternative with or without modification is accepted.

EIA Process

An EIA neither solves problems in itself nor substitutes for the formulation and implementation of appropriate policies. However, the overall process ensures that developments cause minimal environmental damage, do not unnecessarily reduce the productivity of the natural systems and do not impose unwanted costs on other development activities.

EIA consists of the following stages:

(i) Identifications of goals and objectives.

(ii) Survey, forecast and analysis.

(iii) Formulation and evaluation of alternative plans.

(iv) Decision, implementation and monitoring.

(v) Consultation and public participation.

Stage I: Identification of Goals and Objectives

Generally, environmental objectives are not comprehensive mainly due to the absence of clear environmental quality standards. However, as industries have been increasingly adopting ISO 14001 Environmental Management System, the situation has been fast changing and one must clearly identify and formulate goals and objectives of the EIA process.

Stage II: Survey, Forecast and Analysis

Requirement of data with respect to significant parameters of existing environmental conditions is to be identified by using checklists, environmental matrices and other tools. The data obtained through surveys is to be reviewed and monitored continuously. The accuracy of forecasting future environmental conditions is dependent on the quality of population and economic forecasts and other assumptions of impacting parameters. The following actions are to be taken in this stage:

(i) To develop environmental quality standards or targets.

(ii) To predict expected environmental impacts based on available surveys and data.

(iii) To analyse and assess the expected or planned development in terms of likely environmental damage to proceed, modify, change or cancel the project.

Stage III: Formation and Evaluation of Alternative Plans

Alternative plans which satisfy environmental goals and objectives need formulation and critical assessment with respect to ranking of their relative usefulness. Evaluating the environmental impact of each alternative plan involves determining the likely economic, demographic and environmental consequences that would result from their implementation.

Stage IV: Decision, Implementation and Follow-up.

After an environmental plan is chosen, it needs to be implemented and followed up. If evaluation of alternative plans is correct and the right plan is chosen, then there should be no technical problem in implementation and follow-up. However, some times certain deficiencies crop up in implementation and follow-up. Thus environmental impacts of plan implementation should be audited.

Stage V: Public Participation, Consultation and Communication

The content of planning reports, including estimates of environmental impacts should be communicated to the public in a sufficiently clear and comprehensive manner. This is required to get a reliable and representative community assessment of probable impacts.

Environmental Impact Statement (EIS)

It is prepared with the following elements:

(i) Preliminary discussion.

(ii) Writing of EIS.

(iii) Consultation.

(iv) Public participation.
(v) Reviewing.
(vi) Decision.
(vii) Appeal.

Cost-Benefit Analysis

The viability of every project is judged by a single yardstick, a cost-benefit analysis. Cost-benefit analysis is also required for a reliable and representative EIA. Improved communication and timely dissemination of information among the concerned agencies improves cost-effectiveness. The improvement of cost-effectiveness also needs the following:

(i) A close integration of EIA and development planning.
(ii) Clarity of responsibility and continuity of the presence of a responsible person throughout the project.
(iii) Reduction and if possible elimination of overlapping responsibilities and simplification of administrative procedures.
(iv) Laying down an improved system of defining and accounting for environmental costs.

Environmental Risk Assessment (ERA)

ERA begins at the fact-finding/preparation stage when the environmental examination (EIA) indicates a potential hazard. It suggests risk reduction and risk management measures to be incorporated. A typical flowchart highlighting the major components of risk audit management is shown in the Figure 6.8.

Risk Audit/Risk Management Hierarchy

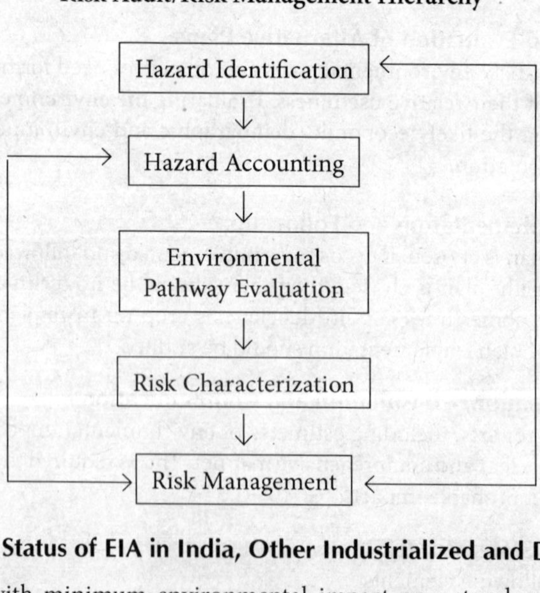

Figure 6.5 Current Status of EIA in India, Other Industrialized and Developing Countries

Sustainable development with minimum environmental impact on natural resources and the world has become an urgent issue in every level and process. Government and non-government environmental groups and associations have actively started putting pressure on the concerned authorities for regulations and enforcement of appropriate actions so as to prevent long-term environmental deterioration. The USA enforced one such Act in 1970, as a result of the Act, EIA became a major component of the National Environmental Policy Act (NEPA) of USA as a planning instrument due to a strong public opinion of its necessity in predicting the impact of actions and coordinate the interests of other groups. Other countries such as Australia, Canada, France, Japan and Germany also adopted the EIA process to be a part of environmental regulations of these respective countries.

India

An expert committee constituted by the Department of Environment and the National Council of Environmental Planning (NCEP) has adopted a preliminary procedure to prepare an EIA. The NCEP and the Planning Commission have also prepared a checklist to be used as a guideline to assess the environmental impact of industrial projects, construction of roads and railways, mining, hydroelectric and irrigation projects. The international standard of Environmental Management System, ISO 14001 is also being increasingly adopted and practiced by many Indian companies. However, EIA needs further development and wider acceptance and implementation in India.

Developed Countries

EIA was adopted in USA in 1970 through NEPA. Public agencies have to prepare a preliminary EIS report before taking any action and federal agencies have to prepare an EIS for the proposed project to determine whether the project affects the environment positively or negatively. The Council of Environmental Quality is the authorized body in the USA to check the implementation of EIA requirement and revise the regulations if required.

In Canada, according to the Canadian Federal Government's environmental assessment and review process, the federal agencies screen the project and its adverse effects followed by the preparation of an initial Environmental Evaluation Report. Then the Department of Environment reviews it thoroughly. Sometimes, the Environmental Assessment Panel, a small group of experts constituted by the Federal Environmental Review Office, reviews the impact of a project and provides guidelines to the environmental agency, collects and reviews public opinions and forwards proposal actions to the government for decision at the level of the Ministry of Environment and the concerned department.

In Japan, the government issued guidelines for project reviews with respect to environmental impact in 1972. This was followed by the formation of the EIA department in the environmental agency and the introduction of the legislation according to which compensations are to be paid to those who are environmentally affected by a project.

In Britain, the Centre for Environmental Management and Planning (CEMP) at Aberdeen University prepared a document on the environment in 1973, with the funding of the Scottish Development Department (SDD) and Department of Environment; this was followed by the adoption of EIA guidelines by the industries.

In Germany, the public is highly aware about environmental concerns, impacts and the short-term and long-term action plans needed. In 1975, the Interior Ministry issued a model procedure for examining public measures for environmental compatibility on the basis of the principles for the EIA of Federal Actions. Though there is no legislation to enforce but the Ministry of Economic Cooperation involves itself in various bilateral programmes regarding implementation of the EIA.

EIA in Developing Countries

Apart from India, Brazil is one of the developing countries which has set up its Secretariats of Environment as long back as 1973. It issued the National Environmental Policy Law in 1981. The permit system, requiring detailed EIA reports from the project developer was introduced by the Government of Rio de Janeiro in 1977. This was followed by the National Environmental Law in 1981 to direct enforcement of EIA implementation. Among developing nations, Brazil is ahead in logistics, legislations and technical manpower in them field of environmental management.

In Indonesia, though the Government passed the Environmental Management Law in 1982 to emphasize the necessity for it, so far, no regulations or procedures have been enforced.

In the Philippines, NEPA which functions under the Ministry of Human Settlement, evaluates projects and issues Environmental Exemption Certificates or refers to the President for approval if the environmental impact is likely to be serious.

The status of EIA in India, a developing country has already been discussed earlier.

6.11 Precautionary Principle

In order to overcome the limitations of the ERA policy and to give a protective cover to human beings and the environment, the Precautionary Principle was introduced for the first time in 1984 in the First International Conference on the protection of the North Sea. The advancement of science and technology has created more and more uncertainties about the after-effects of the applications of these technologies. This cause and effect relationship may take a long time to fully establish or it may not be established scientifically at all, but its after-effects may be disastrous and irreversible for the environment. To protect our environment from damage, strong preventive measures have to be taken at the first sign of damage. For example, in 1970 the dense forests of the then West Germany suddenly started dying. There was no evidence that acid rain generated by emissions of the thermal power plant was responsible for this yet the government ordered cutting down of the emissions of the power plant and the problem was solved. Similarly, the use of mercury in thermometers, freon in refrigerators or even CO_2 exhaust from automobiles and different plants may affect biodiversity. Hence, they are banned. The formal concept of Precautionary Principle was generated from the German socio-legal traditions in the 1930s based on the concept of good household management.

The Precautionary Principle is a moral and political principle which states that, 'If an action or policy might cause severe or irreversible harm to the public, in the absence of a scientific consensus that would not ensue, the burden of proof falls on those who would advocate them to take the action."

In 1998, at the Wingspread Conference on Precautionary Principle convened by the Science and Environmental Health Network the principle was explained as, 'When an activity raises the threat of harm to human health or the environment, precautionary measures should be taken even if some cause and effect relationship is not fully established scientifically.'

In the Rio Declaration (Earth Summit) 1992, in Principle 15 of Agenda 21, the precautionary approach was codified for the first time at the global level to protect the environment. It states that:

(i) Every state should apply the principle according to their capabilities.

(ii) Where there are threats of serious or irreversible damage, lack of full scientific certainty shall not be used as a reason for postponing cost-effective measures to prevent environmental degradation.

The Precautionary Principle is now widely accepted as a fundamental concept of national environmental laws and regulations in management instruments at the international, regional and domestic levels. It is used across diverse areas such as pollution, toxic chemicals, food, fisheries management, species introduction, wildlife, trade and so on. Application of the principle in natural resource management and biodiversity conservation is clearly essential. The Precautionary Principle is elaborated in the Water Law of Israel, in the Environmental Protection Act of the Czech Republic and also in the Environmental Protection Act (1996) of Pakistan.

Polluter Pays Principle (PPP)

This as the name suggests is based on a simple principle that those who pollute the environment must also pay for the damages caused by them. This idea originated in the 1970s when the members of the Organization for Economic Cooperation and Development (OECD) introduced a payment method where pollution control costs are to be financed by polluters alone and not the public in general. Hence, this method of environmental financing gets its name Polluter Pays Principle. Later in 1985, economic instruments such as pollution taxes, user charges and subsidies were also added to OECD environmental guidelines.

This principle, when adopted in a country with a higher level of poverty and economic inequality such as India or Thailand, can hurt the poor and worsen economic inequality. There would be a tendency that low-income families would end up paying the cost of environmental protection while the wealthy would become the major beneficiaries of the improved environment. For instance, insecticide-free vegetables are generally more expensive and are sold at supermarkets instead of local fresh markets. Cheap clone computers usually come in ordinary cardboard boxes while the more expensive Apple machines are packaged in brown recyclable containers. Higher-end luxurious automobiles are now equipped with CFC free air conditioners and the list is endless.

There are some poverty-struck families whose income itself comes from environmental exploitation, either in cash or kind. Landless farmers encroach upon watershed areas in search of insecure cultivated areas to grow suitable cash crops. If one takes the example of poor business people who cannot afford the additional cost of installing a new engine in their pick-up trucks in order to improve exhaust emission and hence have no other option but to pollute clean air. Hence, at this point it would not be an exaggeration to say that environmental conservation or an improved environment is a luxury for the rich and a burden for the poor.

However, the proponents of this principle may argue that it will force only the polluting factories to pay for the environmental damages that they have caused and hence encourage them to install suitable treatment devices to pollute less. Unfortunately, the story does not end here because even if the polluter pays, the principle is targeted at factory owners, the poor will eventually have to end up paying their part of pollution charges through price hikes due to the installation of the treatment devices.

Successful environmental conservation not only incorporates an appropriate method of payment but there is also a need to achieve the objectives of efficiency and cost-effectiveness to promote sustainable development.

PPP cannot ensure efficiency as it determines only **who** and not **how much** one has to invest in pollution control. Even if this principle is adopted, policy makers need to regulate the level of pollution.

Moreover as it only determines the **who** pays factor and not who should be conducting pollution control activities, it is not cost-effective either.

There are some economic instruments such as pollution tax which achieve the dual goals of efficiency and cost-effectiveness and belong to PPP but on the whole this principle hampers sustainable development in a society where both poverty and economic inequality co-exist. Making the polluters pay seems justifiable but when these polluters are poor, then incorporating a principle such as this in the administrative system would not only be morally unjust but it would also be hardly effective in achieving the goals for which it was formed.

Beneficiary Pays Principle (BPP)

In a developing country such as India, the polluters often turn out to be the low-income families who use the environment as a supplement to their tiny income. When these poverty-struck people have to live practically a hand-to-mouth existence to earn two square meals a day then it seems quite logical to think that environmental conservation would not even exist in their list of priorities. At such a stage, the BPP can be convincing when it is used in the context of a developing country because unlike the PPP it assigns environmental rights to those who benefit from environmental improvement than any one who pollutes (more often the poor than the rich).Thus, BPP is used in place of PPP when it is required to ease the economic gap between the rich and the poor. For example, the PPP requires the poor commuters on public buses to pay a higher fee to finance a cleaner and smoke-free bus exhaust emission for everyone to enjoy clean and green air. However, BPP argues that it should be financed by all urban residents and not just the commuters as clean air would benefit everyone.

The BPP suggests that funding for environmental improvement should be obtained from its beneficiaries and hence the name.

This method of financing would generate larger revenue for the government as the rich are willing and capable of paying more for environmental enrichment. Further, a given amount of revenue generated from pollution charges would constitute a greater portion of income of the poor than the rich.

These examples illustrate the validity of BPP as a method of financing environmental improvement in the context of a developing country. The choice among these payment methods depends on considerations of fairness and social justice. There is no readymade formula to dictate which principle should be adopted.

Emissions Trading (Cap and Trade)

This is an administrative approach to promote and control pollution by providing financial incentives for achieving reduction in the emission of pollutants. Here, an administrative body or a central governing body takes the administrative approach. It may be an international body, the Central Pollution Control Board (CPCB) or even the State Pollution Control Board (SPCB). The maximum amount of pollutants that a company or a group can emit has a limit or cap set by a governing body. The company or the group is then issued an emission permit and is required to hold an equivalent number of allowances or credits basing on this cap. At no given time can the cap be less than the total amount of allowances and credits. It is also called 'cap and trade' because emission trading involves both cap and trade. The underlying principle of emission trading is that a company or a group must pay a charge for polluting and this charge is proportional to the quantum of pollution that they emit into the environment.

Several pollutants such as Greenhouse gases and nitrous oxide (that cause acid rain), have active trading programs.

Carbon Trading

The idea of carbon trade watch came into force in 2002. The inclusion of pollution trading policy in the Kyoto Protocol signals a historical proliferation of the free market principle into the environmental sphere. With a focus on emerging Greenhouse gases, carbon trading watch monitors the impact of pollution trading upon environmental, social and economic justice.

CASE STUDY

Powerguda village in Adilabad district of Andhra Pradesh had sold 147 tonnes equivalent of saved carbon credits to the World Bank for US$ 645. According to villagers of Powerguda and Emmanuel D'Silva a farmer, the World Bank staff who are working in the area for creating awareness about their trade have extracted biodiesel from 4,500 Pongamia trees in their village. By using biodiesel instead of petroleum they were able to save 147 MT of CO_2 and were also able to enhance the air quality. The World Bank was buying the carbon credits to balance the aviation fuel burnt by aircrafts carrying bank officials. At present, many other villages of India are following Powerguda and making carbon credit sales.

According to the Kyoto Protocol of 1997, all the countries are required to reduce their Greenhouse gas emission by five per cent from the 1990 levels in the next 10 years or pay a price to those that do. That means if a country is a consumer of any environmental value (clean air) it must pay a producer an equivalent value.

ISO 14000

ISO 14000 is a series of international standards on environmental management tools and systems introduced in 1996 by the International Standards Organization (ISO). The ISO is an apex body on standardizations in the world, formed in 1946 with the national bodies of most of the countries as its members; it has mostly developed technical standards to facilitate international exchange of goods and services. In 1987, it introduced ISO 9000 — the quality management system with emphasis on continual improvement of the process and the quality of goods and services. After the success of ISO 9000 quality management system, the ISO introduced the ISO 14000 series of environmental management system in 1996.

In the past, environmental management systems were reactive, curative and oriented towards solving a problem for compliance of regulations. Increasing irreversible environmental degradation, resource depletion, greater socio-environmental compulsions, stronger regulations, international obligations and global competition have made economic growth without environmental sustainability of an industry temporary, self defeating and disastrous. Industries throughout the world have now been voluntarily adopting quality management systems and environmental management systems to bring down production cost, waste and environmental degradation and achieve continual improvement of process efficiencies to remain globally competitive.

Many countries and regional groupings — UK, Canada, USA, Germany, Japan and the European Union — have their own standards, regulations and eco-labelling programmes to address environmental issues. These standards, regulations and eco-labelling programmes vary from each other. A single standard will ensure that there are no conflicts between regional interpretations of good environmental practice and the same standard is uniformly followed throughout the world. The ISO 14000 series of environmental management system emerged primarily as a result of the Uruguay round of GATT negotiations and the Rio Summit on environment held in 1992.

SAGE, the Strategic Advisory Group on Environment was formed in 1991 and it recommended the creation of TC 207— a new international committee in 1992. TC 207, with its different sub-committees, formulated the new series of ISO 14000 standards which were finally approved in 1996.

ISO 14000 series of environmental management system addresses the following six areas of environmental management:

(i) Environmental Management System (EMS).
(ii) Environmental Auditing (EA).
(iii) Environmental Performance Evaluation (EPE).
(iv) Environmental Labelling (EL).
(v) Life Cycle Assessment (LCA).
(vi) Environmental Aspects in Products Standards (EAPS).

EMS, EA and EPE are related to an environmental management system addressing organizational issues LCL, EL and EPAS are product-oriented standards addressing environmental issues in practice and deal with core business issues such as how products are designed and marketed.

The ISO 14000 series of environmental management system does the following:

(i) Promote a common approach to environmental management similar to quality management.
(ii) Enhance an organization's ability to attain and measure environmental performance.
(iii) Facilitate trade and remove trade barriers.

The ISO 14000 series requires an environmental policy to be adopted by an organization with the commitment from the top management for continuous improvement and compliance of legislations communicated to its employees and the public and audited regularly to assess implementation. It focuses on management rather than on goals. It promotes audit and review of management, not of attainment. It establishes principles and uniform approaches for product evaluation and communication of environmental attributes. It addresses the immediate and long-term impact of the organization's products, processes and services on the environment and provides a consistent system in the allocation of resources, assigning of responsibilities and continual evaluation of practices, procedures and processes. Like ISO 9000, ISO 14000 standards are voluntary management standards and not performance specifications and thus do not establish required environmental performance levels. The interested organizations that have designed and implemented an EMS can apply for certification from the Certification Body accredited by the Accreditation Council which is the national body set up to evaluate the competence of the certification bodies.

Six sub-committees of TC 207 namely SC1 (EMS) and SC2 (EA) developed the family of specifications of ISO 14000 series. Table 6.1 gives the list of specifications.

Table 6.1 Specifications of ISO 14000 Series

Standard	Title/description	Applications/Remarks
ISO 14000	Guide to Environmental Management Principles, Systems and Supporting Techniques	
ISO 14001	Environmental Management Systems-Specification with Guidance for Use	Applicable to any organization producing and/or manufacturing any product or providing a service.
ISO 14004	Guidance to Requirements of ISO 14001	A structured approach to setting environmental objectives and targets and to establishing and monitoring operational controls.
ISO 140010	Guidelines for Environmental Auditing: General Principles	
ISO 140011	Guidelines for Environmental Auditing: Audit Procedures	
ISO 140012	Guidelines for environmental auditing—Criteria for Environmental Auditors.	
ISO 140013 to 14015	Guidelines for environmental auditing: Audit Programmes, Reviews and Assessment	
ISO 14020 to 14025	Environmental Labeling.	Covers labels and declarations.
ISO 140030	Post Production Environment Assessment	
ISO 140031/32	Guidelines on Environmental Performance Evaluation	
ISO 14040 to 14044	Life Cycle Assessment: General Principles and Practices	Discusses pre-production planning and environment goal setting.
ISO 14050	Terms and Definitions	Glossary
ISO 14060	Guide for Inclusion of Environmental Aspects in Product Standards	
ISO 14062	Environmental Impact Goals	Discusses improvements of it
ISO 14063	Further Communications on Environmental Impacts	An addendum to ISO 14020.
ISO 14064–1 to 3:2006	Greenhouse Gases Part 1, 2 & 3	Specification with guidance.
ISO 19011	Audit Protocol for both 14000 and 9000 Series Standards together	Replaces ISO 14011

SUMMARY

Air and water pollution are one of the many consequences of the overuse and abuse of natural resources. Others are depletion of the earth's protective ozone layer, global warming due to increase in carbon dioxide content by burning of fossil-fuels, the Greenhouse effect, acid rain, decrease in forest cover, soil erosion, the increased rate of species extinction just to name a few. The concept of sustainable development strengthens the interdependence of human needs and environmental requirements. Sustainable development maintains a balance between social and economic development along with protection of the environment.

Energy is one of the major pillars of economic development of society. The importance of the development of renewable energy sources as an alternative to fossil fuels for a sustainable energy base was recognized in India since 1970.

Efforts are also being made to meet the water crisis by building dams, reservoirs and digging wells to conserve water. Efforts are made for conservation of water by rain water harvesting and watershed management techniques. These surface water bodies are also responsible for recharging the ground water table of the area. Thus, attempts are made to recharge the aquifer, thereby recharging surface water bodies and thus increase the water table.

Environmental movements

The Constitution of India has provisions to make environmental legislation. Many environmental legislation have already been enacted to protect the environment. In addition to the efforts of the government in preventing and controlling environmental pollution, NGOs play a significant role in environmental protection.

Environmental Impact Assesement (EIA)

The Environmental Impact Assessment (EIA) is a widely recognized study to assess the environmental impact of developmental projects. Environmental Impact Assessment (EIA) is the analysis of any possible alteration of environmental conditions, adverse or beneficial, caused by the project under consideration.

Polluter Pays Principle (PPP)

Polluter Pays Principle (PPP) as the name suggests, is based on a simple principle that those who pollute the environment must also pay for the damages they have caused. This principle, when adopted in a country with higher levels of poverty and economic inequality such as India or Thailand, can hurt the poor and worsen economic inequality.

Beneficiary Pays Principle (BPP)

As the name indicates, the Beneficiary Pays Principle or BPP suggests that funding for environmental improvement should be obtained from its beneficiaries and hence the name.

This method of financing would generate larger revenue for the government as the rich are willing and capable of paying more for environmental enrichment.

Furthermore, a given amount of revenue generated from pollution charges would constitutes a greater portion of the income of the poor than the rich.

Emissions Trading, Carbon Trading

Carbon trading watch monitors the impact of pollution trading upon environmental, social and economic justice. After the success of the ISO 9000 quality management system, the International Standards Organization introduced the ISO 14000 series of environmental management system in 1996. Increasing irreversible environmental degradation, resource depletion, greater socio-environmental compulsions, stronger regulations, international obligations and global competitiveness have now made economic growth without environmental sustainability of an industry temporary, self-defeating and disastrous. The

ISO 14000 series of environmental management system addresses the following six areas of environmental management:

(i) Environmental Management System (EMS).
(ii) Environmental Auditing (EA).
(iii) Environmental Performance Evaluation (EPE).
(iv) Environmental Labeling (EL).
(v) Life Cycle Assessment (LCA).
(vi) Environmental Aspects in Products Standards (EAPS).

EMS, EA and EPE are related to the environmental management system that addresses organizational issues. The ISO 14000 series of environmental management system promotes a common approach to environmental management similar to quality management, enhances an organization's ability to attain and measure environmental performance and facilitates trade and removes trade barriers.

ESSAY TYPE QUESTIONS

1. What do you understand by the concept of sustainable development?
2. How do you propose that development work progress along with sustenance of our environment?
3. 'Urban people have more energy demand than rural people.' Explain the statement.
4. What is a watershed? How can watershed management solve the problem of depletion of the ground water table?
5. Write a short note on the success story of sustainability and environmental restoration at Auroville.
6. Explain how consumerism has affected the present society.
7. Write a note on the Chernobyl disaster.
8. What do you understand by the term Environmental ethics?
9. Write a note on different environmental movements for the conservation of environmental resources.
10. What are the powers provided to the Central Government by the Environmental Protection Act, 1986?
11. What are the Indian laws related to human rights?
12. What are the recent climate changes noticed in the environment?
13. What is acid rain? Explain the steps taken to prevent acid rain.
14. What are the effects of global warming on the environment and what are the means to combat it?
15. Discuss the environmental policies at the national and state level.
16. What are the objectives of Environmental Impact Assessment (EIA)?

SHORT-ANSWER TYPE QUESTIONS

1. Define sustainable development?
2. What is Frontier Mentality?
3. What are the major criteria for a sustainable society in India?
4. Write a short note on the energy problem related to urban people.

5. Write short notes on:
   (i) Rain water harvesting.
   (ii) Watershed management.
   (iii) Biodiesel generation.
6. Explain the role of nuclear energy for the betterment of human civilization.
7. Write short notes on:
   (i) *Chipko* movement.
   (ii) *Appiko* movement.
   (iii) Movements in Orissa.
8. How are the environment and environmental pollution defined in the Environmental Protection Act?
9. Name five important EPAs in India.
10. Write short notes on:
    (i) Greenhouse gases.
    (ii) Global warming and its control.
    (iii) Ozone layer depletion.
11. What are effects of ozone layer depletion?
12. What is acid rain? How is it formed?
13. What are the effects of acid rain?
14. What is Agenda-21?
15. What is Green Bench?
16. Write short notes on:
    (i) The Air (Prevention and Control of Pollution) Act, 1981.
    (ii) The Water (Prevention and Control of Pollution) Act, 1974.
    (iii) The Environment (Protection Act), 1986.
    (iv) The Wildlife (Protection) Act, 1971.
17. Explain the difference between smog and fog.
18. Write short notes on:
    (i) London Smog.
    (ii) Los Angeles Smog.

MULTIPLE CHOICE QUESTIONS

1. Sustainable development
   (a) meets the need of the present without compromising with those of future generations.
   (b) maintains a balance between social and economic development along with protection of nature.
   (c) is consistent with long term development and growth.
   (d) All of the above.

2. Frontier mentality is believed by
   (a) environmentalists and conservationists.
   (b) politicians and bureaucrat.
   (c) economists and technocrats.
   (d) farmers and laymen.

3. A geographic unit that collects, stores and releases water is
   (a) wasteland.
   (b) watershed.
   (c) wetland.
   (d) water divider.

4. Saline and alkaline lands are types of
   (a) barren land.
   (b) fallow land.
   (c) uncultivable land.
   (d) wasteland.

5. Atmospheric pollutants are mainly present in
   (a) ionosphere.
   (b) stratosphere.
   (c) lower mesosphere and whole ionosphere.
   (d) troposphere and lower stratosphere.

6. The most important Greenhouse gas that is being affected by human activities and results in global warming is
   (a) carbon dioxide.
   (b) water vapour.
   (c) methane.
   (d) nitrous oxide.

7. The two acidic oxides mainly responsible for acid rain are
   (a) sulphur dioxide and nitrous oxide.
   (b) sulphur dioxide and sulphur trioxide.
   (c) sulphur trioxide and nitrogen dioxide.
   (d) sulphur dioxide and nitrogen dioxide.

8. Acid rain is any rain with pH
   (a) below 5.6.
   (b) above 5.6.
   (c) equal to 5.6.
   (d) exactly equal to 7.

9. The Taj Mahal at Agra is affected by
   (a) stone disease.
   (b) stone cancer.
   (c) stone tumour.
   (d) marble cancer.

10. Smog is an odd combination of
    (a) smoke and mist.
    (b) mist and fog.
    (c) smoke and fog.
    (d) smoke, fog and mist.

11. Chlorofluorocarbons and freons are
    (a) used as coolants in refrigerators.
    (b) air conditioners.
    (c) responsible for ozone layer depletion.
    (d) All of the above.

12. PAN is a byproduct of
    (a) photochemical smog.
    (b) London smog.
    (c) sulphurous smog.
    (d) reducing smog.

13. The perfect condition for the formation of sulphurous smog is
    (a) humidity.
    (b) subsidence inversion.
    (c) low temperature.
    (d) sunlight to carry out photochemical reactions.

14. The NGOs
    (a) play a significant role in environmental protection.
    (b) mainly work at the grass root level.
    (c) act as the eyes and ears of the government.
    (d) All of the above.

15. The *Appiko* Movement ('tree hugging') was started in
    (a) Karnataka.
    (b) Delhi.
    (c) Orissa.
    (d) Gujarat.

16. The United Nations Conference on Human Environment was organized at
    (a) Rio de Janiero.
    (b) Paris.
    (c) Stockholm.
    (d) Delhi.

17. ________ is a widely recognized study of environmental impacts on developmental projects
    (a) Environmental Impact Assessment (EIA)
    (b) Environmental Risk Assessment (EIA)
    (c) Environmental Impact Statement (EIS)
    (d) Environmental Management System (EMS)

18. The Precautionary principle was codified for the first time at
    (a) Earth Summit in 1992.
    (b) United Nations Framework Convention in 1997.
    (c) The Kyoto Protocol.
    (d) Is yet to be codified.

19. Carbon trading
    (a) is a fundamental concept of national environmental laws and regulations.
    (b) monitors the impact of pollution upon environmental, social and economic justice.
    (c) suggests that funding for environmental improvement should be obtained from its beneficiaries.
    (d) monitors the emission of hydrocarbons from automobiles.

20. The Apex body of standardization in the world formed in 1946 is
    (a) the International Society of Standardization.
    (b) the International Organization of Standardization.
    (c) the International of Standards Organization.
    (d) the Council of International Standardization.

ANSWERS

1 (d)	2 (c)	3) b	4 (d)	5 (d)	6 (a)	7 (d)	8 (a)	9 (b)	10 (c)
11 (d)	12 (a)	13 (b)	14 (d)	15 (a)	16 (c)	17 (a)	18 (a)	19 (b)	20 (c)

UNIT VII
Human Population and the Environment

LEARNING OBJECTIVES

After reading this chapter, students will be able to:

- Describe human population and the environment.
- Explain population growth variation among nations.
- Explain population explosion.
- Describe family welfare programmes.
- Define environmental and human health, human rights and value education.
- Define HIV/AIDS women and child welfare.
- Describe the role of information technology in the environment and human health, with case studies.

7.1 Population Growth Variation among Nations

According to an anthropological estimate, the history of human evolution and civilization is about 3 million years old. About 50,000 years ago, as a result of evolution human beings appeared in their present form (Homo-sapiens erectus). It took the world population several million years to reach one billion (in 1850 AD) and less than 100 years to double that figure.

Decrease in famine-related deaths and infant mortality rates are partially responsible for this rapid population growth. In addition, advances in public health, nutrition, sanitation and medicine have increased the life expectancy of countries worldwide. Industrialization has provided for better economies and has sustained human development. Hence, world population growth has been rapidly expanding at an exponential rate.

The total human population in 3000 BC was estimated to be 0.1 billion or 10 crore. It reached 250 to 300 million around the birth of Christ which nearly doubled to 500 million in 1650 AD that is in 1,650 years. Human population growth was slow till the 17th century. Since then there has been an exponential growth and rapid population explosion. Table 7.1 exhibits the growth of the world population.

Table 7.1 Growth of the World Population

Year	World Population No.	Time taken
3000 BC	100 million	45,000 years to reach 100 million from the origin of Homosapien erectus.
1 AD	250 million	≈3,000 years to grow 2.5 times.
1650 AD	500 million	≈1,650 years to double.
1850 AD	1 billion	200 years to double.
1925 AD	2 billion	75 years to double.
1960 AD	3 billion	35 years to increase by 1 billion.

Year	World Population No.	Time taken
1975 AD	4 billion	15 years to increase by 1 billion or 50 years to double.
1987 AD	5 billion	12 years to increase by 1 billion.
1999 AD	6 billion	12 years to increase by 1 billion.

A study of the geographical distribution of the world population in different continents over the last few centuries as tabulated in Table 7.2 reveals population growth variations among the nations.

Table 7.2 Geographical Distribution of the World Population in Different Continents

Year →	1650	1750	1800	1850	1900	1933	1995
World Population (Millions)	545	728	906	1,171	1,608	2,057	5,716
Percentage Distribution							
Europe	18.3	19.2	20.7	22.7	24.9	25.2	12.7
North America	0.2	0.1	0.7	2.3	5.1	6.7	5.1
Oceania	0.4	0.3	0.2	0.2	0.4	0.5	0.5
Latin America	2.2	1.5	2.1	2.8	3.9	6.1	8.4
Africa	18.3	13.1	9.9	8.1	7.4	7.0	12.8
Asia	60.6	65.8	66.4	63.9	58.3	54.5	60.5

Source: World Population: Past Growth & Present Trends (1936) Carr-Sounders, AM & Demographic Year Book, United Nations, 1995.

Population growth in the developing nations is faster than that in the developed world. The population growth in India, a developing country with a population of 10 billion and 2.2 per cent of the world's land area, is very rapid. In India, 52 births occur every minute and 12 million heads are added every year. The number of births in India adds up to the population of Singapore in five weeks and that of Australia in eight months thereby posing a serious threat to the resource crisis and the ecological balance. Table 7.3 shows the population growth of India since 3000 BC.

Table 7.3 Population Growth of India

Year (AD/BC)	Population (crore)	Time for the population to double (years)	Annual exponential growth (in per cent)
3000 BC	1.3 (estimated)	-	-
1 AD	3.2	>2500	-
1650 AD	6.5	1650	-
1800 AD	20.0	<150	-
1901 AD	24.0	-	-
1911 AD	25.0	-	0.56
1921 AD	25.0	-	−0.03
1931 AD	28.0	-	1.04
1941 AD	32.0	-	1.33

Year (AD/BC)	Population (crore)	Time for the population to double (years)	Annual exponential growth (in per cent)
1951 AD	36.0	-	1.25
1961 AD	44.0	≈150	1.96
1971 AD	55.0	≈40	2.20
1981 AD	68.0	≈40	2.25
1991 AD	85.0	>30	2.50
2000 AD	100.0	≈30	≈2.50

Population Explosion

Population explosion throughout the world and in developing countries in particular, is increasingly threatening the global environmental balance because of the increasing environmental degradation and pollution. Over-population and overuse of resources by the technological societies are the root causes of environmental degradation and pollution. It depends on:

(i) Size of population.

(ii) Per capita consumption of resources.

(iii) Extent of population and degradation per unit of resources consumed.

The world population growth rate is exponentially high and rapid. The United Nations projects the world population to be 7.3–10.7 billion by 2050. The world population of 6.5 billion in 2005 is 380 million more than that of 2000, which is a gain of 76 million annually. Despite the lower fertility levels projected, the world population according to a medium variant is expected to reach 9.1 billion and would still be adding 34 million persons annually by mid-century, with fertility declining to slightly over two children per woman in 2050 from the present rate of 2.6. The projected world population in 2050 and beyond is given in Figure 7.1.

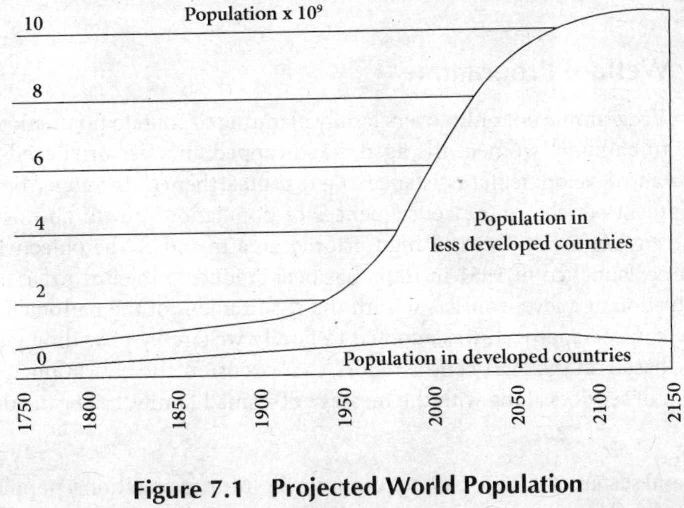

Figure 7.1 Projected World Population

Very rapid population growth is expected to prevail in a number of developing countries, the majority of which are least developed. The population of the 50 least developed countries is projected to more than double, from 0.8 billion in 2005, to 1.7 billion in 2050. Growth in the rest of the developing world is also projected to be fast, though less rapid with its population likely to rise, from 4.5 billion to 6.1 billion between 2005 and 2050. Between 2005 and 2050, the population is projected to at least treble in underdeveloped countries such as Afghanistan, Burundi, Chad, Congo, Liberia and Uganda. During the years between 2005 and 2050, eight countries namely, India, Pakistan, Nigeria, Democratic Republic of the Congo, Bangladesh, Uganda, United States and China listed according to the size of their contribution are likely to account for half the world's population growth.

Because of its low and declining rate of population growth, the population of the developed countries as a whole will virtually remain constant at about 1.2 billion between 2005 and 2050. The population growth in USA at about 1 per cent or nearly 2.7 million people per year is one of the highest among the developed countries. The population of 51 developed countries including Germany, Italy, Japan and most of the successor states of former Soviet Union is expected to be lower in 2050 than it was in 2005.

The fertility rate in developed countries is projected to increase slowly from the present level of 1.56 per woman to 1.84 children per woman, between 2045 and 2050. In the least developed countries the current fertility of children per woman will dip to 2.57 children per woman by 2045–50. In the rest of the developing world, the same will be 1.92 from the current 2.57.

Global life expectancy at birth was 46 years in 1950–55 which rose to 65 years in 2000–2005 and is likely to increase to 75 years in 2045–50. In the developed countries, it is now 75 years and likely to be 82 years by 2045–50. Among the least developed countries, the life expectancy today, is less than 50 years which is likely to increase to 66 years, by 2045–50. The primary consequence of fertility decline especially in combination with increasing life expectancy is population ageing whereby the share of older persons in a population increases vis-a vis that of younger persons.

During the period 2005–2050 the net number of international migrants to more developed regions is projected to be 98 million which will result in an increase in population despite the deaths that are projected to be 73 million more than births there. In terms of annual averages projected, the major net receivers of international migrants are United States (1.1 million), Germany (204,000), Canada (201,000), UK (1, 33,000), Italy (1, 20,000) and Australia (100,000). The major donor countries are China (333,000), Mexico (304,000), India (245,000), Philippines (180,000), Pakistan (173,000) and Indonesia (168,000).

7.2 Family Welfare Programme

The family welfare programme not only covers family planning to contain population growth but also extends welfare measures to children, women, the aged, handicapped and less privileged to provide food, shelter, education, medical and developmental assistance so as to protect them from hunger, poverty, undernourishment and underdevelopment which are the consequences of population growth and overexploitation of natural resources and the environment. This is a high priority area in India. The objective of the National Family Welfare Programme, launched in 1951 in India has been 'reducing the birth rate to the extent necessary to stabilize the population at a level consistent with the requirement of the national economy.' In the first two Five Year Plans a 'clinical' approach was adopted to family welfare by providing facilities of services. In the subsequent plans, based on the 1961 census report, the 'extension and education approach' of family welfare through expansion of services along with the message of a small family, has been adopted.

Family Planning

To save the earth and sustain its fast-depleting resources for future generation's, population control is extremely important. While global environmental change cannot be stopped the pace can be slowed and we may be able to reduce environmental damage and risk markedly, by prudent policy actions. Family planning programmes provide educational and medical services which help couples choose the size of their family and the timing of starting one. The size of the family is smaller in educated and economically better-off families. Illiteracy and lack of education are associated with high rate of population growth in developing and underdeveloped countries including India.

Population can be effectively controlled with sex education, information on methods of birth control, distribution of contraceptives, information on sterilization and abortions and other family planning methods. Family planning has become very successful in China, Indonesia, Singapore, Hong Kong, Sri Lanka, Cuba, Thailand, South Korea and some other developing countries while it achieved moderate results in the countries of the Indian sub-continent and in Brazil, Nigeria and other developing and underdeveloped countries. The success of population control programmes depends on committed leadership, wide availability of contraceptives and family planning services, suitable implementation methods specific to social and cultural characteristics, the role of print and electronic media to propagate sex education, level of education and wealth in the society.

7.3 Environment and Human Health

Human activities in every sphere of life have continuously been changing our environment which in turn is affecting human health. Health is an outcome of the interactions between people and their environment. Millions of people mainly children, suffer from malnutrition and poor health due to parasitic infections such as amoebiasis and worms occurring from infected food or water. About three million children die each year across the world from water-borne diseases such as diarrhoea. In India, it is estimated that every fifth child below the age of five dies due to diarrhoea. Contaminated water results in gastrointestinal diseases in the communities and even sporadic large epidemics.

Oral Rehydrating Salts

Diarrhoeal diseases are not only very common among young children but are also a leading child-killer. About 2.2 million children die from dehydration occurring due to diarrhoeal diseases every year. Diarrhoea is the discharge of watery stool. This means body fluids and salts are quickly drained out of the body and the child becomes dehydrated. This is very dangerous and may even be fatal.

Diarrhoea is caused by bacteria or viruses. The main factors are unclean water and dirty hands and malnourished children are more likely to suffer. Diarrhoea can be prevented through exclusive breast feeding, improved weaning practices, washing hands, keeping food and water clean and the sanitary disposal of stool.

Oral Rehydration Therapy (ORT) is the process of continuous replacement of essential body fluids and salts in proper quantities during an attack of diarrhoea. By replacing lost body fluid, ORT prevents dehydration and nurses the child back to health.

ORT is best administered with the use of pre-packed formulae called Oral Rehydration Salts (ORS) available in a sachet to make a litre of solution. According to UNICEF, these sachets should be stocked in each household to prevent or treat dehydration. Oral rehydration salts are a combination of salt electrolytes and sugar meant to be taken orally. The combination of electrolytes and sugar stimulates water and electrolyte absorption from the gut and this prevents or reverses dehydration and replaces lost salts in conditions such as diarrhoea and vomitting.

WHO and UNICEF recommend the following single formula of glucose-laced ORS to treat and/or prevent dehydration in persons of any age:

Table 7.4 Formulation of ORS

Reduced osmolarity ORS	gms/litre	Reduced osmolarity ORS
Sodium chloride	2.6	Sodium chloride
Glucose, anhydrous	13.5	Glucose anhydrous

Reduced osmolarity ORS	gms/litre	Reduced osmolarity ORS
Potassium chloride	1.5	Potassium
Trisodium citrate dihydrate	2.9	Citrate
Total weight	20.5	Total osmolarity

ORS is used for the following purposes:

(i) Dehydration.

(ii) Fluid and salt replacement in conditions such as sudden diarrhoea and vomitting.

(iii) Severely low concentration of salts in the blood (severe electrolyte depletion).

While using ORS we should keep the following 10 points in mind:

(i) Prior to solution preparation, hands should be washed with soap and water.

(ii) Clean water should be added up to the mark of the bottle, which is equal to one litre.

(iii) Empty the contents of one packet of ORS into the water and stir it.

(iv) Give the sick child as much of the solution as it needs, frequently and in small quantities.

(v) Give other fluids, such as breast milk and juices, alternately.

(vi) Continue to give solids if the child is four months or older.

(vii) If the child still needs ORS after 24 hours, make a fresh solution.

(viii) ORS does not stop diarrhoea. It prevents the body from drying up. The diarrhoea will stop by itself.

(ix) If the child vomits, wait for 10 minutes and give ORS, again. Usually vomitting will stop.

(x) If diarrhoea increases and/or vomitting persist take the child to a health clinic.

ORS has no significant reactions and harmful effects. However, it should be carefully used in case of diabetes, prolonged illness, infant patients, kidney or liver problems, shocks, allergies to any of the ingredients and so on. According to some experts, ORS/ORT is lifesaving.

The following are some of the comments about ORT:

'ORT can rightly be called the medical miracle of this century… For people in Asia, Africa and Latin America, ORT holds the promise of healthier childhoods and more productive adult lives.' (Shamsul Hag, Minister of Health and Population Control, Government of Bangladesh)

'Today the high success rate and low cost of ORT makes it the preferred treatment for mildly or moderately dehydrated children. According to current American Academy of Pediatrics practice guidelines, ORT is effective as a therapy for mild to moderate dehydration.' (Mayo Clinic)

'Nearly 90 per cent mortality from diarrhoea is due to fluid loss. Accurate replacement of that loss is lifesaving.' (Dr. Norbert Hirsclorn, Nutrition Reviews, Volume 40, page 87, 1982)

Every year millions of people in tropical countries get malaria when mosquitoes breed in stagnant water. This also causes several casualties. Hundreds of millions of people suffer from serious respiratory diseases such as lung cancer, tuberculosis and so on due to the fumes of industries, motor vehicles and smoke from fires, kitchens and smoke of tobacco. Millions of people are exposed to hazardous chemicals and radioactive rays resulting in ill-health and sometimes even death. Thousands of people die or get injured in road accidents. About 1,000 million people live in abject poverty, with no clean water, air, proper shelter and nutrition. Population growth threatens the environmental balance and the health of nearly all individuals due to the way resources are exploited and wasted and the inequalities of distribution of wealth and resources.

Human health and quality of life depend on the physical, chemical, biological, social and psychological factors in the environment. Climate and weather too, affect human health. Public health depends on adequate, safe and quality food, water and shelter. Floods, storms, hurricanes and tsunamis kill many people every year. Unprecedented drought and rainfall cause water-borne diseases; global climate change and global warming cause serious health implications. The transmission of vector-borne diseases such as malaria depends on the weather and climate.

Acquired Immune Deficiency Syndrome (AIDS) due to Human Immunodeficiency Virus (HIV) and Severe Acute Respiratory Syndrome (SARS) and Bird Flu were practically unknown earlier. Many infectious diseases related to the environment such as plague, dengue and tuberculosis reappear due to changes in the environment. Cholera, typhoid, diarrhoea, dysentery, polio, meningitis, hepatitis A and E are caused due to impure drinking water. Arsenic poisoning due to presence of arsenic in water; cancer, infertility and neurological diseases due to the harmful pesticides that contaminate water, and blood disorders due to excessive nitrate content in water are all examples of water-borne diseases which claim thousands of lives every year. Aquatic organisms such as guinea worm which affect the feet and roundworm which affects the small intestine cause a variety of water-based diseases. Malaria and filariasis are water-related vector diseases spread by insects that breed in stagnant water. They cause illness and death in many people. Tuberculosis, tetanus and leprosy which take many lives every year are caused due to scarcity of water and consequent unhygienic conditions.

The environment and human health are interdependent. A sustainable environment cannot be achieved without proper health care. Clean potable drinking water and nutrition are absolutely essential for human health and thus for a sustainable environment. Industrial pollution and transport emissions cause ill-health and need to be reduced. The use of renewable non-conventional energy instead of thermal power generation, contribute to improving human health. Factors like population control and reduction in the consumption of natural resources are instrumental towards this cause. We must aim for a sustainable environment, which ultimately will result in 'health for all.'

Human Rights

Man gets his sustenance from the environment which in turn is greatly affected by humans. Therefore, human rights and environmental issues are closely interlinked. They are equity in the distribution of environmental resources, utilization of resources, intellectual property rights, tribal people's rights in forest areas and conflicts with wildlife, particularly in protected areas, development, displacement and resettlement issues and accessibility to nutrition and health care to prevent diseases mainly those caused by negligence to the environment.

The foundation of freedom, justice and peace in the world, depends on the recognition of the inherent dignity and equal rights of all members of the human family. The General Assembly of the United Nations, on December 10, 1948 adopted and proclaimed the Universal Declaration of Human Rights which needs to be disseminated, displayed, read and expounded principally in schools and other educational institutions. The full text of the Declaration is appended at the end of the chapter.

In spite of the widening gap between the rich and the poor, components of human rights such as equality and social justice are essential for men and women, irrespective of their caste, creed and nationality. The right to the use of natural resources is a human right. For example, tribal people should have access to fuel woods and forest products for their livelihood and water should be available to the villagers, irrespective of their caste or wealth.

The right to life and thus proper nutrition and health are fundamental human rights. Poverty, hunger, malnutrition and a poorly managed environment affects human health as well as the socio-economic development of the society.

Traditional medicines, complementary and alternative medicines, herbal products, yoga and naturopathy are widely used in our country since ancient times. Now, the western world is taking patents for the same because of lack of seriousness on our part. This may result in denial of this type of medication or payment of royalties to Multinational Companies (MNC) as per the patent laws, intellectual property rights and so on. Over one-thirds of the population in developing countries does not have access to essential allopathic medicines. The provision of safe and effective traditional medicines and complementary and alternative medicines could increase health care. Therefore, there is a need for serious and proper handling of intellectual property rights.

Value Education

Values are one's own beliefs, principles, perceptions, feelings and behaviour to judge what is right or wrong. Value education in the context of a sustainable environment must transmit environmental values, values for nature's bounty, cultures, human heritage, equitable use of resources and common property resources, social justice, protection of environment and so on. Value education must give weightage to sustainability, equitable use of resources, avoiding consumerism, wastefulness and overexploitation of non-renewable natural resources. Environmental values make us understand the importance of conservation of nature and encourage us to look beyond the physical values. Rivers, forcsts and seas are not to be merely seen as sources of water, wood and fish, respectively. In the years gone by, killing a tiger or other wild animals was considered a profession or a sport, but in current times it is a crime against biodiversity. This is what environmental values and value education teach us.

Till the 18th century, value education remained the core of our education system in institutions and *gurukuls*. In ancient times we were taught ethical and moral values. The economic and technical development of society has now changed the social structure, education system and also shifted the emphasis of to economic and secular values. Presently, value-based education in schools at all levels has again taken the centrestage. The National Resource Centre on Value Education was set up at National Council of Educational Research and Training (NCERT) by the Government of India to cultivate good values in the future citizens of India.

Environmental values are also emphasized in the Constitution of India in the following articles:

Article 51A(g): The constitution expects citizen of the country to 'protect and improve the natural environment including forests, lakes, rivers and wildlife and to have compassion for all living creatures'.

Article 48A: The state shall endeavour to protect and improve the environment and to safeguard the forests and wildlife in the country.

Value education in the context of the environment must teach us the following:

(i) Environmental values.
(ii) Values for nature.
(iii) Values for culture.
(iv) Values for social justice.
(v) Values for human heritage.
(vi) Values for equitable use of resources.
(vii) Values for sharing common property resources.
(viii) Values to stop/minimize environmental degradation.

7.4 HIV/AIDS

AIDS is one of the most serious deadly diseases in human history. The first cases of AIDS in the world were reported in New York and San Francisco, USA in 1981. In India, the first HIV infection report came from

Chennai in 1986. Now, there are an estimated 42 million people living with HIV or AIDS worldwide. This includes about four million people in India. It has spread worldwide threatening all of mankind so much so that it is estimated that 7,000 adults and 500 babies are infected with HIV or AIDS every day. It causes the death of more than three million people every year throughout the world and the cumulative number of deaths due to HIV/AIDS is around 30 million.

HIV stands for Human Immunodeficiency Virus. A person infected with HIV is referred to as a HIV positive person. Only a blood test can establish one's status. HIV causes AIDS; it destroys a type of defence cells in the body called CD_4 helper lymphocyte cells (CD_4 cells). Under normal circumstances these cells help the immune system to function normally and fight-off certain kinds of infections by acting as messengers to other immune system cells telling them to become active and fight against an invading germ. The HIV attached to these CD_4 cells infects them and uses them to multiply resulting in loss of ability of CD_4 cells to do their job of fighting infections. The name for this condition is Acquired Immuno Deficiency Syndrome (AIDS). Because their immune systems are weakened, people who have AIDS are unable to fight-off many infections particularly tuberculosis, pneumonia, meningitis, encephalitis and cancers. AIDS is fatal. It is essential therefore to detect AIDS/HIV at an early stage through blood tests.

An HIV positive person may not necessarily have AIDS. Such a person remains infected for the rest of his life but could remain asymptomatic for about three to 12 years, or longer. The amount of time it takes for the AIDS symptoms to appear varies from person to person. It is possible to infect others with HIV even if the person with the virus has absolutely no symptoms. In a person infected with AIDS symptoms can include extreme weakness or fatigue, rapid weight loss, frequent long fevers, heavy sweating at night, minor infections that cause skin rashes and sores in the mouth, anus or genitals, swollen lymph glands, white spots in the mouth or throat, chronic diarrhoea, loss of memory and consistent cough.

HIV can spread from one infected person to another in the following ways:

(i) **Unprotected sex:** HIV is transmitted mostly through semen and vaginal fluids during unprotected sex.

(ii) **Contaminated blood:** HIV can be transmitted through transfusion of blood and blood products.

(iii) **Using contaminated syringes and needles:** Sharing of syringes and needles among intravenous drug users can transmit HIV and the chances of transmission are the highest via this.

(iv) **From an infected mother to her baby:** HIV can be transmitted to the baby from an infected mother before birth, during birth and after birth. Breast milk can also transmit HIV infection to the infant.

Chances of infection are highest in blood transfusion (90–95 per cent) followed by prenatal (from mother to baby, 20–40 per cent). While unprotected sex has a low efficacy of 0.1–1 per cent of HIV transmission, yet it contributes the highest infection globally (75 per cent). The efficacy of HIV transmission by different routes is shown in Table 7.5.

Table 7.5 HIV Transmission through Different Routes

Mode of transmission	Percentage of Efficacy	Percentage of Infection (Globally)	Percentage of Infection in India
Unprotected sex	0.1–1.0	75	79
Blood transfusion	90–95	5	7
Prenatal	20–40	10	0.15
Injecting drug use	0.5–1.0	10	6

Mode of transmission	Percentage of Efficacy	Percentage of Infection (Globally)	Percentage of Infection in India
Needle type exposure (health care settings, tattooing needle stick injury and so on)	0.5	0.1	Data not available
Others (not specified)			8

HIV is not transmitted by casual contact such as touching, holding hands, body contacts in crowded places, shaking hands, working or playing together, sharing food vessels and clothes or eating food, cooked or handled by an infected person. HIV/AIDS does not spread by donation of blood, by mosquito or insect bites. Swimming pools and toilet-sharing also are not known to spread HIV infection. HIV has been found in small amounts in body fluids such as saliva, faeces, urine and tears but no evidence has been found that it could spread through these body fluids.

Prevention is the only solution against HIV/AIDS. Awareness and proactive action only can ensure a HIV/AIDS-free society. Some of the protective actions are safe sex (use of condoms), having a faithful monogamous sexual relationship with an uninfected partner, use of sterilized disposable syringes, sterilization of needles, transfusion of unaffected blood and blood products after proper test. Children and adolescents have to be made aware about HIV infection and its mode of transmission before they become sexually active. A tolerant, compassionate and non-discriminatory attitude towards people with HIV infection is very important, as ill-based fear of HIV and AIDS affects the patient's environment adversely. Proper medical care for HIV positive pregnant women can prevent HIV infection to the newborn. Every woman should know about HIV and AIDS to protect herself and her family. According to a World Health Organization estimate, approximately 44 per cent of the 40 million adults living with HIV/AIDS are women. Women are more affected because they are more vulnerable. Moreover, women in India and in most of the countries invariably occupy a subservient status within the family and society which further heightens their vulnerability to infection. In a developing country such as India, HIV affects people in the productive age groups and leads to premature deaths. In most of the families the male member is also the earning member and his death/incapacitation affects the whole family. Besides, no cure for AIDS has been found till date and the medicines used in the treatment are very costly therefore beyond the reach of many people. This severely affects the economic condition of the whole family and in turn the community and the country.

7.5 Woman and Child Welfare

Woman and child welfare are very important for a sustainable environment and a healthy society. There are strong connections between several environmental factors and woman and child welfare. Women in the developing world invariably occupy a subservient status within the family and society. Women, especially in lower income group families both in towns and villages work for longer hours than men. Their work pattern also differs and is more prone to health hazards. They are often the last to eat and fed mostly on leftovers and may not get enough nutrition. The girl child is given few facilities for education. This socio-environmental divide is a major concern. Each year about 11 million children die due to diseases and malnutrition in the world. Pneumonia, diarrhoea, measles, malaria and malnutrition are the main causes of such deaths. In most of the underdeveloped countries one in five children die before they are five years old. The WHO estimates that 1.6 billion early deaths occur annually from the pollution caused by the cooking oven (*chulha*). Oven smoke is the third highest cause of diseases and death after dirty water and lack of sanitation. Around the globe about 4, 40,000 women suffer from HIV/AIDS. There is always a close relationship between poverty, a degraded environment and malnutrition. As per the 1991 census, India has 150 million children constituting 17.5 per cent of the population. Many of them are deprived of primary health care, elementary education, sanitation, proper food and nutrition. Under these circumstances, woman and child welfare becomes an absolute necessity for the development of a society.

In India, the Department of Women and Child Development was set up in 1985 as a part of the Ministry of Human Resource Development to give the much needed impetus to the holistic development of women and children. As the national machinery for the advancement of women and children, the Department formulates plans, policies and programmes, enacts/amends legislation, guides and coordinates the efforts of both governmental and non-governmental organizations working in the field of women and child development. Besides playing a nodal role, the Department implements certain innovative programmes for women and children. The programmes for children emphasize on nutrition, pre-school education and various aspects of health, while those for women cover welfare and support services, training for employment and income generation, awareness generation and gender sensitization. These play a supplementary and complementary role to other developmental programmes in the sectors of health, education, rural development and so on. All these efforts are directed to ensure that a sound foundation is laid for the physical, social and psychological development of children and towards the economic and social empowerment of women.

The Department has three autonomous organizations namely National Institute of Public Cooperation and Child Development (NIPCCD), *Rashtriya Mahila Kosh* (RMK) and the Central Social Welfare Board (CSWB) working under its aegis. A Minister of State with an interdependent charge heads the Department. Four Joint Secretaries, who head Child Development, Child Welfare and Nutrition, Women's Development and Micro-credit Development Bureau, assist the Secretary.

Some of the women welfare programmes are:

(i) Providing assistance to women welfare organizations such *as mahila samajams/mandals* by giving equipment and raw material for manufacture of garments, handicrafts and tailoring products.

(ii) Establishment of self-reliant women's self help groups.

(iii) Conducting seminars/workshops/exhibitions on mother and child care.

(iv) Organizing medical awareness camps on the health aspects of women.

(v) Organizing meeting of mothers on health and nutrition especially for pregnant and lactating mothers.

(vi) Providing additional central assistance for foodgrains to undernourished adolescent girls, pregnant and lactating mothers.

(vii) Organizing special programmes on nutrition and *swayam sidha* through establishment of self-help groups.

7.6 Role of Information Technology in Environment and Human Health

The development of information technology has caused knowledge explosion and our society has become a technical society where IT is used in all spheres in mega-cities and remote villages in the field of arts, advanced sciences, environment management, weather forecasting, agriculture, banking, industry and medical profession. Information technology, Internet and satellite communication enhance interaction between people and institutions and allow people to communicate with the government more easily and instantaneously and share information globally.

Environmental changes are taking place constantly, due to man-made activities and natural factors. Accurate and timely information of various environmental changes are of great significance. A timely forecast of a tsunami on December 26, 2004 or of the super cyclone on November 29, 1999, in Orissa could have saved a lot of loss of human life and property. An environmental information system is an essential part of governance at the local, national or international level and can be done with the support of modern information technology so as to acquire, store, manage, disseminate information and awaken and warn people and the society of any eventualities. Linking together citizens, schools, libraries, newspapers, local authorities, institutions and people will create a powerful knowledge sharing/managing infrastructure. Information technology has an important role in planning, developing, policymaking, statistics generation and maintenance, natural resource management, census demographic research analysis and forecasting and in other spheres of governance.

Institutions in Uganda, an underdeveloped country in Africa shared information with other African countries about the development of plans of actions in various sectors. *Bhoomi*, Karnataka's GIS scheme has revolutionized the way farmers access their land records. Karnataka has brought computers to 6.7 million farmers of the state. A farmer can get a copy of the records of rights, tenancy and crops from a computerized information kiosk, without harassment or bribes. Information technology has enabled us to access information from regional and international organizations such as Network for Environmental and Sustainable Development in Africa (NESDA), United Nation's Environment Programme (UNEP), World Meteorological Organization (WMO) and World Bank for weather forecasts, natural resource management, policymaking and planning and development. Information technology is a powerful tool of sustainable development through improved decision making, quick, transparent and effective communication of reliable information to more people and more areas. The application of information technology in the environment and human health ranges from IT-based medical diagnosis to the operation of sending satellites to space to explore the universe. Some of the applications of IT in the environment and human health are:

(i) Telemedicine, advanced diagnostic, health care and operation equipment.

(ii) Statistics based on IT for planning, decision making and developmental activities, family welfare, population control and health care.

(iii) Use of radio sensing technology in wildlife tracking census and study.

(iv) Environmental pollution monitoring and weather forecasting through remote sensing and geographical information system.

(v) Communication, collaboration and coordination among environmental scientists for decision-making.

(vi) Computer-based modelling and simulation of environmental scenarios for analysis and prediction.

SUMMARY

Human population growth was slow till the 17th century, followed by a rapid population explosion. Population growth in developing nations is faster than that in the developed world. The population growth of India, a developing country, with a population of 10 billion, living on 2.2 per cent of the world's land area, is extremely rapid. Moreover, the decrease in famine-related deaths and the infant mortality rate are partially responsible for this rapid population growth. In addition, advances in public health, nutrition, sanitation and medicine have increased the life expectancy of countries, worldwide.

The population explosion throughout the world and in the developing countries in particular, is increasingly threatening the global environmental balance due to the fast depletion of resources and energy sources due to the increasing demand. This is leading to rapid environmental degradation and pollution.

The population growth in the USA is one of the highest among the developed countries at about 1 per cent or nearly an addition of nearly 2.7 million people per year. The fertility rate in developed countries is projected to increase slowly to 1.84 children per woman in 2045–2050 from the present level of 1.56 per woman. In the least developed countries the fertility of children per woman will dip to 2.57 children per woman by 2045–50.

Family Welfare Programme

The family welfare programme not only covers family planning to control population growth but also extends to welfare measures for children, women, the aged, the handicapped and the less privileged. These welfare measures aim to meet the needs for food, shelter, education, medicine and development, of the most vulnerable sections of society so as to protect them from hunger, poverty, undernourishment and underdevelopment which are the consequences of population growth and overexploitation of the natural resources and the environment.

Family planning

Family planning programmes provide educational and medical services which help couples to choose their family size and when to have children. The size of the family is smaller in educated and economically better-off families. Illiteracy and lack of education are associated with a high rate of population growth in developing and underdeveloped countries, including India.

Environment and Human Health

Human activities in every sphere of life have continuously been changing our environment which in turn is affecting human health. Millions of people, mainly children are malnourished and in poor health due to parasitic infections such as amoebiasis and worms, occurring due to infected food or water. About three million children die each year across the world from water-borne diseases such as diarrhoea. These are caused by bacteria or viruses. The main factors are unclean water and dirty hands and malnourished children suffer the most. Diarrhoea can be prevented through exclusive breastfeeding, improved weaning practices, washing hands, keeping food and water clean, sanitary disposal of stool and following Oral Rehydration Therapy. Oral Rehydration Therapy (ORT) is the process of continuous replacement of essential body fluids and salts in proper quantities during the attacks of diarrhoea. This prevents dehydration and nurses the child back to health.

Disasters such as floods, storms, hurricanes and tsunamis kill many people every year. Unprecedented drought or rainfall also leads to water-borne diseases. Symptoms of global climate change such as global warming have serious health implications. The environment and human health are interdependent. Clean potable drinking water and nutrition are absolutely essential for human health and thus for sustainable environment.

Value Education

Value education in the context of the environment must transmit environmental values to all members of the society. These values include value for natures, cultures, human heritage, equitable use of resources and common property resources, social justice and protection of the environment, among others. The National Resource Centre on Value Education was set up at NCERT by the Government of India to realize value-based education at the school stage in the country.

HIV/AIDS

Today, there are an estimated 42 million people living with HIV or AIDS worldwide. This includes about four million people in India. A person infected with HIV is referred to as an HIV positive person. AIDS is caused by the HIV. AIDS is fatal. However, HIV positive persons may not have AIDS. HIV-contaminated blood can transmit the virus through transfusion of blood and blood products. Breast milk can also transmit HIV infection to the infant.

Prevention is the only solution available against HIV/AIDS. Awareness and proactive action can ensure a HIV/AIDS-free society. Proper medical care to a HIV positive pregnant lady can prevent the transmission of the HIV infection to the new born. Every woman should know about HIV and AIDS to protect herself and her family. According to a World Health Organization estimate, approximately 44 per cent of the 40 million adults living with HIV/AIDS are women. In a developing country such as India, HIV affects people in the productive age groups and leads to premature deaths.

Women and Child Welfare

Women and child welfare are very important for a sustainable environment and a healthy society. Women in the developing world invariably occupy a subservient status within the family and society. Oven smoke is the third highest cause of diseases and death after dirty water and lack of sanitation. About 44 per cent of the 40 million HIV/AIDS affected adults are women. India, as per the 1991 census, has 150 million children, constituting 17.5 per cent of the total population. In India, the Department of Women and Child Development was set up in the year 1985 as a part of the Ministry of Human Resource Development to give the much needed impetus to the holistic development of women and children. Besides playing a nodal role, the Department implements certain innovative programmes for women and children. The secretary is assisted by four joint secretaries, who head Child Development, Child Welfare and Nutrition, Women's Development and Micro Credit Development Bureau.

Role of Information Technology in environment and human health

New technologies such as information technology, the internet and satellite communication improve the interaction between people and institutions. Information Technology, or IT, makes it easier for people to interact with the government, as well as share information globally and instantaneously. The development of information technology has caused a knowledge explosion. Our society is increasingly becoming a technological society where IT is used in all spheres of life. Accurate and timely information about various environmental changes are of great significance. Information technology has also been supporting planning, developmental policymaking, statistics generation and maintenance, natural resource management, census demographic research analysis and forecasting and also in different spheres of governance, linking together citizens, schools, libraries, newspapers, local authorities, institutions and people. In doing so, IT has created a powerful knowledge sharing/managing infrastructure.

ESSAY TYPE QUESTIONS

1. What are the contributing factors of world population growth?
2. Sketch a brief history of population growth in developing countries and developed countries of the world.
3. 'Population explosion threatens the global environmental balance.' Justify the statement.
4. 'Human health and diseases are generally caused by the environment.' Justify the statement with suitable examples.
5. Write a note on human rights and the environment.
6. What are the objectives of the Universal Declaration of Human Rights by the United Nations?
7. Discuss the process of HIV infection and its mode of transmission.
8. What is the difference between HIV Positive and AIDS and how can they be prevented?
9. Describe various roles of information technology in the protection of human health and the environment.
10. What is value education, how does it help to maintain a sustainable environment?

SHORT-ANSWER TYPE QUESTIONS

1. Define the following:
   (i) Population ageing.
   (ii) Fertility rate of reproductive women.
   (iii) Birth rate.
   (iv) Growth rate.
2. 'Population growth in developing countries is more compared to developed countries.' Justify the statement.
3. What are the different family welfare and family planning programmes?
4. What is the major objective of family welfare and family planning programmes undertaken by India?
5. Write short notes on the effect of population growth on:
   (i) Energy Resources.
   (ii) Food Resources.
   (iii) Water Resources.
6. Write short notes on:
   (i) Water-borne diseases.
   (ii) Air-borne diseases.
   (iii) Value education.
7. Differentiate between HIV & AIDS.
8. Why are women considered the worst victims of environmental hazards?
9. Describe how environmental values are emphasized in the Indian Constitution.
10. Name a few applications of information technology in environment and human health.

MULTIPLE CHOICE QUESTIONS

1. The history of human evolution and civilization is about
   (a) 3 million years old.
   (b) 100 million years old.
   (c) 1 billion years old.
   (d) only 50,000 years old.
2. Population growth in developing countries as compared to developed countries is
   (a) negligible.
   (b) slower.
   (c) faster.
   (d) at almost the same growth rate.

3. Family welfare programme
   (a) covers family planning only.
   (b) covers welfare measures of children, women, aged and the handicapped.
   (c) Both a and b.
   (d) None of the above.

4. ORS is used in case of
   (a) dehydration.
   (b) fluid and salt replacement due to diarrhoea and vomitting.
   (c) severe electrolyte depletion.
   (d) All of the above.

5. HIV destroys a kind defence cells in the body called
   (a) CD_4 helper lymphocyte cells.
   (b) white blood cells.
   (c) nerve cells.
   (d) red blood cells.

6. HIV does not spread through
   (a) contaminated blood.
   (b) unprotected sex.
   (c) from infected mother to her baby.
   (d) sharing food vessels and eating food cooked by the infected person.

7. The process of continuous replacement of essential body fluids is
   (a) rehydration therapy.
   (b) hydration therapy.
   (c) oral hydration therapy.
   (d) oral rehydration therapy.

8. Which of the following is not a woman welfare programme?
   (a) Establishment of a self-reliant women's self-help group.
   (b) Conducting workshop on mother and child care.
   (c) Conducting medical awareness camps on health aspect of women.
   (d) None of the above.

9. The Department of Women and child development was set 1985 as a part of
   (a) Ministry of Women and Child Welfare Department.
   (b) Ministry of Human Resource Development.
   (c) Both of the above.
   (d) None of the above.

10. Which of the following is an autonomous organization of the Department of Women and child development?
   (a) *Rashtriya Mahila Kosh* (RMK).
   (b) Central Social Welfare Board (CSWB).
   (c) Both of the above.
   (d) None of the above.

11. Telemedicine and weather forecasting are examples of
   (a) computer-based modelling for analysis of weather.
   (b) use of radio sensing technology.
   (c) diversification of the Healthcare Department.
   (d) application of IT in environment and healthcare.

12. UNEP stands for
   (a) United Nations Environmental Policy.
   (b) United Nations Environmental Programme.
   (c) United Nations Environmental Protection.
   (d) None of the above.

13. The threat to global environmental balance and fast depletion of natural resources are all outcome of
   (a) overuse of natural resources.
   (b) population explosion.
   (c) pollution.
   (d) All of the above.

14. An environmental information system is an essential part of any government at
   (a) only local level.
   (b) local and international level.
   (c) national, international and local levels.
   (d) national and local levels.

15. The first cases of AIDS in the world was reported at
   (a) Stockholm and Italy.
   (b) Beijing and Taiwan.
   (c) Italy, San Francisco and Beijing.
   (d) New York and San Francisco.

16. One's own belief, principles, perceptions, feelings and behaviour to judge what is right and wrong is called
   (a) morality.
   (b) perception.
   (c) life science.
   (d) value education.

17. The fundamental human rights include
    (a) the right to eat and access to livelihood.
    (b) the access to livelihood and health.
    (c) the right to eat and nutrition.
    (d) proper nutrition and health.

18. SARS stands for
    (a) Severe Acute Respiratory Syndrome.
    (b) Sudden and Acute Respiratory Syndrome.
    (c) Severe and Acquired Respiration State.
    (d) None of the above.

19. Equity and social justice are essential components of
    (a) value education.
    (b) conservation of biodiversity.
    (c) human rights.
    (d) environmental ethics.

20. Diarrhoea is caused by
    (a) bacteria and protozoa.
    (b) viruses and fungi.
    (c) viruses only.
    (d) bacteria and viruses.

ANSWERS

1 (a)	2 (c)	3 (c)	4 (d)	5 (a)	6 (d)	7 (d)	8 (d)	9 (b)	10 (c)
11 (d)	12 (b)	13 (b)	14 (c)	15 (d)	16 (d)	17 (d)	18 (a)	19 (c)	20 (d)

UNIT VIII

Field Work

LEARNING OBJECTIVES

After reading this chapter, students will be able to:

- Attempt field work.
- Visit a local area to document environment problems in forest, grassland, hill or mountain.
- Visit a local polluted site in an urban, rural, industrial or agricultural environment.
- Study common plants, insects and birds.
- Study simple ecosystems such as ponds, rivers and hill slopes.
- Demonstrate a few experiments.

8.1 Introduction

Human actions are always changing the environment they live in. Man loves nature aesthetically, he wants to exploit it gainfully and also retain its beauties and bounties for his successors. We study our nature and environment knowingly or unknowingly, by observing it through travel, sports, sightseeing, visits, excursions, mountaineering, boating, fishing, hunting and so on.

After going through environmental studies theoretically, students should visit nature for acquiring practical knowledge of their studies. This requires understanding of some basics of physics, chemistry and biology of land, air, and water. Certain environmental actions that need to be understood are:

(i) Determination of concentration of pollutants in air, water and land.

(ii) Diffusion and spread of pollutants through air, water, and land.

(iii) Transfer of pollutants from one media to another and their rate of transfer.

(iv) Rate of chemical reactions associated with pollution.

(v) Physical, chemical and biological methods to clean up the pollutants.

(vi) Understanding the natural mechanism of pollution control and using those principles for waste treatment plants.

To understand environmental studies, not only observations are required but also experimentations, especially during a visit to a forest, grassland, hills, mountains or a polluted place such as urban/rural/industrial/agricultural sites or to study plants, insects, birds or even simple ecosystems such as ponds and rivers. Some important experiments are included in this unit for the benefit of students and the readers.

8.2 Experiments

Experiment 1

Aim

To study the physical characteristics such as, temperature, colour and texture of soil.

Requirements

Different types of soil, soil thermometer, Munsell's soil colour chart, sieves of different number of meshes.

Procedure

(i) **For measurement of soil temperature**: Soil temperature is measured with the soil thermometer, which has a vertical arm with a bulb at one end and a dial with a deflection needle, at the other. The bulb of the thermometer is buried at different depths, such as 1″, 5″, 10″, 15″, 20″ of soil and the respective temperatures in °C are noted from the dial. A comparative study thus denotes whether or not any variation of temperature exists along the depth of soil.

(ii) **Study of colour of the soil:** The soil sample is spread uniformly over a cardboard chart and the colour of the soil particles is matched with the chips of Munsell's soil colour chart to identify the colour of that particular soil sample.

(iii) **Detection of texture of the soil:** Primarily, rubbing the soil between the thumb and fingers can help determine the texture of the soil. But with the help of mesh sieves of graded sizes, the exact diameter range of the soil particles can be determined.

The soil sample is examined under a hand lens. A pinch of the soil, in dry, as well as, moist condition, is rubbed between the thumb and fingers to understand its texture. Table 8.1 describes how the texture is determined.

Table 8.1 Determination of soil texture

Examination of soil as felt between thumb and fingers.	Texture
Individual grains may be seen or felt, form a cast when moist soil is squeezed.	Sand
Individual grains may be seen or felt, form a cast, but fall apart, when dry soil is squeezed but casts formed from moist soil does not break.	Sandy loam
Mellow with a gritty feel, when dry soil is squeezed, it forms a cast that is stable with careful handling, but with moist form, the cast is more stable.	Loam
Dry, appearing cloddy, soft to touch, the cast of both dry and moist sample can easily be handled without breaking.	Silt loam
Breaks into clods or lumps, hard when dry, thin ribbons can be made from moist soil but it breaks easily, cast from moist sample is more stable than dry ones.	Clay loam
Forms very hard lumps when dry, sticks when wet, if moist sample is squeezed long flexible ribbons are formed.	Clay

For measuring the diameter of the soil, the sample is passed through different sized mesh sieves, which is a standard method. The soil sample is first passed through a maximum mesh sieve then gradually through decreasing mesh sieves. Finally the texture is determined as per Table 8.2.

Table 8.2 Types of soil texture

Diameter (mm) of the particle	Texture
< 0.002	Clay
0.002–0.02	Silt
0.02–0.20	Fine Sand
0.2–2.00	Coarse Sand

By calculating the proportion of each particle as it occurs in the sample under examination, the texture as well as the composition of the soil can be determined.

Experiment 2

Aim

To examine the chemical characteristics of soil such as:

(i) Moisture content.
(ii) Carbonate content.
(iii) Nitrate content.
(iv) Base deficiency.
(v) pH of the soil sample with the help of rapid tests.

Requirements

Different soil samples, hot air oven, china-clay tiles, test tubes, capillary tubes.

Chemicals Required

Barium sulphate, diluted hydrochloric acid, concentrated sulphuric acid (0.2 per cent), saturated solution of ammonium thiocyanate, hydrogen peroxide, universal indicator, and diphenylamine.

Procedure

(i) **Moisture content**: A small amount of the soil sample is taken, weighed, and then kept in the hot air oven for drying at 105–110°C for 24 hours. Then the sample is weighed again. The difference (loss) in the weights is due to the evaporation of the moisture in the soil.

Calculation

Let weight of the soil prior to heating = a gm
weight of the soil after heating = b gm
∴ The amount of moisture present in the given sample = (a–b) gm
Thus:
moisture content (in per cent) = $\frac{(a-b) \times 100}{b}$

(ii) **Carbonate content**: The presence of carbonate can be instantly determined by the evolution of carbon dioxide gas (effervescence) by adding a drop or two of diluted HCl to the soil sample. For a comparative study, a small amount of each of the different soil samples is taken in test tubes, a drop or two of diluted HCl is added to each of the test tubes. From the extent of effervescence in each test tube, the soil samples can be graded as A, B, C or D. The sample showing maximum effervescence contains the highest amount of carbonate and so on.

(iii) **Nitrate content:** This is a wet test, where a clear solution of the soil sample in water is made in a ratio of 1:5. With the help of a dropper, two to three drops of diphenylamine are added followed by two drops of 2 per cent concentrated H_2SO_4 to about 1 ml of this clear solution. A blue colour confirms the presence of nitrate in the soil. The test is performed with all the different soil samples. An increase in the intensity of the blue colour indicates an increase in the presence of nitrate content in the soil.

(iv) **Base deficiency:** A pinch of different soil samples is taken in separate clean and dry test tubes. To each test tube, saturated alcoholic solution of ammonium thiocyanate is added and the test tube is shaken thoroughly. It is allowed to stand for some time so that the solids settle down. Then a drop of hydrogen peroxide is added to the supernatant liquid in each test tube. A red colour indicates the deficiency of exchangeable bases in the supplied soil sample. The sample with a faint red colour indicates minimum deficiency whereas a deep red colour indicates maximum deficiency of exchangeable bases in the supplied soil sample.

(v) **pH value:** The pH value of the soil is tested with a universal indicator. A pinch of each soil sample is taken in different test tubes. To each of these test tubes 15 ml of water and an equal amount of barium sulphate is added. The test tubes are allowed to stand for some time to allow the mixture to settle. Then to the clear supernatant liquid in each test tube 1 ml of universal indicator solution is added. The colour that shows up is matched with the colour chart provided on the bottle for the different pH values of the soil samples.

Experiment 3

Aim

Measurement of particulate air pollutants (from various sites such as urban, rural, industrial and agricultural) namely:

(i) Dust fall and

(ii) Suspended Particulate Matter

Although, various sophisticated methods/instruments are available for this type of study, here are a few methods which can be performed in all laboratories, including those with limited facilities.

Procedure

(i) **Dust fall**: This is a rough estimate of air pollution and can be measured through dust drops. A large polyethylene container with a tapered cylinder of a minimum diameter of 15 cm and two to three times its height is used as a dust trap. It is fitted with a suitable guard frame and sieves of number 18 or 20. The traps are half-filled with distilled water along with some algaecide or fungicide. Then, the trap is kept at least 3 metres above ground level or on a rooftop in such a way that a line drawn from the trap to the nearest point of the highest edge of any building is not at an angle of more than 30° horizontally. Traps are exposed to the atmosphere for a minimum period of one month for proper sampling.

At the end of the set time, the trap is closed and brought to the laboratory, for analysis. The sample is sieved to remove extra material. The water insoluble portion is weighed and also the supernatant liquid is evaporated to restore the insoluble portion. The combined weights give the total dust particle weight.

The dust fall rate is then calculated as:

$$\text{Dust fall rate} = \frac{w}{a} \times \frac{30}{t} \text{ g/m}^2\text{/month}$$

Where w = Weight of the dust collected
a = Total open area of sample container.
t = Total days of exposure.

With the help of a dust trap, total dust fall of any place, be it rural, urban, industrial or agricultural, can be obtained.

(ii) **Suspended Particulate Matter (SPM):** Suspended Particulate Matter are smaller than dust particles, hence a dust trap cannot trap them. For their sampling, special instruments such as a high volume sampler is used. It is placed inside a plywood or aluminium box. Air is passed through a previously weighed filter paper inside the box, where the suspended particulate matter gets retained and clean air passes out. Air is sucked in and passed out at a rate of 1.13 m³/min and sampling can be done for 6, 8, 12 or 24 hours depending upon the SPM load of the area. After sampling, the filter paper is taken to the laboratory where it is kept at 105°C for two hours, then it is cooled and weighed.

Calculation

The air volume sampled

$$X = \frac{(Q_1 + Q_2) \times T \text{ m}^3}{2}$$

Where Q_1 = Initial air flow rate m³/min.
Q_2 = Final air flow rate m³/min.
T = Sampling time in min.

$$\text{Thus, SPM} = \frac{(W_2 - W_1)}{X} \times 106 \text{ mg/m}^3$$

Where W_1 = Initial weight of filter paper
W_2 = Final weight of filter paper

Experiment 4

The use of air samples for monitoring the air quality. The two main parameters responsible for air quality are (i) Suspended Particulate Matter and (ii) Gaseous pollutants.

(i) **Suspended Particulate Matter:** Very sophisticated high volume air samplers are available these days that are suitable for air quality monitoring within the work space environment and in the ambient air outside. Air is passed through a high efficiency filter paper, which retains the particles at a very high flow rate of 0.9 to 1.4 m³/min. A blower also has a free air flow rate of 1.4 to 1.8 m³/min. Thus the instrument measures the volume of air sampled per minute for a given period of time. The increase in the weight of the filter paper determines the amount of total suspended particulate matter collected in the above-mentioned condition.

(ii) **Gaseous pollutants:** For gaseous pollutants such as SO_2, NO_2, NH_3, O_3 and so on, suitable absorbing reagents are used instead of filter paper. Sampling is done for eight to 24 hours depending on the type of sampler. Different gaseous pollutants are simultaneously absorbed in separate glass impingers, using suitable absorbing reagents kept in ice trays so as to allow maximum absorption and to reduce any loss due to evaporation.

Each of the absorbed solutions is analysed either volumetrically or spectrophotometrically to find out the total amount of gaseous pollutants in the air sample.

Experiment 5

Aim

Measurement of the following water qualities in the supplied sample

(i) Hardness of water.

(ii) Total alkalinity.

(iii) Total solids.

(iv) Total dissolved solids.

Chemicals Required

NH_4Cl-NH_4OH buffer, Na_2S solution, Eriochrome black T indicator, EDTA solution, phenolphthalein indicator, methyl orange indicator and 0.1(N) HCl.

Procedure

(i) **Hardness of water:** The hardness of water is due to the presence of bicarbonates of Ca and Mg in water. To about 50 ml of water sample in a conical flask, add 1 ml of NH_4Cl-NH_4OH buffer followed by 1 ml of Na_2S solution. To this add about 100 mg of Eriochrome Black T indicator. The colour of the solution changes to wine red. Then, it is titrated with EDTA solution till the colour of the solution changes to blue, to mark the end point.

Calculation

Hardness (mg/l) = (EDTA in ml × 1000)/ ml of sample

(ii) **Total alkalinity:** Alkalinity of water is due to the presence of excess hydroxyl ions in water. This can be easily analysed with the addition of different indicators.

To about 100 ml of water sample in a conical flask add a few drops of phenolphthalein indicator. If the colour of the solution changes to pink then the solution is titrated with 0.1(N)HCl to find phenolphthalein alkalinity (PA). But if the solution remains colourless, then the PA is zero (that is carbonate is absent). In that case, the alkalinity is determined with methyl orange indicator. On addition of methyl orange (two to three drops), the colour of the solution becomes yellow, it is titrated with 0.1(N)HCl till the colour of the solution changes to pink (end point). This is total alkalinity (TA).

Calculation

$$\text{PA} = \frac{(\text{X} \times \text{N}) \text{ of HCl} \times 1000 \times 50}{\text{ml of sample}}$$

$$\text{TA} = \frac{(\text{Y} \times \text{N}) \text{ of HCl} \times 1000 \times 50}{\text{ml of sample}}$$

Where X = ml of HCl used only with phenolphthalein

Y = ml of HCl used only with methyl orange (that is, total HCl used with both the indicators)

(iii) **Total solids:** Total solids are all solid substances suspended, dissolved and volatile which are present in a water sample. Thus it can only be measured as the residue left after evaporation of the unfiltered sample.

Thus, in a previously weighed evaporating dish, about 250–300 ml of unfiltered and well-shaken water is taken and evaporated. After evaporation, it is kept in a hot air oven at 105°C for one hour and then cooled in a desiccator. The final weight is taken only after it is properly cooled.

Calculation

$$\text{Total solids mg/l} = \frac{(W_2 - W_1) \times 1000 \times 1000}{V}$$

Where W_1 = Weight of empty evaporating dish.

W_2 = Weight of evaporating dish with residue.

V = Volume of sample evaporated.

(iv) **Total dissolved solids:** It is the solids present in a water sample in a fully dissolved condition. So, the procedure for the calculation of total dissolved solids is the same as the previous one. Only in this case, a filtered water sample is taken.

About 250–300 ml of water sample is filtered through Whatman filter paper so that no turbidity is left. Then the sample is evaporated, cooled and weighed as was done in the previous experiment.

Calculation

Thus,

$$\text{Total dissolved solids mg/l} = \frac{(W_2 - W_1) \times 1{,}000{,}000}{V}$$

Where W_1 = Weight of the empty evaporating dish in gm.

W_2 = Weight of the empty evaporating dish with sample in gm.

V = Volume of sample evaporated in ml.

Experiment 6

Aim

Measurement of dissolved oxygen and free carbon dioxide levels in the water sample.

Chemicals Required

$MnSO_4$ solution, alkaline KI solution, concentrated H_2SO_4, $Na_2S_2O_3$ solution, starch solution, NaOH solution and phenolphthalein.

(i) **Dissolved oxygen:** In a glass stoppered bottle, a known volume of water sample is taken so that no air bubble remains trapped inside. Add to it 1 ml of $MnSO_4$, 1 ml of alkaline KI solution and shake it thoroughly. The appearance of a brown precipitate indicates the presence of oxygen in the sample. To dissolve the precipitate, 2 ml of concentrated H_2SO_4 is added and shaken again. Then the total contents of the bottle are titrated with $Na_2S_2O_3$ solution (within one hour of the dissolution of the precipitate) using starch as an indicator. The dissolved oxygen is calculated as:

Calculation

$$\text{Dissolved oxygen (in mg/l)} = \frac{(V_1 \times 8 \times 1000)}{V_2}$$

Where V_1 = Volume of (N) $Na_2S_2O_3$ solution used

V_2 = Volume of water sample used

(ii) **Free carbon dioxide:** To about 100 ml of water sample in a conical flask, add a few drops of phenolphthalein indicator. The colour change to pink indicates the absence of free CO_2. If the sample remains colourless it is titrated with 0.05 (N) NaOH solution. The end point is indicated by the appearance of a pink colour.

Calculation

$$\text{Free } CO_2 \text{ (mg/l)} = \frac{V_1 \times 1000 \times 44}{V_2}$$

Where V_1 = Volume of NaOH used in ml.

V_2 = Volume of sample taken

Experiment 7

Aim

Analysis of pollutants like oil and grease in water.

Requirements

500 ml separatory funnel, water bath, filter paper, sulphuric acid, petroleum ether and ethyl alcohol.

Procedure

The total sample is taken in a separatory funnel. To this sample, 10 ml of H_2SO_4 and 25–50 ml of petroleum ether is added and shaken well. A small amount of ethyl alcohol is added to it to dissolve any suspended particles present. On standing, the liquid separates into two distinct layers. The lower layer is discarded while the upper layer is filtered and collected in a previously weighed crucible. The filter paper is also moistened with petroleum ether. After filtration, the sides of the filter paper are rinsed with a little more petroleum ether to collect any residual oil or grease attached to it. The crucible is heated carefully on a water bath and finally it is cooled in a dessicator and weighed again.

Calculation

$$\text{Oil and grease (mg/l)} = \frac{(W_2 - W_1) \times 1{,}000{,}000}{V}$$

Where W_1 = Weight of empty crucible

W_2 = Weight of crucible with residue

V = Volume of sample taken in (ml.)

Experiment 8

Aim

To study the abiotic components like pH, turbidity, temperature and light intensity of water in a pond ecosystem.

Requirements

Secchi disc, long thread, metre scale, thermister, photometer, universal indicator.

Procedure

Pond water samples are collected from different places, depths and seasons. A few drops of Universal indicator are added into the test tubes containing a few ml of each sample, and the colour developed is compared with the colour chart to find out the approximate pH value.

A Secchi disc is a circular metallic disc of 20 cm diameter and painted radially white and black alternately. It is lowered and raised in pond water to measure turbidity. The degree of indistinction between the two colours in water determines the index of turbidity of water.

The temperature of pond water is measured by the thermister, an instrument based on the principle of thermo couples.

A Photometer, with the photocell sealed in a transparent water-proof case, is placed at different sites and depths of pond water to measure variations in light intensity values.

Experiment 9

Aim

To study the biotic components of a pond ecosystem.

Requirements

Collection nets with different mesh sizes, metallic chains with hooks, specimen tubes, glass jars, hand lens, meter scale, scissors, forceps, centrifuge, microscope and suitable stains.

Procedure

The organisms of plants and animals are collected in polythene bags/bottles from different depths of the pond with various collection means. The organisms are brought to the laboratory and cultivated on suitable culture media, if required. The concentration of the presence of different biotics namely submerged, free-floating, rooted, floating, amphibious and phytoplanktons type of green plants, zooplanktons, insects, frogs, fishes, birds, bacteria, and fungi are studied.

Experiment 10

Aim

Qualitative and quantitative estimation of planktons in fresh water.

Requirements

Plankton nets/plankton samplers, plankton traps, pumps, tubes haemocytometer, cover glass, microscope and filter paper.

Procedure

Water samples are collected from different depths of water bodies at suitable intervals.

Plankton nets/samplers/traps are used to collect planktons, the selection being determined by the accuracy required and the equipment available. Phytoplanktons are counted by several methods. One common method is Lackey's drop method, where a known volume of water, which fits below a 22 mm cover glass, is placed over a glass slide. The volume is one drop taken by a dropper. Organisms in this drop are counted through a high-powered micro field of a compound light microscope.

Experiment 11

Aim

To determine the potability of water through the following three essential tests:

(i) Jar test.
(ii) Chlorine demand.
(iii) Residual chlorine.

(i) **Jar test**: This test is carried out to find out the optimum dose of alum and lime used for coagulation and flocculation of potable water.

Requirements

Alum stock solution, lime stock solution and 1.5 litre beakers fitted with multiple stirrers with continuous speed variation from 20 to 150 rpm.

Procedure

The colour, turbidity, alkalinity, pH of the water sample is determined. Then to one litre of the water sample taken in different beakers, different concentrations of alum and lime are added to each and shaken thoroughly. Time required for the appearance of the first flock is noted. Then the beakers are allowed to settle for one hour. After an hour a little water is pipetted out from each beaker and re-examined for colour, turbidity and alkalinity. More alum and lime are added, if required. In this way, the optimum dose of alum and lime for a particular water sample is determined.

(ii) **Chlorine demand**: This is the amount of chlorine required for the organic matter and other substances present in water. It is estimated as the difference between the amount of chlorine added and the amount of residual chlorine available after a specific interval of time. For the purpose, 200 ml of sample water is taken in at least 10 amber bottles to which increasing amount of chlorine solution (350 mg of 30 per cent bleaching powder in 1 litre water) is added and allowed to stand for 10 minutes. After 10 minutes have passed, free available chlorine and combined available chlorine is estimated.

$$\text{mg/l chlorine demand} = \text{mg/l chlorine added} - \text{mg/l residual chlorine.}$$

(iii) **Residual chlorine**: Since chlorine is added to water, it is necessary to estimate the amount of residual chlorine in water, for obtaining the optimum amount of chlorine required for the treatment of water. This is done either through iodometry or the ortho-toluidine method.

Iodometry: This method is sensitive when residual chlorine concentration is more than 1 mg /l. To about 500 ml of the sample in a conical flask, add 0.5–1 gm of KI crystal and 5 ml of acetic acid. Titrate the resulting solution with 0.025 (N) $Na_2S_2O_3$ solution till the brown colour of iodine disappears. To this, add 1 ml of starch solution and titrate till the blue colour disappears.

$$\underset{(\text{mg/l})}{\text{Residual chlorine}} = \frac{\text{ml } 0.025\ (\text{N})\ Na_2S_2O_3 \times 0.0885 \times 100}{\text{ml of sample taken for estimation}}$$

Ortho-toluidine method: This method is used when residual chlorine concentration is less than 1 mg/l.

Requirements

Ortho-toluidine solution (1.35 gm ortho-toluidine hydrochloride dissolved in 500 ml distilled water + 350 ml distilled water and 150 ml concentrated HCl), sodium arsenite solution, Nessler tubes or comparator tube, standard colour glass dishes.

Procedure

To about 10 ml of the water sample, add 0.5 ml of ortho-toluidine solution and 0.5 ml of sodium arsenite. Within three to five minutes, a colour may develop (due to free available chlorine). The colour is compared with the standard colour disc and noted. Let the reading be FC (or free chlorine). For total chlorine the same procedure is followed and the colour is recorded after five minutes. Let this be TC (or total chlorine). For interference, the same quantity of sodium arsenite solution is taken in a Nessler tube and the same volume of sample solution and ortho-toluidine 0.5/10 ml of sample is added. The developed colour is compared immediately. Let it be I_1, and after five minutes the colour is compared and noted as I_2.

Calculation

Total available residual chlorine = $TC - I_2$

Free available residual chlorine = $FC - I_1$

Combined available residual chlorine = $(TC - I_2) - (FC - I_1)$

Conclusion

The optimum amount of coagulant (lime), chlorine and residual chlorine estimation is very important in order to declare whether a water sample is potable or not. These are the fundamental tests performed by all supply persons for confirming the purity and suitability of potable water.

Model Question Papers

These are a few model question papers in which 70 marks are assigned for the theory paper while 30 marks are assigned to project/field work.

MODEL QUESTION PAPER – 1

Total marks – 70
Time – 3 hours

1. Answer the following questions: 1 x 10
   - (i) What are the components of environment?
   - (ii) What is soil?
   - (iii) What does the term biological resource mean?
   - (iv) In which year was the Earth Summit held?
   - (v) What is a food web?
   - (vi) What is the permissible limit of lead in drinking water?
   - (vii) What is the full form of SPCB?
   - (viii) What is fly ash?
   - (ix) What is the name of the disease caused in human beings due to asbestos?
   - (x) Which gas is both harmful and beneficial to life on earth?

2. Answer the following questions: 2 x 5
   - (i) What do you mean by habitat?
   - (ii) What is soil?
   - (iii) What is biodiversity?
   - (iv) What is biomagnification?
   - (v) What is sustainable development?

3. Answer any four of the following questions: 5 x 4
   - (i) What is acid rain?
   - (ii) What is noise pollution?
   - (iii) What is *Chipko* movement?
   - (iv) What is organic farming?
   - (v) What is eutrophication?
   - (vi) What are environmental ethics?

4. Answer any three of the following questions: 3 x 10
   (i) Write the composition of atmosphere.
   (ii) What is Greenhouse Effect? What are the major gases responsible for the same?
   (iii) What is waste? How can it be converted to wealth?
   (iv) Write a note on sanitary landfilling of solid waste explaining its merits and demerits.
   (v) Differentiate between Polluter Pays Principle and Beneficiary Pays Principle.

MODEL QUESTION PAPER – 2

Total marks – 70
Time – 3 hours

1. Answer the following questions: 1 x 10
   (i) Who wrote *Silent Spring*?
   (ii) Name the components of the biosphere.
   (iii) Name the different types of ecological (Eltonian) pyramids.
   (iv) Which layer of the atmosphere is also known as ozonosphere?
   (v) What is PAN?
   (vi) What is lotic water?
   (vii) What is a wetland?
   (viii) What is ex-situ conservation?
   (ix) What are fossil fuels?
   (x) What are the major components of soil?

2. Answer the following questions: 2 x 5
   (ii) What is surface water?
   (iii) What is meant by the word 'resource'?
   (iii) What is BOD?
   (iv) What is trophic level?
   (v) Name the two main causes of eutrophication.

3. Answer any four of the following questions: 5 x 4
   (i) What is marine water pollution?
   (ii) What are the characteristic structural changes noticed in the human body affected from fluorosis?
   (iii) Differentiate between HIV and AIDS.
   (iv) Explain the need for value education.
   (v) Write a short note on *Narmada Bachao Andolan*.
   (vi) Write a short note on mushroom cultivation.

4. Answer any three of the following questions: 3 x 10
   (i) Explain the impact of environment on human health.
   (ii) What is ISO 14000?
   (iii) Explain the necessity of Environmental Impact Assessment.
   (viii) What is smog? Explain in detail.
   (ix) What is rain water harvesting? Explain the main techniques of rain water harvesting.

MODEL QUESTION PAPER – 3

Total marks – 70
Time – 3 hours

1. Answer the following questions: 1 x 10
   (i) What is biofertilizer?
   (ii) Explain the term afforestation.
   (iii) What is organophosphorous insecticide?
   (iv) Define biopesticide.
   (v) Name the gases that cause ozone depletion.
   (vi) Define the term dendrothermal energy.
   (vii) What is hazardous waste?
   (viii) Where and when did the last tsunami occur?
   (ix) Name two ecological hotspots of India.

2. Answer the following questions: 2 x 5
   (i) What is the role of blue-green algae in the nitrogen cycle?
   (ii) What are the diseases caused by heavy metal poisoning?
   (iii) Name the landslide prone areas of India?
   (iv) Write two points of difference between urban environment and natural environment?
   (v) What is tsunami?

3. Answer any four of the following questions: 5 x 4
   (i) Explain the different methods of timber extraction.
   (ii) Explain the impact that dams have on forests as well as tribal people.
   (iii) What do you mean by green manuring?
   (iv) State the importance of wetlands.
   (v) Write a short note on generation of electricity from waste.
   (vi) Explain how an individual can contribute in the prevention of pollution.

4. Answer any three of the following questions: 3 x 10
   (i) What do you mean by the term biomedical waste? Write a note on the options that are recommended for the treatment of different varieties of biomedical wastes.

(ii) Write a short note on recycling of plastic.

(iii) Give a brief description of the useful functions of forest resources.

(iv) Differentiate between ex-situ and in-situ conservation.

(v) What are endemic species? Write a note on measures taken to preserve endemic species.

MODEL QUESTION PAPER – 4

Total marks – 70
Time – 3 hours

1. Answer the following questions: 1 x 10

   (i) What do you understand by the term contaminant?

   (ii) What is biofuel?

   (iii) What is a biome?

   (iv) What is ocean thermal energy?

   (v) What is a pollutant?

   (vi) What is the full form of DDT?

   (vii) Name a source of fly ash.

   (viii) Where is the Silent Valley situated?

   (ix) What is MIC?

   (x) What is temperature inversion?

2. Answer the following questions: 2 x 5

   (i) Name three main effects of disaster.

   (ii) What are the effects of acid rain?

   (iii) What is BOD?

   (iv) Define secondary producer in a food chain.

   (v) What does the term endangered species mean?

3. Answer any four of the following questions: 5 x 4

   (i) Briefly discuss about the alternative sources of energy.

   (ii) What is carrying capacity?

   (iii) What is bioremediation?

   (iv) Briefly discuss about environmental protection laws.

   (v) What is an inverted pyramid?

   (vi) Differentiate between a wasteland and a wetland.

4. Answer any three of the following questions: 3 x 10

   (i) What is earthquake? What steps should be followed for mitigating the damage caused by earthquakes?

   (ii) How does energy flow through an ecosystem?

   (iii) Write a short note on arsenic pollution.

(iv) Write a short note on different methods that can be employed for control of noise pollution.

(v) Write a short note on photochemical smog.

MODEL QUESTION PAPER – 5

Total marks– 70
Time – 3 hours

1. Answer the following questions: 1 x 10
   (i) What is Gross Primary Productivity?
   (ii) What is the full form of SPCB?
   (iii) What are the present day numbers of sanctuaries of India?
   (iv) What is the percentage of tribal people displaced as a result of dam-building?
   (v) Name two pollutants of automobile exhaust.
   (vi) When did the supercyclone occur in Orissa?
   (vii) What do you mean by the term 'Human Resource'?
   (viii) What is PAN?
   (ix) Name two tiger projects of India.

2. Answer the following questions: 2 x 5
   (i) Mention two sources of lead pollution.
   (ii) What are the basic differences between in-situ and ex-situ cultivation?
   (iii) What is albedo?
   (iv) What is an aquifer?
   (v) What was the main objective of the Earth Summit?

3. Answer any four of the following questions: 5 x 4
   (i) Define ecological niche.
   (ii) What is Chernobyl disaster?
   (iii) What are the main causes of drought?
   (iv) Define population explosion.
   (v) Write a short note on EIA.
   (vi) Classify different types of fresh water ecosystem.

4. Answer any three questions of the following: 3 x 10
   (i) Write about the role that biogeochemical cycles play in nature.
   (ii) What do you mean by environmental ethics? What role do they play in getting rid of environmental hazards?
   (iii) Write a note on fluoride pollution.
   (iv) What do you mean by emission trading?
   (v) Write a note on consumerism and waste products.

Self Assessment Questions

MULTIPLE CHOICE QUESTIONS (SOLVED)

1. The mean temperature of the earth is approximately
   (i) 10^{0}C.
   (ii) 15^{0} C.
   (iii) 20^{0}C.
   (iv) 29^{0}C.

2. Soil erosion takes place due to
   (i) loss of vegetation cover.
   (ii) excessive activities of ploughing and irrigation.
   (iii) natural phenomenon.
   (iv) All of the above.

3. Control measures of soil erosion caused by man is/are
   (i) intense plantation.
   (ii) less use of chemicals (herbicide).
   (iii) taking preventive action to reduce run off of soil with water.
   (iv) All of the above.

4. Incineration is the disposal method of
   (i) water pollutants.
   (ii) air pollutants.
   (iii) solid waste.
   (iv) All of the above.

5. The term environment denotes
   (i) only living beings.
   (ii) only non-living beings.
   (iii) both living and non-living that effect an individual organism or population at any point in the life cycle.
   (iv) None of the above.

6. The conservation of nature and natural resources
   (i) were not encouraged in ancient India.
   (ii) were encouraged in ancient India.

(iii) are being encouraged recently in India.
(iv) None of the above.

7. Which of the following is a Greenhouse gas?
   (i) Nitrogen.
   (ii) Methane.
   (iii) Oxygen.
   (iv) All of the above.

8. Water in the ocean comprises of:
   (i) 77 per cent of the total surface water of the Earth.
   (ii) 67 per cent of the total surface water of the Earth.
   (iii) 87 per cent of the total surface water of the Earth.
   (iv) 93 per cent of the total surface water of the Earth.

9. Water in the ocean comprises of:
   (i) 77 per cent of the total water of the Earth.
   (ii) 67 per cent of the total water of the Earth.
   (iii) 87 per cent of the total water of the Earth.
   (iv) 93 per cent of the total water of the Earth.

10. Which of the following is not a fossil fuel?
    (i) Coal.
    (ii) Wood.
    (iii) Petrol.
    (iv) LPG.

11. The interaction of animals and plants in any well defined area like pond is called
    (i) ecosystem.
    (ii) biosphere.
    (iii) eco-community.
    (iv) biome.

12. The metal associated with ground water pollution in the present day is
    (i) mercury.
    (ii) lead.
    (iii) cadmium.
    (iv) arsenic.

13. Water pollution may be reduced by the biological treatment with the help of
    (i) fungi.
    (ii) chlorine.

(iii) phytoplanktons.

(iv) ammonia.

14. Which of the following is non-biodegradable?

    (i) Cotton.

    (ii) Wool.

    (iii) Nylon.

    (iv) Silk.

15. There are two samples of waste water. Sample-I has BOD 200mg/l and Sample-II has BOD 40mg/l.

    (i) The degree of pollution is same in both the samples.

    (ii) Sample-I is more polluted than Sample-II.

    (iii) Sample-II is more polluted than Sample-I.

    (iv) No inference can be drawn on degree of pollution.

16. The coldest region of atmosphere is

    (i) troposphere.

    (ii) stratosphere.

    (iii) mesosphere.

    (iv) thermosphere.

17. The biotic factor of the ecosystem is

    (i) soil.

    (ii) wind.

    (iii) producer and consumer.

    (iv) sunlight.

18. The temperature range of troposphere is

    (i) –2 to –920^0 C.

    (ii) 15 to –560^0 C.

    (iii) –56 to –2^0 C.

    (iv) –92 to 1200^0 C.

19. Ozone is not a pollutant if it does not reside in

    (i) troposphere.

    (ii) stratosphere.

    (iii) mesosphere.

    (iv) ionosphere.

20. Which one of the following unit measures the intensity level of noise?

    (i) Met.

    (ii) Candle.

(iii) Decibel.
(iv) Newton.

21. Which is the unit dose of ionization radiation?
    (i) ROM.
    (ii) RAD.
    (iii) CAD.
    (iv) CNCED.

22. The dengue fever is caused by the mosquito
    (i) anopheles.
    (ii) culex.
    (iii) aedes.
    (iv) All of the above.

23. India is a megadiversity country with
    (i) 4 ecological hotspots.
    (ii) 3 ecological hotspots.
    (iii) 2 ecological hotspots.
    (iv) 1 ecological hotspot.

24. Which of the following is a geothermal source of energy?
    (i) Tidal waves.
    (ii) Ocean current.
    (iii) Gases and hot springs.
    (iv) All of the above.

25. Electrostatic precipitator and filters are used in factories to minimize
    (i) air pollution.
    (ii) water pollution.
    (iii) noise pollution.
    (iv) pesticide pollution.

26. The wildlife is beneficial to mankind in several ways, because it provides beneficial substances such as
    (i) rubber.
    (ii) tannin.
    (iii) drugs.
    (iv) All of the above.

27. The Great Indian Bustard is a
    (i) vulnerable species.
    (ii) endangered species.

(iii) rare species.
(iv) flourishing species.

28. The two largest dam-reservoirs which were planned to be constructed on river Narmada are
(i) Sardar Sarovar and Narmada Sagar.
(ii) Sardar Sarovar and Nagarjuna Sagar.
(iii) Narmada Sagar and Nagarjuna Sagar.
(iv) Narmada Dam and Sardar Dam.

29. The names of Baba Amte and Medha Patkar are associated with which of the protest movements of India?
(i) Silent Valley Movement.
(ii) *Narmada Bachao Andolan.*
(iii) *Chipko* Movement.
(iv) All of the above.

30. *Chipko* Movement is
(i) a movement of political strength.
(ii) an environmental movement.
(iii) a movement for independence.
(iv) None of the above.

31. Environmental planning organization is
(i) ICAR.
(ii) NEERI.
(iii) NCO.
(iv) NPO.

32. Which one of the following is not a Ramsar site?
(i) Bhitarkanika Mangroves.
(ii) Chilika Lake.
(iii) Deepor Beel.
(iv) None of the above.

33. United Conference on Environment and Development (UNCED) was held in
(i) June 1992.
(ii) December 1993.
(iii) December 1995.
(iv) November 1996.

34. There will be a limit to economic growth, set by the available stocks of exhaustible resource. This is known as
(i) waste availability limit to growth.
(ii) resource availability limit to growth.

(iii) raw material availability limit to growth.
(iv) All of the above.

35. In *Chipko* Movement
(i) women were protesting against replacement of mixed forest by commercially valuable pine tree.
(ii) women were protesting against alcoholic husbands.
(iii) *Garhwali* women were protesting against atrocities of contractors.
(iv) it is a movement against corrupt political leaders.

36. Variation of genes within the same species (single population) and also among geographically separated populations is called
(i) biodiversity.
(ii) species diversity.
(iii) genetic diversity.
(iv) None of the above.

37. The rich biodiversity of India is under severe threat of extinction on account of
(i) habitat loss/degradation due to agriculture; extraction, etc.
(ii) fragmentation and over exploitation of resources.
(iii) poaching and international trade of wild species and products.
(iv) All of the above.

38. The BOD of a pond is high, the pond is
(i) highly polluted.
(ii) not polluted.
(iii) without any green vegetation.
(iv) None of the above.

39. DDT is a
(i) biochemical pollutant.
(ii) biodegradable pollutant.
(iii) non-biodegradable pollutant.
(iv) non-pollutant.

40. Soil is a
(i) collection of organic materials.
(ii) collection of non-organic materials.
(iii) top layer of land surface.
(iv) None of the above.

41. What type of energy resource is Petroleum?
(i) Synthetic.
(ii) Inconvenient.

(iii) Non-renewable.
(iv) Renewable.

42. Ramsar Convention 1971 aimed at the conservation of
    (i) wasteland.
    (ii) wetland.
    (iii) desert.
    (iv) All of the above.

43. Which one the following is a world heritage site:
    (i) Kaziranga National Park.
    (ii) Nandadevi National Park.
    (iii) Sundarban National Park.
    (iv) Keoladeo National Park.

44. The principal pollutant from automobile exhaust is
    (i) mercury.
    (ii) lead.
    (iii) chromium.
    (iv) cadmium.

45. Agenda 21 refers to
    (i) Earth Summit at Rio de Janeiro.
    (ii) Montreal Protocol.
    (iii) Stockholm Conference.
    (iv) None of the above.

46. Which of the following does not contain biomass?
    (i) Municipal garbage.
    (ii) Sewage.
    (iii) Plastic materials.
    (iv) All of the above.

47. Stone cancer is caused by
    (i) UV rays.
    (ii) ozone layer depletion.
    (iii) acid rain.
    (iv) Greenhouse gases.

48. CFC is also known as
    (i) Chlorofluoro Carbon.
    (ii) Center for fuel control.

(iii) Carcinogenic fluoride compound.
(iv) None of the above.

49. Source of oxygen in atmosphere is
    (i) respiration.
    (ii) excretion.
    (iii) photosynthesis.
    (iv) transpiration.

50. Acidity in rain is measured using a
    (i) Barometer.
    (ii) Hygrometer.
    (iii) Ammeter.
    (iv) pH meter.

51. Rapid urbanization and industrialization
    (i) leads to environmental pollution.
    (ii) is responsible for economic growth all over the world.
    (iii) is responsible for the population explosion.
    (iv) All of the above.

52. Environmental Impact Assessment (EIA)
    (i) should be made compulsory for starting a developmental project.
    (ii) should not be made compulsory for starting a developmental project.
    (iii) should be done after completion of a developmental project.
    (iv) None of the above.

53. The pesticide that directly attacks the nervous system is
    (i) Aldrin.
    (ii) DDT.
    (iii) Organic Phosphates.
    (iv) None of the above.

54. The most polluted city in the world is
    (i) Kolkata.
    (ii) Tokyo.
    (iii) New York.
    (iv) Mexico.

55. The major effect of global warming is
    (i) animals will die.
    (ii) aquatic plants will die.

(iii) crop production and yield will be increased.
(iv) glaciers will melt and water level on earth will increase.

56. The main cause of global warming is
    (i) trapping of high wavelengths of sunrays by SO_2.
    (ii) trapping of shorter wavelengths of sunrays by CO_2.
    (iii) trapping of longer wavelengths of sunrays by CO_2.
    (iv) All of the above.

57. The main chemical responsible for ozone depletion is
    (i) PAN.
    (ii) Hydrocarbon.
    (iii) Freon.
    (iv) CFC.

58. The most powerful eye irritant present in the smog is
    (i) NO.
    (ii) O_3.
    (iii) PAN.
    (iv) SO_2.

59. The main cause of ozone layer depletion in the upper atmosphere is
    (i) SO_2.
    (ii) CH_4.
    (iii) NO_x.
    (iv) NH_3.

60. Which of the following is a source of air pollution?
    (i) CO.
    (ii) CO_2.
    (iii) SO_2.
    (iv) N_2.

61. Thickness of ozone layer is measured in which of the following units?
    (i) Meters.
    (ii) Centimeters.
    (iii) Dobson units.
    (iv) Decibel units.

62. The recent air pollution due to radiation that causes hazards to the world and disturbed the ecological balance
    (i) MIC gas tragedy in Bhopal.
    (ii) Chernobyl explosion.

(iii) Challenger explosion in USA.
(iv) All of the above.

63. The most possible cause of tension, headache, high blood pressure, loss of concentration and loss of hearing abilities is
(i) water pollution.
(ii) air pollution.
(iii) noise pollution.
(iv) pesticide pollution.

64. Biological Oxygen Demand (BOD) determines the health of
(i) air.
(ii) water.
(iii) soil.
(iv) All of the above.

65. Recent changes in the climate is due to
(i) environmental pollution.
(ii) different natural phenomenon in recent years.
(iii) natural cyclic change of climates.
(iv) Chernobyl explosion.

66. First International convention on environment was held in
(i) New Delhi.
(ii) New York.
(iii) Stockholm.
(iv) Geneva.

67. *Chipko* Movement was held in
(i) Uttar Pradesh.
(ii) Assam.
(iii) Karnataka.
(iv) Kerala.

68. In India, Control of air Pollution is protected by
(i) Forest Act.
(ii) Police Act.
(iii) Environmental Pollution Act.
(iv) Wildlife Act.

69. Acid rain is caused due to mixing of rain water with
(i) Carbon monoxide.
(ii) Carbon dioxide.

(iii) Sulphur dioxide and some oxides of nitrogen.
(iv) All of the above.

70. The pyramid which can not be inverted in a stable ecosystem, is a pyramid of
(i) number.
(ii) food.
(iii) energy.
(iv) biomass.

71. The logical sequence of carbon cycle is
(i) decomposer-photosynthesis-consumer.
(ii) photosynthesis-decomposer-consumer.
(iii) consumer-photosynthesis-decomposer.
(iv) photosynthesis-consumer-decomposer.

72. According to ecological food chain, man is a
(i) decomposer.
(ii) producer.
(iii) consumer.
(iv) All of the above.

73. The term trophic level is used for
(i) an organism's position in a food chain.
(ii) an organism's position in the ecosystem.
(iii) an organism's position in a biome.
(iv) All of the above.

74. A food web consists of
(i) many independent food chains.
(ii) a single food chain.
(iii) many interconnected food chains.
(iv) All of the above.

75. The transfer of food energy from the producer, through a series of organisms, with repeated eating and being eaten is called
(i) food web.
(ii) food habit.
(iii) food chain.
(iv) food roll.

76. A food chain consists of
(i) producers, carnivores and decomposers.
(ii) producers, herbivores and carnivores.

(iii) producers, consumers and decomposers.
(iv) producers and primary consumers.

77. At the earth summit held in Rio de Janeiro, a globalization plan was adopted with aim of integrating environmental imperatives and developmental aspirations. The Global Action Plan is known as
(i) Agenda 22.
(ii) Agenda 32.
(iii) Agenda 21.
(iv) Agenda 31.

78. Minamata disease is caused due to
(i) intake of cadmium.
(ii) intake of arsenic.
(iii) intakc of fish contaminated with mercury.
(iv) intake of fluoride contaminated water.

79. In 1984, the worst gas tragedy in India took place in
(i) Bengaluru.
(ii) Mumbai.
(iii) Bhopal.
(iv) Patna.

80. The Silent Valley is a
(i) place in Himachal Pradesh.
(ii) densely forested valley in Palghat district of north Kerala.
(iii) valley in Garhwal Himalayan region.
(v) valley in Shivalik range.

81. Which one of the following is not a gaseous cycle?
(i) Carbon cycle.
(ii) Phosphorous cycle.
(iii) Nitrogen cycle.
(iv) Oxygen cycle.

82. Which one of the following is capable of entrapping solar energy?
(i) Producer.
(ii) Decomposer.
(iii) Primary consumer.
(iv) None of the above.

83. What is smog?
(i) Dense fog.
(ii) Fog plus smoke.

(iii) A special type of smoke.
(iv) None of the above.

84. Which one of the following is not a renewable source of energy?
(i) Wood.
(ii) Petroleum.
(iii) Tidal energy.
(iv) Wind energy.

85. The land mass of earth is known as
(i) biosphere.
(ii) hydrosphere.
(iii) lithosphere.
(iv) stratosphere.

86. Richter scale is used for measuring
(i) velocity of light.
(ii) intensity of sound.
(iii) amplitude of seismic waves.
(iv) All of the above.

87. The name Sundarlal Bahuguna is related to
(i) *Chipko* movement.
(ii) *Narmada Bachao Andolan.*
(iii) Silent Valley.
(iv) All of the above.

88. Which of the following is not a marine pollutant?
(i) Oil spillage from tankers.
(ii) Oil reaching the marine water from land.
(iii) Radioactive waste.
(iv) Industrial sludge.

89. The safest method for biomedical waste disposal is
(i) incineration.
(ii) autoclaving.
(iii) sharp pit encapsulation.
(iv) Both (ii) and (iii).

90. HIV is not transmitted by which of the following:
(i) Sharing needles and injection equipments.
(ii) Blood transfusion.

(iii) Breast feeding.
(iv) Mosquito bites.

91. The endangered species Brown Palmcivet (*Paradoxurus Jerdoni*) is found in
(i) Western Ghats.
(ii) Eastern Ghats.
(iii) Eastern Himalaya.
(iv) Sundarbans.

92. In 1978, world's most notorious and infamous soil pollution took place in
(i) Chernobyl.
(ii) Bhopal.
(iii) Love Canal.
(iv) Smatler.

93. Which of the following virus is responsible for the Avian Influenza (Bird Flu)?
(i) H_5N_1 virus.
(ii) NH_3 virus.
(iii) $H_2 N_{10}$ virus.
(iv) None of the above.

94. Which of the following is an advantage of forests:
(i) They bring rainfall in the monsoon.
(ii) They act as watersheds.
(iii) They prevent rainfall in monsoon.
(iv) None of the above.

95. Oral Rehydration Therapy (ORT) is the process of
(i) continuous replacement of essential body fluids and salts in proper quantities during the attacks of diarrhoea.
(ii) providing mineral water and fruit juice by railway during journey.
(iii) taking saline injection.
(iv) None of the above.

96. The best way of disposal of non-biodegradable plastic is
(i) incineration.
(ii) landfilling.
(iii) recycling.
(iv) All of the above.

97. ISO 14000 is a series
(i) of international standards on environmental management tools and systems.
(ii) of international standards on quality management tools and systems.

(iii) of international standards on environmental and quality management tools and systems replacing ISO 9000 series.
(iv) All of the above.

98. Lubrication and proper maintenance of machine reduces
(i) air pollution.
(ii) water pollution.
(iii) soil pollution.
(iv) noise pollution.

99. Use of jack-hammers and dynamic blasting are related to
(i) soil pollution.
(ii) noise pollution.
(iii) air pollution.
(iv) Both (ii) and (iii).

100. In Kolkata, major air pollution is caused by
(i) fungal spores.
(ii) algae.
(iii) hydrocarbons.
(iv) carbon monoxide.

FILL IN THE BLANKS

1. ________ is the main contributor of global warming.
2. Earthquake is a ________ hazard.
3. The convention to safeguard ozone layer depletion was held at ________, Canada.
4. ________ June is known as World Environmental Day.
5. *Chipko* Movement was initiated by local tribal people in ________ in Uttar Pradesh.
6. The permissible level of sound in and around silence zone in daytime is ________ decibel.
7. Acid rain is caused by the presence of ________ acid and ________ acid in rain water.
8. Ozone depletion causes ________ cancer.
9. ________ is the best way of protecting soil erosion.
10. Biogas is a ________ type of energy source.
11. Asbestos causes ________ disease in human body.
12. ________ of the atmosphere protects us from the UV rays of the Sun.
13. A sequence of transfer of food energy from organisms in one trophic level to another is called ________.
14. On Earth, autotrophic components can directly fix ________ energy.
15. Greening ________ environmental degradation.

16. HIV stands for ________.
17. The innermost layer of earth is known as ________.
18. In 1986, nuclear plant disaster took place in ________.
19. Two non-conventional energy sources are ________ and ________.
20. El Nino is a ________ phenomenon connected with ________ of ocean surface.
21. Ozone is a pollutant gas on the surface of the ________ but high up, it ________ life below.
22. Itai-Itai disease is associated with ________ pollution.
23. The zone where fresh water mixes with marine water is called ________.
24. The names of two endangered mammals in India are ________ and ________.
25. ________ is responsible for Greenhouse Effect and ________ is responsible for ozone hole formation.
26. CO is a ________ pollutant and PAN is a ________ air pollutant.
27. ________ led the movement on Tehri Dam.
28. Heavy rain on vegetation free slopes causes ________ erosion.
29. India is a megadiversity country with two ________ hotspots.
30. The animals that live at the bottom of the sea are called ________.
31. ________ causes sea level rises.
32. Bhopal gas disaster occurred in the year ________.
33. Stone cancer of Taj Mahal is caused by ________.
34. The two important Greenhouse gases are ________ and ________.
35. All environmental changes take place in lower and ________ atmosphere.
36. The ________ of the atmosphere protects us from the UV rays of the Sun.

ANSWERS OF MULTIPLE CHOICE QUESTIONS

1 (i)	2 (iv)	3 (iv)	4 (iii)	5 (iii)	6 (ii)	7 (ii)	8 (i)	9 (iv)	10 (ii)
11 (iv)	12 (iv)	13 (i)	14 (iii)	15 (ii)	16 (i)	17 (iii)	18 (ii)	19 (i)	20 (iii)
21 (ii)	22 (iii)	23 (iii)	24 (iii)	25 (i)	26 (iv)	27 (ii)	28 (i)	29 (ii)	30 (ii)
31 (ii)	32 (iv)	33 (i)	34 (ii)	35 (i)	36 (iii)	37 (iv)	38 (i)	39 (iii)	40 (iii)
41 (iii)	42 (ii)	43 (i)	44 (ii)	45 (i)	46 (iii)	47 (iii)	48 (i)	49 (iii)	50 (iv)
51 (i)	52 (i)	53 (iii)	54 (iv)	55 (iv)	56 (iv)	57 (iv)	58 (iii)	59 (iii)	60 (i)
61 (iii)	62 (ii)	63 (iii)	64 (ii)	65 (i)	66 (iii)	67 (i)	68 (iii)	69 (iii)	70 (iii)
71 (iv)	72 (iii)	73 (i)	74 (iii)	75 (iii)	76 (iii)	77 (iii)	78 (iii)	79 (iii)	80 (ii)
81 (ii)	82 (i)	83 (ii)	84 (ii)	85 (iii)	86 (iii)	87 (i)	88 (ii)	89 (iv)	90 (iv)
91 (i)	92 (iii)	93 (i)	94 (ii)	95 (i)	96 (iii)	97 (i)	98 (iv)	99 (iv)	100 (iv)

ANSWERS OF FILL IN THE BLANKS

1	CO_2	2	natural	3	Montreal
4	Fifth	5	Uttarakhand	6	50
7	Sulphuric, nitric	8	skin	9	Afforestation
10	renewable	11	silicosis	12	Ozone layer

13	food chain	14	light	15	prevents
16	Human Immuno Virus	17	core	18	Chernobyl
19	Solar energy, wind energy	21	Earth, protect	22	mercury
23	estuary	24	tiger, one horned rhinoceros	25	CO_2, CFC
26	primary, secondary	27	Medha Patekar	28	Soil
29	two	30	benthic	31	Global warming
32	1984	33	air pollution	34	Carbon dioxide, methane
35	middle	36	ozone layer		

Appendix I

National River Action Plan

Rivers are one of the pillars for the survival and development of mankind. All major civilizations of the world such as the Indus valley civilization, Mesopotamian civilization and Chinese civilization have grown on the banks of perennial flowing rivers.

Rivers are thus considered as both the source and the sink for human civilization. While water from the rivers is drawn for drinking, household use, irrigation and industrial use, waste water, industrial effluents and sewage are also discharged into the same river. In India and other countries, the rivers are named after gods and goddesses. People also worship the rivers as their mother. Inspite of this, the same people do not think twice before polluting the rivers.

In India, particularly this is the fate of every river. As water from the rivers is drawn for irrigation, industry and other developmental works (dams, hydroelectric projects), the water in the rivers has reduced to a great extent. Therefore, it can no longer dilute the loads of sewage, industrial effluents and other pollutants that are being continuously discharged into it. A study conducted by the Central Pollution Control Board reveals that sewage and municipal effluents constitute about 75 per cent of the pollutants while the remaining 25 per cent is from industrial effluents.

The rivers were always a boon for human beings and other biotic societies in this world. Now the rivers have turned into a bane for them. Thus, the need of the hour is to create public awareness and implore participation to check further pollution of the rivers and also to carry out cleaning steps to restore the rivers to their previous condition. The Government of India came up with some actions plans to prevent pollution of rivers and to improve the water quality of different rivers.

Ganga Action Plan (GAP)

In June 1985, the Government of India launched the Ganga Action Plan to improve the water quality of the river Ganga by using a multi-pronged strategy. This project was sponsored by the Central Government. The plan envisaged interception and diversion of waste water reaching the Ganga with the installation of sewage treatment plants for treatment, management of solid waste, installation of crematoriums, installation of low cost sanitation and development of the riverfront. Thus, it is clear that the total project is based on public awareness and participation.

An apex body called the Central Ganga Authority was set up with the Prime Minister of India as the Chairman to oversee the implementation of the programme and to coordinate the efforts of various agencies involved.

A steering committee under the Chairmanship of the Secretary, Ministry of Environment and Forests was formed to consider the approval of schemes, annual allocation of funds and review the progress of the schemes. A monitoring committee was also constituted to monitor the progress of the scheme. A Ganga Project Directorate was set up as a part of the Ministry of Environment and Forests to service the steering committee, the monitoring committee and the Central Ganga Authority and also help to coordinate the implementation of the Ganga Action Plan.

The Ganga Action Plan has been taken up in two phases with an allocation of Rs 462.04 crore. Out of this, an amount of Rs. 451.70 crore has been released to and utilized by the participating states namely Uttar Pradesh, Bihar and West Bengal. The Ganga Action Plan Phase I was declared closed since March 31, 2000.

Also, an amount of about Rs. 488 crore has been utilized so far out of the allocated funds of Rs 1276.26 crore in the second phase of the Ganga Action Plan which was approved in stages between 1993 and 1996.

Under the GAP, a total of 242 out of 261 sanctioned schemes have been completed till date.

The pollution in Ganga with respect to pollution parameters, namely, Dissolved Oxygen (DO) and Biochemical Oxygen Demand (BOD) during the critical periods of summer in 1986 and 1999 was measured as follows:

1986–1999

Location	DO (mg/l)	BOD (mg/l)	DO (mg/l)	BOD (mg/l)
Haridwar	8.1	1.8	8.6	1.2
Allahabad	6.6	15.5	7.9	3.2
Patna	8.1	2.2	7.8	2.4
Kolkata at Uluberia	6.9	2.1	6.6	2.2

The desired standards fixed by the Central Pollution Control Board of the Ganga for these two parameters are 3 mg/l (maximum) for BOD and 5 mg/l (minimum) for DO.

The Ganga river can be expected to be pollution free only after Phase II of GAP is completed and a minimum flow in the river is maintained to allow the necessary dilution of treated waste water discharged into the river.

The evaluation of Phase I of GAP was done by experts from IIT Roorkee, Aligarh Muslim University, IIT Kanpur and All India Institute of Hygiene and Public Health, Kolkata. According to the report submitted by them, the primary step in restoring the water quality of the river has been achieved as there is a fair reduction of organic matter discharge into the river. They also feel that GAP as conceived and implemented is an appropriate programme for replication in other river basins also. The team also recommended appropriate intervention to reduce the microbial pollution of the river, rigorous qualitative and quantitative characterization of the sewage by adopting the most appropriate technology for treatment and resource recovery from the treated waste waters.

Ganga Action Plan (Phase-II) (Main Stream)

After the successful implementation of GAP Phase-I, the Government of India approved GAP Phase-II in 1991 for the abatement of pollution in the tributaries of Ganga.

In the GAP Phase-II plan with an outlay of Rs 378.00 crore, pollution abatement works are being done in 29 towns of which 10 are in Uttar Pradesh and eight in West Bengal. About 618 mld of sewage is targeted to be intercepted, diverted and treated. The plan includes substantial outlays for scheme implementation at Allahabad, Kanpur and Varanasi in Uttar Pradesh. Schemes in Bihar are only being sanctioned after due scrutiny. The Yamuna Action Plan, Gomti Action Plan and Damodar Action Plan were also put into action.

Ganga Action Plan (Supreme Court Cases)

Under this, pollution abatement works are being taken up in 30 towns. Of these 12 are in Uttar Pradesh, three in Bihar and 15 in West Bengal. About 162 mld of sewage is to be intercepted, diverted and treated. Out of the project outlay of Rs 209.90 crore, the funds released by the Government of India amount to Rs. 5.18 crore.

Yamuna Action Plan

Under this plan, pollution abatement works were being done in six towns of Haryana and six towns of UP and Delhi respectively. Six other towns of Haryana were also included in the Yamuna Action Plan following the Supreme Court of India's order. About 744 mld of sewage was proposed to be intercepted, diverted and treated under the plan. The approved cost for the project was Rs 509.45 crore with an external assistance of 17.77 billion yen. The deadline for the completion of the project was fixed as March 2000.

However, under the National River Conservation Plan, an extended phase for the Yamuna Action Plan was approved with a cost of Rs 222.60 crore and an external assistance of 8 billion yen by the Japan Bank of International Corporation. Of this total sanctioned amount, nearly 75 per cent was spent on the construction of a Community Toilet Complex (CTC) in Delhi, 1,146 CTCs were constructed in resettlement colonies where proper sanitation facilities were lacking. The remaining amount was spent for the states of Haryana, UP and Bihar. The project was completed in March 2002. The project resulted in significant health benefits, particularly to the people of Delhi.

Gomti River Action Plan

Under the Gomti River Action Plan, pollution abatement works are being done along the Gomti River in Lucknow, Sultanpur and Jaunpur in UP with an approval of Rs 58.11 crore. The United Kingdom Government also agreed to fund the pollution abatement work in Lucknow town but only in two phases. The first phase comprising certain emergency works and master plan preparation was completed in March 1999. However, the UK Government delayed the decision for funding the main action programme and finally in August 1997 refused to fund it at all. Thus, a major component of the work under the Gomti Action Plan could not be started even after years of approval from the CCEA. Subsequently, the UP Government submitted a revised cost estimate for Rs 237.40 crore for the same. The project target is 269 mld of sewage for interception, diversion and treatment.

Damodar Action Plan

The pollution abatement works for the Damodar river targets 12 towns of which eight are in Jharkhand and four in West Bengal. A number of schemes were taken up for priority sanction. About 68 mld of sewage is targeted to be intercepted, diverted and treated. The main towns to benefit from this action plan are Bokaro, Kangali, Chinchuda, Dugdha, Jharia, Ramgarh, Sindri, Sudamdih and Telunochu in Bihar/Jharkhand and Raniganj, Andal, Asansol and Durgapur in West Bengal.

National River Conservation Plan

The National River Conservation Plan was launched in 1995 at an estimated cost of Rs 37.13 crore to cover 18 major rivers in 10 states of India. About 1,928 mld of sewage is targeted to be intercepted, diverted and treated in 46 towns of Andhra Pradesh, Bihar, Gujarat, Karnataka, Madhya Pradesh, Maharashtra, Orissa, Punjab, Rajasthan and Tamil Nadu. The rivers involved are Brahmani, Mahanadi, Wain Ganga, Betwa, Tapti, Kshipra (Shipra), Khan, Kaveri, Tungabhadra, Sutlej, Chambal, Godavari, Krishna, Suvarnarekha, Sabarmati and Narmada.

National Lake Conservation Plan

A proposal for the conservation and management of three lakes namely Powai (in Mumbai), Kodaikanal and Ooty was approved by the Union Government of India in 2001 under the National Lake Conservation Plan. The estimate approved was Rs 14.90 crore out of which Rs. 6.62 crore was for Powai, Rs 6.63 crore for Kodaikanal and Rs 1.95 crore for Ooty. The plan envisaged the use of low cost sustainable bioremediation processes for the conservation of these lakes. On completion, the project will not only improve the (i) biodiversity and ecology of the lakes but will also help in (ii) pisciculture, (iii) improvement of overall hygiene and sanitation, (iv) boosting recreation and tourism and (v) producing biological manure from composting of water hyacinth.

Appendix II

Global Summits on Environment

In order to move towards an environmentally sound planet, the global community must address the issue of global development through international cooperation, bilateral and multilateral programmes and regional initiatives. Removal of poverty is an integral part of the environmental goal of the world. International conventions normally attempt to build international consensus on environmental issues through:

- Discussions, convergence of decisions and signatures by the participant nations in agreement of the decisions.
- Ratification of the above at a national level by the participant countries.
- Enforcement of the convention decisions at international levels by member nations after the same are ratified by a minimum number of nations (to be decided during negotiations). The environmental calamities during the 1950s and 1960s in USA, UK and Japan generated widespread awareness about the dangers of environmental pollution. Today, the air we breathe in, is full of toxic gases, the water we need is either scarce or contaminated and the land we inhabit is also polluted. Much of the pollution is caused by the pattern of production and consumption in industrialized countries. The earth's resources are mostly enjoyed by just 20 per cent of the world's population. The first UN Conference on Human Environment was held in Stockholm (Sweden) from June 5–10, 1972 to express concern over the depletion of forest resources, mineral wealth, marine life and other natural wealth, and also to share a common concern over air and water pollution. Since then, the world has moved together to chalk out specific environmental action plans, concrete agendas, framing time, necessary standards, conservation strategies and developmental priorities. The UNCED (United Nations Conference on Environment and Development) or the Earth Summit-1992 was held at Rio de Janeiro, the capital of Brazil from June 3–14, 1992 to ensure the relationship between environment and development on a global partnership level.

The major international conventions/events/protocols on environment are listed below:

A The conventions on oil, sea and hazardous waste disposals are listed as follows:

- International Convention for the Prevention of Pollution of the Sea by Oil tankers, 1954 (London).
- International Convention of Civil Liability for Oil Pollution Damage (as amended), 1969 (Brussels).
- International Convention for the Prevention of Pollutions from Ships, 1973 (as modified by the Protocol of 1978).
- UN Convention on the Law of the Sea 1982 (Montege Bay). It came into force in 1994 and established the rules governing all uses of the oceans and seas and their resources.
- Basal Convention on the Control of Transboundary Movement of Hazardous Wastes and their Disposal, 1989 (Basal, Switzerland).

In the 1980s, some African countries suffered from environmental pollution caused by wastes moved from developed European states. To deal with this environmental issue, the convention was held to properly administer the transboundary movement of waste, mostly by ships.

B The conventions on biodiversity and other related issues are listed as follows:

- CITES (Convention on International Trade in Endangered Species of wild fauna and flora)

  A meeting of members of the IUCN (the world conservation union) took a resolution in 1963 which ultimately drafted CITES and agreed to be the representative of 80 countries in Washington DC in 1973. This became effective from July 1, 1975 and regulates international trade in specimens of selected species and bars import, export, re-export, and collection/disturbance of species in seas and flora/fauna banned or covered under licence. Today, CITES accords varying degrees of protection to more than 30,000 species of animals and plants, whether they are traded as live specimens, leather goods, food products, curios, timbers, medicines, fur coats or dried herbs while the international wildlife trade is estimated to be worth billions of dollars.

- International Conference on Preservation of Marshes and Waterfowl, 1971 (Ramsar, Iran).

  This Convention was held to recognize the importance of marshes for animals, plants and migratory birds and the ecological system as a whole and to promote the conservation of marshes.

- Convention on the Conservation of Migratory Species of Wild Animals, 1979 (Bonn).

  It came into force on November 1, 1983.

- Convention on Protection of Migratory Birds between India and USSR (former), 1986.

- CBD (Convention on Biological Diversity) was adopted in 1992 in Rio de Janeiro at the Earth Summit. It aims for the conservation of biological diversity. The sustainable use of its components and the just and equitable sharing of benefits arising out of the utilization of genetic resources.

- Cartagena Protocol on Biosafety, 2000.

  Taking into account transboundary movement and risks to human health, the Convention ensures protection in the safe transfer, handling and use of living modified organisms resulting from modern biotechnology that may have adverse impacts on the conservation and sustainable use of biological diversity. Biotechnology is an emerging science that enables man-made techniques of change in DNA/genetic materials in plants, animals, microbial systems, leading to useful products and technologies.

- International Plant Protection Convention (IPPC).

  It is a treaty under the United Nations Food and Agricultural Organisation (FAO) to control pests, with the more specific purpose of adopting common and effective actions to prevent the spread and introduction of pests of plant and plant products and promoting measures for their control.

C Conventions on issues such as climate change, ozone layer depletion are listed as follows:

- Convention for the Protection of Ozone Layer, 1985 (Vienna).

  The ozone layer around the globe absorbs most of the ultraviolet (UV-B) rays that harms creatures. Depletion of the ozone layer is therefore, harmful to both the human body and the ecological balance due to the increase in ultra violet rays on the earth. Chlorofluorocarbons (CFC) and some other substances destroy the ozone layer and thus substances such as CFCs are to be avoided.

- Montreal Protocol on Substances that Deplete the Ozone Layer, 1987.

  The parties to the Montreal Protocol established a fund with contributions by developed countries to provide financial resources to developing countries to reduce ODS (Ozone Depleting Substances), namely CFCs, halons, carbon tetrachloride, methyl chloroform and so on. Ultimately, these substances are to be phased out.

- UNFCCC (United Nations Framework Convention on Climate Change), 1992.

  It was adopted at the Earth Summit at Rio de Janeiro with the objective of stabilizing the density of greenhouse gases and reducing the limits of emissions of these gases.

- Kyoto Protocol, 1994.

  It is the protocol of the UNFCCC for the purpose of stabilizing Greenhouse gas concentration in the atmosphere at a level that would prevent dangerous man-made climate changes. It became effective on March 21, 1994.

D International protocols on regional development/protection are listed as follows:

- UNCCD (United Nations Convention to Control Desertification).

  Developing countries such as African countries, which are experiencing serious drought and/ or desertification need to evolve national and regional initiatives to stop desertification. This Convention provides the scope for the affected developing countries to initiate such actions and the developed countries to support such efforts.

- Antarctica Treaty.

  This Treaty provides a protocol on environmental protection to the Antarctica by the countries exploring the possibilities of using its resources.

E International conventions on pesticides, chemicals, hazardous wastes are listed as follows:

- Stockholm Convention, 1998–2000.

  The Stockholm Convention aims to reduce and eliminate 12 Persistent Organic Pollutants (POPs) that may affect our future generations. They are dioxin, furan, DDT, and so on. These chemicals remain intact in the environment for a prolonged period, get widely distributed geographically and accumulate in the fatty tissues of living organisms. They are toxic and harmful to human beings and wildlife. This convention was completed in five inter-governmental meetings during 1998–2000.

- Rottendam Convention on the Prior Informed Consent (PIC) Procedures for Certain Hazardous Chemicals and Pesticides in International Trade.

  Toxic pesticides and other hazardous chemicals are very harmful to human beings, wildlife and the environment and are affecting many of them every year. Various governments started a voluntary PIC procedure in the 1980s in their trading to minimize the harmful effects. Rottendam convention in 1998 wanted to enforce and legalize it. It is applicable to banned or severely restricted chemicals and severely hazardous pesticides formulation. It became effective on February 24, 2004.

F Convention on Access to Information, Public Participation and Decision-making and Access to Justice in Environmental Matters, June 15, 1998 (Aarhus, Denmark).

This is a new form of environmental agreement related to environmental rights.

G Earth Summit-1992

The United Nations Conference on Environment and Development, that is the Earth Summit-1992 was held in Rio de Janeiro, the capital of Brazil. Held from June 3–14, 2002, the Summit was the largest ever historic assembly of world leaders in which representatives of more than 170 countries, including 115 heads of states and governments participated. The Rio Earth Summit marked the 20th anniversary of the Stockholm Conference and the founding of UNEP. The main objective of the summit was:

- Scientific understanding of global climate changes phenomena, namely, ozone layer depletion, acid rain and global warming.
- Biodiversity.
- Conservation of the national ecosystem including forests.

The following four crucial documents were adopted in the summit:

(i) Rio Declaration or Draft Earth Charter with 27 guiding principles.
(ii) Convention on climate change and biodiversity.
(iii) Convention on forests.
(iv) Agenda-21 or action plan for achieving sustainable development in the 21st century.

The 27 guiding principles of the Earth Charter are as follows:

(i) Human beings are the core beneficiaries of sustainable development. Therefore, they must live a healthy and productive life in harmony with nature.

(ii) In accordance with the UN charter on the principles of international law, the states have a sovereign right to exploit their own resources as regards their own environmental and development policies, and will be responsible to ensure that those activities are within their jurisdiction or control and do not cause damage to the environment of other states or of areas beyond the limits of national jurisdiction.

(iii) The right to development must be fulfilled to ensure fulfillment of equitable developmental and environmental needs of the present and future generations.

(iv) For sustainable development, environment protection is an integral part of the development process and cannot be considered in isolation from it.

(v) All states and people shall be cooperative in the essential task of eradication of poverty as an indispensable requirement for sustainable development and decreasing the disparities in standards of living and fulfilment of the needs of the majority of the people of the world.

(vi) The special situation and needs of developing countries, particularly the least developed and those most environmentally vulnerable shall be given special priority. International actions should also address the interests and needs of all countries.

(vii) States shall cooperate in the spirit of global partnership to conserve, protect and restore the health and integrity of the earth's ecosystem. In view of the different contribution to global environment degradation, states have common but differentiated responsibilities.

(viii) To achieve sustainable development and better quality of life, the states should reduce and eliminate unsustainable patterns of production and consumption and promote appropriate demographic policies.

(ix) States should strengthen endogenous capacity building for sustainable development by improving scientific understanding and enhancing development by adaptation, diffusion and transfer of technologies including new and innovative technologies.

(x) Environmental issues are best handled with the participation of all concerned citizens, at the relevant level. Each individual shall have appropriate access to information concerning the environment and the opportunity to participate in decision-making processes. States shall facilitate and encourage public awareness and participation by making information widely available. Effective access to judicial and administration proceedings including redressal and remedy shall be provided.

(xi) States shall enact effective environmental legislation. Environment standards, management objectives and priorities should reflect the environmental and development context to which they apply.

(xii) States shall cooperate to promote a supportive and open international economic system that would lead to economic growth and sustainable development in all countries, to better address the problem of environmental degradation. Environmental measures addressing transboundary or global environment problems should as far as possible be based on an international consensus.

(xiii) States shall develop a national law regarding the liability and compensation for the victims of pollution and other environmental damages and also contribute to the expenditure in a more determined manner to develop future international laws regarding the liability and compensation for adverse effects of environment damage.

(xiv) States should effectively cooperate to discourage or prevent the relocation and transfer of any activities and substance harmful to human health and environment to other states.

(xv) In order to protect the environment, the precautionary approach shall be widely applied by states in accordance to their capabilities.

(xvi) National authorities should endeavour to promote the internalization of environmental cost and the use of economic instruments, keeping in mind that the polluter should, in principle bear the cost of pollution.

(xvii) Environmental Impact Assessment shall be undertaken for proposed activities that are likely to have a significant adverse impact on the environment and are subject to the decision of a competent national authority.

(xviii) States shall immediately notify other states of any natural disaster or other emergencies that are likely to produce sudden harmful effects on the environment of those states and the international community to help the afflicted states.

(xix) States shall provide prior and timely notification, relevant information and consult with potentially affected states on activities that may have significant adverse transboundary environmental effects.

(xx) Women have a vital role to play in environmental management and development. Their total participation is therefore essential to achieve sustainable development.

(xxi) The creativity, ideas and courage of the youth of the world should be mobilized to forge a global partnership in order to achieve sustainable development and a better future for all.

(xxii) Indigenous people and their communities and other local communities have a vital role to play in environmental management and development because of their knowledge and traditional practices. States should recognize and duly support their identity, culture and interests and enable their effective participation in the achievement of sustainable development.

(xxiii) The environment and natural resources of the people under oppression, domination and occupation shall be protected.

(xxiv) Warfare destroys sustainable development. Therefore, there shall be respect for the international law providing protection for the environment in times of armed conflict.

(xxv) Peace and development are interdependent and indivisible for environmental protection.

(xxvi) States shall resolve all their environmental disputes peacefully and by appropriate means in accordance with the chapter of the United Nations.

(xxvii) States and people shall cooperate in good faith and in spirit of partnership in the fulfilment of the principles embodied in this declaration and in the future development of international law in the field of sustainable development.

Agenda-21 (Action Plan)

Agenda–21 envisages implementation of the programme with target and cost estimates in all the areas of environment and development. Targets are to be linked to availability of resources. The crucial question was how much funds should be provided and by whom. There are 150 items of which nine depend on future mechanisms. According to a UNCED estimate, $625 billion was required by the end of 2000 to implement this action plan. But developing countries felt that they would need 'new and additional' resources over and above the estimated levels for sustainable development. Hence, the developing countries should not be asked to pay the sum they can ill-afford. The developed world, led by USA has been talking of a common responsibility. However, inclusion of the word 'differential responsibility' to the document has been an achievement for the developing countries. The US was more concerned with the Dunkel draft over patent laws and Intellectual Property Rights and shunned its obligation of providing new and additional funds and transfer of technology to needy nations. The summit also failed to raise enough funds for GEF for launching Agenda-21 Plan of action for the 21st century.

Though the US and other industrialized countries, including EEC, signed a convention on climate change, they refused to make a specific time-bound commitment to bring down the levels of emission of CO_2 and other gases. The only achievement of the summit is that it has put environmental issues including protection of the 'Only One Earth', on the world agenda, with a new added consciousness to protect it against any future degradation.

The full implementation of Agenda-21, the programme for further implementation of Agenda-21, and the commitments to the Rio principles were strongly reaffirmed at the World Summit on Sustainable Development (WSSD) held in Johannesburg, South Africa from August 26 to September 4, 2002.

Appendix III

Important World Dates

January 1	World Peace Day
February 2	World Wetland Day (In 1971, on this day, the Ramsar Convention on Wetlands of International Importance was signed)
March 8	United Nations Day for Women's Rights & International Peace also known as International Women's Day.
March 21	World Forestry Day and International Day for racial discrimination.
March 22	World Water Day (UNDP)
March 23	World Meteorology Day (WMO)
April 7	World Health Day (In 1948, on this day, the World Health Organization (WHO) was constituted).
April 22	Earth Day
May 1	Labour Day
May 3	World Press Freedom Day
May 15	International Day of Families
May 31	World No Tobacco Day (WHO)
June 5	World Environment Day (In 1972, on this day the Stockholm Conference on Human Environment was held in Sweden)
June 17	World Combat Day against Desertification and Drought
June 20	World Refugee Day
July 11	World Population Day
July 28	World Nature Conservation Day
August 6	Hiroshima Day
August 9	International Day of World's Indigenous People (observed every year during the decade 1994 to 2003).
September 8	International Literacy Day
September 16	World Ozone Layer Preservation Day (In 1987, on this day, the Montreal Protocol was signed.)
Last Week of September	World Ocean Week
October 1	World Elders' Day
1st Monday of October	UNESCO Day
October 1–7	Wildlife Week
October 13	International Reduction on the Natural Disaster Day
October 16	World Food Day (FAO)
October 17	International Day for Eradication of Poverty
December 1	World AIDS Day
December 2	International Day for Abolition Of Slavery
December 3	International Day for Disabled Persons
December 10	Human Rights Day
December 29	International Biodiversity Day

Appendix IV

Important Abbreviations

CAZRI	Central Arid Zone Research Institute
CEC	Council on Environment Quality
CFC	Chlorofluoro carbon
CITES	Convention on International Trade in Endangered Species
CPCB	Central Pollution Control Board
DNES	Department of Non-Conventional Energy Sources
EIA	Environmental Impact Assessment
ENVIS	Environmental Information System
EPA	Environmental Protection Agency
EPC	Environmental Protection Council
FAO	Food & Agricultural Organization
GEF	Global Environment Facility (Funds)
GEMS	Global Environmental Monitoring System
IAEA	International Atomic Energy Agency
ICRAF	International Council for Research in Agro-Forestry
IIED	International Institute for Environment & Development
IUCN	International Union for Conservation of Nature & Natural Resources
MAB	Man & Biosphere Programme
MINAS	Minimal National Standards
NAWDF	National Afforestation & Wastelands Development Fund (For Greener India)
NCEP	National Committee on Environmental Planning
NCEPC	National Committee on Environmental Planning & Coordination
NEPA	National Environmental Policy Act (1969)
NLUWDC	National Land Use & Wasteland Development Council
NWDB	National Wasteland Development Board
SACEP	South Asia Cooperative Environmental Programme
SCOPE	Scientific Committee on Problems of the Environment
SPCB	State Pollution Control Board
UNCED	United Nations Conference on Environment & Development, also called 'Earth Summit 1992'
UNEP	United Nations Environment Programme
WCED	World Commission on Environment & Development
WCS	World Conservation Strategy
WHO	World Health Organization
WMO	World Meteorological Organization
WWF	World Wide Fund for Nature-India
WRI	World Resources Institute (Washington DC, USA)

Appendix V

Albedo (Latin word meaning white)

Albedo is a measure of the reflectivity of a surface or body. It is the ratio of total electromagnetic radiation reflected to the total amount of incident light on it. The albedo is unitless and is sometimes expressed in percentage, from 0% to 100%. It is an important concept in climatology and astronomy. The ratio depends on the frequency of radiation considered unqualified. Fresh snow albedos are high, upto 90% whereas the ocean surface has low albedo. The average albedo of the earth is about 30% and of the moon is about 7%, respectively. Enceladus, a moon of Saturn, has the highest known albedo of any body in the solar system, that is 99%.

Human activities such as deforestation, farming and so on have changed the albedo in various areas around the globe. However, quantification of this is different on a global scale and the relationship of this change in albedo with global warming is yet to be established. However, the albedo effect can be observed clearly through snow temperature feedback. If a snow covered area warms and the snow melts, the albedo increases, more sunlight is absorbed, the temperature tends to increase and vice versa. Earth observation satellite sensors, such as NASA's MODIS instruments on board the 'Terra' and 'Aqua' satellites, regularly measure the albedo of the earth's surface.

Ambergris

Ambergris is a wax-like substance originating as a morbid concentration in the intestine of the sperm whale. Being lighter than water, it is found floating on tropical seas or cast up on the shore in yellow, grey, black or variegated masses, usually a few ounces in weight. Large pieces weighing about several hundred pounds are also not very rare. Ambergris is used as a fixative in perfumes. Its chemical composition is found to be crystalline alcohol with empirical formula of $C_{30}H_{51}OH$ named Ambrein.

Aquifer

Aquifer is a term used for a body of rock that contains water and releases it in significant quantities for use. The rock contains water-filled pores that are sufficiently connected to the water flow through the rock matrix to wells and springs.

Catalytic Converter

A catalytic converter is an anti-pollution device which runs exhaust gases of automobiles through a bed of allumina pellets coated with platinum or palladium catalyst. When this device is attached to the exhaust system of an automobile, it converts carbon monoxide, hydrocarbons and nitrogen oxides into carbon dioxide, water and nitrogen gas, respectively. But the catalyst gets poisoned with the presence of lead. Hence, automobiles fitted with catalytic converters must use lead-free petrol.

Ecological Niche

Ecological niche refers to the habitat in an ecosystem, which supplies every requirement and factors essential for the existence and survival of a species or an organism.

El Nino and La Nina

El Nino is an inter-annual fluctuation in the ocean atmosphere system. The local phenomenon of warming of the ocean surface near South America is called El Nino. In this phenomena, the western tropical Pacific cools, while the eastern tropical Pacific warms. It occurs in December just before Christmas every two years. It is also known as Southern oscillation. Although the exact cause of this phenomena is not known yet, but it takes place along the coast of Peru and Ecuador. The Peruvial coast is a place of strong high tides, which brings large quantities of nutrients to the sea surface and thus maintains a very high production of Anchovy fishes. During El Nino, warm equatorial waters, which are low in nutrients, move southwards along the Peruvian coast and replace the cold nutrient-enriched water. Thus, the fish industry is greatly affected by the El Nino effect.

El Nino also affects the weather conditions around the Pacific Ocean. It brings heavy rain and floods to countries like Peru and Ecuador in South America but drought to Australia. A severe El Nino occurred in 1982–83 when the bush fires reached the suburbs of Melbourne.

The return of the warm water back to the western Pacific is known as La-Nina. In the 1980s, La Nina was responsible for the intense storm that hit Australia and parts of Eastern Asia and brought droughts to North America, floods in Bangladesh and Sudan, where El Nino brings drought as the weather pattern gets reversed.

Fly Ash

Fly Ash is the low density, finely divided non-combustible particles present in a flue gas. These fine particles of the non-combustible portion of a solid burning fuel like coal are carried away by air drift and spread over a large region. Fly ash is widely used for brick-making and other purposes.

GAIA Hypothesis

British Biochemist James Lovelock developed the GAIA hypothesis. The hypothesis considers the whole biosphere as a single functional unit, where all the flora, fauna, all biotic and abiotic factors act as a functioning whole in the development of a lasting environment that would preserve itself irrespective of anthropogenic activities and interferences.

Green Bench

In popular response to combat the different facets of environmental abatements, and monitor and supervise industry-related environmental problems, the Supreme Court of India directed all High Courts of India to set up a Green Bench. The role of the Green Bench is to maintain the ecological balance of the environment.

Menagerie

This is a term representing the historical form of keeping wild and exotic animals in human captivity (predecessor of modern zoological gardens). As per the Encyclopedie Methodique of 1782, menagerie is defined as an establishment of luxury and curiosity. In fact, in France around the 17th century, the term menageries was meant for the management of household (domestic) stock and later for royal animals (aristocratic) collection. Later, the term was used to refer to even traveling animal collections that exhibited wild animals at fairs across Europe and North America.

Minimata incident

The Minimata incident of mercury pollution took place in Japan during the 1950s. From 1953–1960, many villagers had been suffering from unknown diseases in Japan. The worst situation occurred in a village facing the bay wherein 15 per cent of the villagers were either killed or paralyzed and about 50 babies were born with genetic diseases. The scientific investigation revealed that the toxic effect of mercury caused the hazard. The effluents from a vinyl chloride plant of a Minimata chemical company containing mercury was discharged into the sea thereby causing mercury pollution in water. The mercury pollution affected the villagers facing the bay when they consumed fish, which were living on the aquatic ecosystem contaminated with mercury pollution.

Following the incident, the Environmental Protection Agencies of USA and Sweden recommended the following measures to prevent mercury pollution:

(i) All Chloralkali plants must stop using Hg-electrodes and switch over to a new technology.
(ii) All alkyl mercury pesticides must be banned.
(iii) All other mercurial pesticides must be restricted to some selected areas.

Odour Pollution

Air pollution due to odorous substances is called odour pollution. Various odorous substances are generated in the environment due to anthropogenic causes in daily life Some sources of odour pollution are:

(i) Municipal organic and inorganic wastes.
(ii) Uncovered drains, wells, manholes.
(iii) Public urinals, toilets.
(iv) Industries.
(v) Livestock market.
(vi) Polluted water bodies (ponds, lakes).
(vii) Crematorium and morgue.

Ozone Hole

In 1987, British scientists observed from Halley Bay, the depletion of ozone above the Antarctica from single Dotson instrument. This was confirmed by the observation made by the V. S. Nimbus-7 satellite and the measurements made at the Japanese station Swoya. All the measurements concluded the presence of a ozone hole at an altitude of 15–20 km above the Antarctica. The reason for this large scale depletion of ozone was due to extensive use of CFCs, and (i) the very low temperature of Antarctica (-80^{0}C or less), (ii) presence of stratospheric clouds, (iii) extension of a polar vertex that produces a containment wall.

Silent Valley Movement

The Silent Valley is a densely forested valley situated in Palakkad district of Kerala. As the forest is very deep, dark and peaceful, valley got the name Silent Valley. The river Kunithpara, that flows across the valley from north to south, is not navigable. Following a report by a British technical expert team, the Kerala state government planned to generate hydroelectricity from this flowing stream by constructing a dam on the river Kunithpara. The Planning Commission also approved the proposal to generate 120 MW electricity initially and 240 MW subsequently. However, later in 1976, a Task Force of the WWF advised against the implementation of the project as it would cause ecological imbalance in the area and suggested the project be banned. The *Kerala Shastra Sahitya Parishad* (KSSP), the popular science organization of Kerala, opposed the project through a mass signature campaign. Many environmentalists, Salim Ali, IUCN along with KSSP, rallied and protested against the implementation of the project. Finally, in December 1980, the Kerala Government announced the rejection of the project and Silent Valley was declared a National Park.

Wetland

Wetlands are generally defined as areas where during a large part of the year, water stands at 2.5 cm to around 300 cm. Wetlands include swamps, bogs, fresh water lakes and marshes, estuaries and deltas, salt marshes, mangroves, rivers and streams, rice fields, fish ponds, reservoirs, and so on.

Glossary

A

Abatement To minimize the intensity of pollution through reuse or waste treatment.

Acid rain Precipitation (rain, snow fall, hail storm etc.) having pH less than 5.6.

Activated sludge A sewage treatment process in which the aerobic bacteria are continuously circulated and put in contact with biodegradable waste to increase the rate of decomposition of the waste.

Adaptation Process of adjustment of an organism in a set of habitat conditions different from its own.

Adsorption Process of accumulation of a gas or liquid particles on the surface of a solid or a liquid (adsorbent) forming a film of molecules or atoms. This process is often used for the removal of pollutants. Activated carbon, silica gel and alumina are used as adsorbent for the purpose.

Aeration Process of circulating/bubbling air through a solution. The process is used in purification of waste water.

Aeolin Sediment deposits carried by wind.

Aerobe An organism having oxygen-based metabolism. During cellular respiration, aerobes use molecular oxygen to oxidize substrate (carbohydrates) to obtain energy.

Algae (singular alga) A type of typical autotrophic organism.

Algal bloom The sudden spurt of algal growth responding to the great increase in the nutrient (nitrates and phosphates) content entering the aquatic ecosystem. Algal bloom decreases the dissolved oxygen level of the water body causing death of all aquatic life and resulting in eutrophication.

Ambient air Any unconfined (open) portion of the atmosphere. Open air or outdoor air.

Anaerobic digestion The process of decomposition of biodegradable waste with micro-organisms in the absence of air (oxygen), used often for the treatment of sludge.

Anaerobes Micro-organisms that cannot use molecular oxygen but obtain their oxygen from inorganic compounds of nitrates, sulphates and phosphates. This type of micro-organisms are used in anaerobic digestion of biodegradable waste.

Anthropogenic Effects or processes that are caused by human activities.

Atmosphere The envelope of air surrounding the earth.

Aquifer The underground layer of water-soaked sand and rock that acts as a source of useable water.

Autecology Ecology of individual species.

Autotroph A group of organisms, capable of producing complex organic compounds from simple inorganic compounds like carbonates (from soil), using energy from light or oxidizing inorganic compounds (hydrogen sulphides, ammonium, ferrous ion). They are producers in a food chain. Plants, algae, some bacteria are examples of autotrophs.

Altitude The elevation of a point or object from a known level (sea level).

B

Benthos Aquatic life living on, in, or at the bottom of a water body like oceans, rivers etc.

Biodegradable Organic substances that can be broken down by micro-organisms.

Biodegradation The process of decomposition of organic matter by enzymes secreted by microorganisms.

Biochemical cycles (Biogeochemical cycle) The pathway (cycle) through which chemical elements (or molecules) move through biotic (bio) and abiotic (geo) components of an ecosystem.

Biogeosphere Outer layer of the earth's crust (lithosphere) where life is found.

Biodiversity The variety and variability among various life forms (flora and fauna) found in a definite geographical area. Biodiversity is the measure of the health of an ecosystem.

Biota The flora and fauna of a region.

Biotic Related to life or caused by living beings.

Biological Oxygen Demand Biochemical Oxygen Demand or Biological Oxygen Demand (BOD) is a chemical procedure for determining how fast biological organisms use up oxygen in a body of water. It is used in water quality management and assessment. It is not an accurate quantitative measurement and can only indicate the quality of a water body.

Biomagnification Biomagnification, also known as bioamplification, or biological magnification is the cumulative increase of a persistent chemical substance in the successive trophic levels in a food chain.

Biomass (a) The total amount of living matter present at a given time in an area.
(b) It refers to living and recently dead biological material that can be used as fuel or for industrial production.

Biome A climatically and geographically defined area of ecologically similar communities of plants, animals and soil organisms, often referred to as ecosystems.

Biotechnology Application of technology to industrialize the biological processes for the benefit of human beings or for any other specific use. It is used in the fields of medicine, DNA study, tissue culture, etc.

Biomethanation The process of enzymatic decomposition of organic matter by micro-organisms to get methane rich biogas.

C

Canopy The leafy part of a shrub or a tree.

Capillary water Water present between pores of soil particles.

Carcinogenic Cancer causing substances.

Carrying capacity The supportable population of an organism, given the food, habitat, water, and other necessities available within an ecosystem, is known as the ecosystem's carrying capacity for that organism.

Catchment area The area from where water drains to a water body like river, lake etc.

Chemical Oxygen Demand It is a scientific method to determine the quantity of oxygen required to oxidize the waste materials (both biodegradable and non-biodegradable) in the waste sample under specific conditions of oxidizing agent, temperature and time.

Chemotroph Organisms that obtain energy by the oxidation of electron-donating molecules in their environment.

Colloids A type of mechanical mixture where one substance is dispersed evenly throughout another. Due to this dispersal, some colloids have the appearance of solutions.

Comminution Mechanical shredding of waste in both solid and liquid state.

Compactor A machine or mechanism used to reduce the volume of waste material or soil through compaction.

Composting The aerobic decomposition of biodegradable organic matter, producing compost, ie manure.

Conservation The conservation movement is a political, social and, to some extent, scientific movement that seeks to protect natural resources including plant and animal species as well as their habitat for the future.

Consumer Organisms which cannot convert solar energy into food and depend on autotrophs to obtain their energy for survival are called consumers or heterotrophs.

Contaminate Contaminants are potentially undesirable substances. These are usually man-made and human induced substances like DDT, Toxaphene etc. They have a long life and cause harmful effects to human beings and the environment at high concentration.

Chlorofluorocarbons(CFCs) Compounds containing chlorine, fluorine and carbon only, that is, they contain no hydrogen. They are a class of chemical compounds that deplete ozone.

Cybernatics It is the science of systems of control in an ecosystem.

D

Decomposers Organisms that use deceased organisms and non-living organic compounds as their food source to get energy.

Deforestation The conversion of forested areas to non-forest land for use such as arable land, pasture, logged area, or wasteland.

Denitrification A microbially facilitated process of dissimilatory nitrate reduction that may ultimately produce molecular nitrogen (N_2) through a series of intermediate gaseous nitrogen oxide products.

Denudation The process by which the removal of material, through means of erosion and weathering, leads to a reduction of elevation and relief in landforms and landscapes.

Desalinization Any of several processes that remove excess salt and other minerals from water or soil.

Dissolved oxygen Oxygen gets into water by diffusion from the surrounding air, by aeration (rapid movement), and as a waste product of photosynthesis. Dissolved oxygen is a measure of the amount of gaseous oxygen (O_2) dissolved in an aqueous medium. Amount of dissolved oxygen present is a water quality parameter of an aqueous system.

E

Ecology The scientific study of the distribution and abundance of life and the interactions between organisms and their environment

Ecosphere Global ecosystem of the earth.

Ecosystem A natural unit consisting of all plants, animals and micro-organisms (biotic factors) in an area functioning together with all of the non-living, physical (abiotic) factors of the environment.

Effluent An outflowing of water from a natural body of water, or from a man-made structure (factories).

Endemism In a broad sense it means native to, or characteristic of a particular geographical area.

Energy flow The passage of energy through the trophic levels of a food chain.

Environment The physical and biological factors along with their chemical interactions that affect an organism.

Erosion The carrying away or displacement of solids by the agents of currents such as, wind, water, or by downward or down-slope movement in response to gravity or by living organisms.

Estuary A semi-enclosed coastal body of water with one or more rivers or streams flowing into it and with a free connection to the open sea.

Euphotic zone The uppermost sunlit layer of the oceans is called the sunlit zone or the euphotic zone.

Eutrophic lakes A highly productive lake due to high nutrient content. These lakes are subject to excessive algal bloom resulting in poor water quality. The water is of such lakes commonly deficient in dissolved oxygen, which causes aquatic lives to gradually perish.

Evapotranspiration A term used to describe the cumulative effect of evaporation and plant transpiration from the earth's land surface to atmosphere.

Exosphere The uppermost layer of the atmosphere, extending from 500 to 1000 km to about 10,000 km above the earth's surface, from where atmospheric gases, atoms, and molecules can, to any appreciable extent, escape into space.

F

Facultative Bacteria that can live under both aerobic and anaerobic conditions.

Fauna The sum of all animal life of any particular region or time.

Flora All plant life occurring in an area or time period, especially the naturally occurring or indigenous plant life.

Fly ash The mineral residue of coal found as a fine glassy powder that comes primarily from the exhaust of gases of coal fired power-plants. It closely resembles volcanic ashes.

Food chain Describes the feeding relationship between species in an ecosystem (plants, animals and human beings).

Food web Depicts how plants and animals are interdependent in many ways (in terms of food, energy, etc.) for their survival.

Fossil fuel The fossil source of fuels, that is, hydrocarbons found within the top layer of the earth's crust. Coal, oil and natural gas are the three main fossil fuels.

Fresh water Bodies of water such as ponds, lakes, rivers and streams containing low concentrations of dissolved salts and other total dissolved solids.

H

Habitat An ecological or environmental area that is inhabited by a particular species.

Half-life The time required for an original amount of substance to reduce to its half.

Halons A group of chemical compounds that contain alkanes with one or more halogens linked to it.

Hazardous waste These comprise all toxic chemicals, radioactive materials and biologic or infectious waste.

Heavy metals Elements like transition metals, some metalloids, lanthanides, and actinides.

Heterotrophic An organism that requires organic substrate to get its carbon for growth and development. Also called a consumer in the food chain.

High Density polyethylene (HDPE) A polyethylene thermoplastic made from petroleum that has little branching giving it stronger intermolecular forces and tensile strength than lower-density polyethylene. It is also harder and more opaque and can withstand somewhat higher temperature.

Humus The organic material in soil having a dark brown or black colouration.

Hydrocarbons An organic compound consisting entirely of carbon and hydrogen.

Hydrological cycle The movement of water within the hydrosphere. Also called the water cycle.

Hydrosphere The collective mass of water found on, under, and over the surface of a planet.

Hygroscopic water Water held by surface forces of soil particles.

I

Incineration A waste treatment technology that involves the combustion of organic materials.

Ionosphere The uppermost layer of atmosphere extending from 80 km to 300 km above sea level. Gaseous atoms get ionized by solar radiation here and hence the name.

L

Land filling Disposal of residual solid wastes on land in a facility designed with protective measures against pollution of ground water, surface water and air fugitive dust.

Land reclamation Restoring an area to a more natural state (usually after being made unusable due to pollution, salinization, etc).

Latitude Gives the location of a place on earth. Lines of latitude are the horizontal lines shown running east-to-west on maps.

LDPE Low density polyethylene is a thermoplastic made from ethylene through free radical polymerization. It is defined by a density range of 0.910–0.940 g/cm^3. It is non-reactive at room temperature, except by strong oxidizing agents and can withstand temperatures of 80°C continuously and 95 °C for a short time.

Leachate Liquid that seeps through solid or other medium and has extracted, dissolved or suspended material in it.

Lentic water Still water, for example, ponds, basins, marshes and ditches.

Limnetic zone The well-lit, open surface water in a lake, away from the shore.

LLDPE Linear low-density polyethylene is a polyethylene with significant numbers of short branches, commonly made by co-polymerization of ethylene with longer-chain olefins.

Lotic water Flowing water, for example, creeks, streams, rivers and springs.

Litter Waste disposed in the wrong place by unlawful human action, which can vary in size of incident, occurrence or items.

Littoral The coast of an ocean or sea, or the banks of a river, lake or estuary.

M

Macrophytes An emergent, submerged or floating type of aquatic plant.

Marshland Treeless land where the water table is just at, above, or below the ground and it is dominated by grasses, reeds, sedges and reed mace.

Marsh tand A type of wetland which is subject to frequent or continuous inundation.

Mesosphere A layer of earth's atmosphere situated just above the stratosphere extending from 50 km to about 80–90 km above the surface of the earth.

Migration Refers to the movement of a person or animal leaving its own home to go to a new place of residence.

Monitoring To observe a system or situation for any change that may occur over time.

Municipal solid waste A type of waste predominantly made up of domestic and commercial wastes. Includes commercial and residential waste (solid or semi-solid form excluding hazardous waste but it may contain treated biomedical waste) generated in a municipal or notified area.

N

Nannoplankton The smallest phytoplankton.

Niche Ecological niche is a term describing the relational position of a species or population in its ecosystem.

Nitrogen fixation A process by which nitrogen is taken from its natural, relatively inert molecular form (N_2) in the atmosphere and converted into nitrogen compounds.

Non-Government Organization (NGO) A legally constituted organization created by private persons or organizations with no participation or representation of any government.

Non-point source Sources of pollution that arise over a comparatively large area rather than from a single point.

Non-renewable source An energy source that has taken too long to form, is exhaustible and cannot be re-created in a short period of time.

Nutrients Chemicals (food) that an organism needs to live, grow and develop or substances consumed by an organism from its environment and used in an organism's metabolism.

O

Oligotrophic An ecosystem with very little nutrient level and thus offers very little to sustain life.

Organic farming A form of agriculture to maintain soil productivity and to control pests without the use of synthetic fertilizers or pesticides.

Osmoregulation Active regulation of osmotic pressure of bodily fluids to maintain homeostasis of the body's water content.

Oxidation pond Artificial pond in which wastes and effluents are discharged.

Ozone layer A layer in the earth's atmosphere which contains relatively high concentrations of ozone (O_3).

P

Particulates Tiny particles of solid or liquid suspended in a gas.

Pelagic Derived from Greek, the word 'pelagos' means open sea. Any water in the sea that lies not too close to the bottom is referred to as pelagic and that region as the pelagic zone.

Pelletisation Preparation of pellets in the form of cubes or cylindrical pieces from solid waste, including fuel pellets termed as refuse-derived fuel.

Pesticides Chemical substance or a biological agent used to kill a pest.

PET Polyethylene Terepthalate, PET, is a thermosetting plastic, a common plastic material.

Phytoplankton The autotrophic component of plankton (drifting organisms that occupy the pelagic zone of oceans).

Plume Column of one fluid moving through another.

Point source A single, identifiable localized source of pollution.

Polymer A large molecule (macromolecule) composed of repeating structural units connected by covalent chemical bonds. It is derived from two Greek words ('poly' meaning many and 'meros' meaning parts).

PP Polypropylene, a common polymer.

Precipitation Refers to rain, sleet, hail, snow and other forms of water falling from the sky.

Predator Organism that feeds on other living organism(s).

Producer (Autotroph), a living being that synthesizes energy-rich organic compounds.

Profundal zone The profundal zone is the deep zone of a water body, such as an ocean or a lake located below the range of effective light penetration.

PS Polystyrene, a common type of plastic.

PVC Polyvinyl chloride, (IUPAC Polychloroethene), commonly abbreviated as PVC, is a widely used thermoplastic polymer.

R

Recycling Using old materials to make new products with the aim of reducing the consumption of energy and raw materials and to prevent the wastage of potentially useful materials.

Red data book The data book containing the red list of all threatened species of fauna and flora listed as per IUCN specification.

Resource management The efficient and effective deployment of an organization's resources when they are needed.

Risk assessment A common first step in a risk management process. Risk assessment is the determination of quantitative or qualitative value of risk related to a concrete situation and a recognized threat.

Run-off The combination of surface runoff and interflow. It is also equivalent to quick flow.

S

Sanitary landfill An engineered operating design and a method of disposing wastes on land without creating nuisance or hazard to public health and safety.

Scrap Anything that is left over.

Segregation Separation of municipal solid waste into organic, inorganic, hazardous and recyclable waste.

Sewage Mainly liquid waste containing some solids produced by humans, for example, faeces, urine, laundry waste and other materials, which go down drains and toilets from households and industries.

Sewer An artificial conduit (or pipe) or system of conduits used to carry and remove sewage (human liquid waste) and to provide drainage.

Smog A mixture of smoke and fog that may also cause air pollution.

Solid waste The unwanted, discarded solid waste material generated through various activities in houses, society, commercial places, industries and agricultural fields, also known as refuse in the cities and litter in the countryside. Depending upon the locality, the nature of solid waste varies.

Species A particular group in the plant or animal kingdom, which is capable of interbreeding among its own and has certain distinct characteristic features.

Stratosphere Part of the earth's atmosphere lying between troposphere and mesosphere.

Sub-surface water Sub-surface water is ground water present in the pores of soil and rocks, as well as water present in aquifers.

Succession (ecological) More or less predictable and orderly changes in the composition or structure of an ecological community.

Sustainable development A pattern of resource use that aims to meet the needs of the present (while preserving and protecting the environment) without compromising those of the future.

Swamp A wetland that features temporary or permanent inundation of large areas of land by shallow bodies of water and covered by aquatic vegetation.

T

Thermosphere Layer of the earth's atmosphere lying between mesosphere and exosphere. Within this layer, UV radiation causes ionization.

Tickling filter Tickling filters are used for biological treatment of domestic sewage and industrial wastes with the help of large gravels, crushed rocks, etc as the filtering media.

Toxic A substance is said to be highly toxic if it contains enough toxin (or poisonous substances) to highly damage an exposed organism.

Transpiration Evaporation of water from aerial parts of plants (root, leaf, stem and flowers).

Trash Unwanted or undesired waste material.

Trophic level The position that an organism occupies in a food chain – what it eats and what eats it.

Troposphere The lowest portion of the earth's atmosphere containing 75 per cent of its mass and nearly all its water vapour.

Turbidity The haziness in a fluid caused by suspended solid particles, invisible to the naked eye.

V

Vegetation Plant life of a region or the ground cover of a place provided by plants.

Vermicomposting Process of using earthworms for conversion of biodegradable waste to compost.

Volatile compounds Chemical compounds having high vapour pressure to significantly vapourize and enter the atmosphere under normal conditions.

Volume reduction (compaction) The process of compacting waste, that is, to reduce the size and volume of waste materials.

W

Waste exchange In waste exchange, the product of one process becomes the raw material for the second process.

Weathering Decomposition of earth, rocks, soil and minerals through direct contact with the earth's atmosphere.

Wetland An area of land saturated with moisture, for example, swamp, marsh and bog.

Z

Zooplankton Heterotrophic component of plankton that drift in oceans, seas, rivers and fresh water bodies.

Bibliography

1. Dey A. K., Environmental Chemistry, Willey Eastern Pvt. Ltd, New Delhi.
2. Dr Bharucha Erach, Text Book of Environmental Studies for UG Course, University Press (India) Pvt. Ltd.
3. Dr Bharucha Erach, The Biodiversity of India, Mapin Publishing Pvt. Ltd, Ahmedabad – 380 013, India.
4. Katyal Timi & Satake M., Environmental Pollution, Anmol Publication Pvt. Ltd, New Delhi.
5. G .R. Chhatwal, M. C. Mehra, M. Satake, T. Katyal & Mohan V., Environmental Radiation and Thermal Pollution and their control, Anmol Publications, New Delhi.
6. R. C. Brunner, Hazardous Waste Incineration, Mc Graw Hill Inc.
7. K. C. Agarwal, Environmental Biology, Nidi Publishing Ltd, Bikaner.
8. R. N. Basu (Editor), Environment Calcutta University, Kolkata.
9. G. S. Dhaliwal, G. S. Sangha & P. K. Ralhan, Fundamentals of Environmental Science, Kalyani Publishers, New Delhi.
10. P. D. Sharma, Ecology and Environment, Rastogi Publications, Meerut.
11. B. K. Sharma & Kaur, An introduction to Environmental Pollution, Goel Publishing House, Meerut.
12. M. Satoke, Y. Mido, S. A. Iqbal & M. S. Sethi, Environmental Chemistry, Discovery Publishing House, New Delhi.
13. S. M. Khopkar, Environmental Pollution Analysis, New Age International (P) Ltd, New Delhi.
14. G. S. Sodhi, Fundamentals Concepts on Environmental Chemistry, Narosa Publishing House, New Delhi.
15. Benny Joseph, Environmental Studies, Tata Mc Graw Hill Publishing Co. Pvt. Ltd, New Delhi.
16. Bindu N., Mohan Alsair & M. North, Environmental Quality Management, South Asian Publishing Pvt. Ltd, New Delhi.
17 Clark R. S., Marine Pollution, Clanderson Press, Oxford.
18. Odum, E. P., 1963, Ecology, Holt Rinehart & Winston, New York.
19. Odum, E.P., 1971, Fundamentals of Ecology, W. B. Saunders Co., USA.
20. Odum, E. P., 1983, Basic Ecology, Holt-Seunders Intl. ed., Japan.
21. Gleick, H. P., 1993, Water in crisis, Pacific Institute for Studies in Dev., Environment & Security, Stockholm Env. Institute, Oxford University Press.
22. Hawkins R. E., Encyclopedia of Indian Natural History, Bombay Natural History Society, Bombay.
23. Heywood, V. H. & Waston, R. T., 1995, Global Biodiversity Assessment, Cambridge University Press.
24. Jadhav, H. & Bhosale, V. M., 1995, Environmental Protection and Laws, Himalaya Pub. House, Delhi.
25. Mckinney, M. L. & School, R. M., 1996, Environmental Science systems & Solutions, Web enhanced edition.
26. Mhaskar A. K., Matter Hazardous, Techno-Science Publication.
27. Miller T. G. Jr., Environmental Science, Wadsworth Publishing Co.

28. Down to Earth, Centre for Science and Environment.
29. Rao M. N. & Datta, A. K., 1987, Waste Water Treatment, Oxford & IBH Publications Co. Pvt. Ltd.
30. Sharma B.K., 2001, Environmental Chemistry, Geol Publishing House, Meerut.
31. Survey of the Environment, The Hindu.
32. Townsend C., Harper J., and Michael Begon, Essentials of Ecology, Blackwell Science.
33. Trivedi R. K., Handbook of Environmental Laws, Rules Guidelines, Compliances and Standards, Vol I and II, Enviro Media.
34. Trivedi R. K. and P. K. Goel, Introduction to Air Pollution, Techno-Science Publication.
35. Wanger K. D., 1998, Environmental Management, W. B. Saunders Co. Philadelphia, USA.
36. Bolin, B. and Cook, R. B. (Eds)., 1983, The Major Biogeochemical Cycles and their Interactions, Wiley, Chicester, England.
37. Swaminathanan, M. S., 1991, From Stockholm to Rio de Janeiro: The Road to Sustainable Agriculture, Monograph No 4, M. S. Swaminathan Research Foundation, Madras.
38. Singh, Samar, 1986, Conserving India's Natural Heritage, Natraj Publishers, Dehradun.

Index

H

I

J

K

L

M

N